Rolf Berndt

Representations
of Linear Groups

Rolf Berndt

Representations of Linear Groups

An Introduction Based on Examples from Physics and Number Theory

Bibliografische information published by Die Deutsche Nationalbibliothek
Die Deutsche Nationalbibliothek lists this publication in the Deutschen Nationalbibliografie;
detailed bibliographic data is available in the Internet at <http://dnb.d-nb.de>.

Prof. Dr. Rolf Berndt
Department of Mathematics
University of Hamburg
Bundesstraße 55
D-20146 Hamburg
Germany

berndt@math.uni-hamburg.de

Mathematics Subject Classification
20G05, 22E45, 20C25, 20C35, 11F70

First edition, July 2007

Editorial Office: Ulrike Schmickler-Hirzebruch | Susanne Jahnel

Vieweg is a company in the specialist publishing group Springer Science+Business Media.
www.vieweg.de

Cover design: Ulrike Weigel, www.CorporateDesignGroup.de
Printing and binding: MercedesDruck, Berlin
Printed on acid-free paper

ISBN 978-3-8348-0319-1

Preface

There are already many good books on representation theory for all kinds of groups. Two of the best (in this author's opinion) are the one by A.W. Knapp: "Representation Theory for Semisimple Groups. An Overview based on Examples" [Kn1] and by G.W. Mackey: "Induced Representations in Physics, Probability and Number Theory" [Ma1]. The title of this text is a mixture of both these titles, and our text is meant as a very elementary introduction to these and, moreover, to the whole topic of group representations, even infinite-dimensional ones. As is evident from the work of Bargmann [Ba], Weyl [Wey] and Wigner [Wi], group representations are fundamental for the theory of atomic spectra and elementary physics. But representation theory has proven to be an inavoidable ingredient in other fields as well, particularly in number theory, as in the theory of theta functions, automorphic forms, Galois representations and, finally, the Langlands program. Hence, we present an approach as elementary as possible, having in particular these applications in mind.

This book is written as a summary of several courses given in Hamburg for students of Mathematics and Physics from the fifth semester on. Thus, some knowledge of linear and multilinear algebra, calculus and analysis in several variables is taken for granted. Assuming these prerequisites, several groups of particular interest for the applications in physics and number theory are presented and discussed, including the symmetric group $\mathfrak{S}_n$ as the leading example for a finite group, the groups $SO(2)$, $SO(3)$, $SU(2)$, and $SU(3)$ as examples of compact groups, the Heisenberg groups and $SL(2,\mathbf{R})$, $SL(2,\mathbf{C})$, resp. the Lorentz group $SO(3,1)$ as examples for noncompact groups, and the Euclidean groups $E(n) = SO(n) \ltimes \mathbf{R}^n$ and the Poincaré group $\mathcal{P} = SO(3,1)^+ \ltimes \mathbf{R}^4$ as examples for semidirect products.

This text would not have been possible without the assistance of my students and colleagues; it is a pleasure for me to thank them all. In particular, D. Bahns, S. Böcherer, O. v. Grudzinski, M. Hohmann, H. Knorr, J. Michaliček, H. Müller, B. Richter, R. Schmidt, and Chr. Schweigert helped in many ways, from giving valuable hints to indicating several mistakes. Part of the material was treated in a joint seminar with Peter Slodowy. I hope that a little bit of his way of thinking is still felt in this text and that it is apt to participate in keeping alive his memory. Finally, I am grateful to U. Schmickler-Hirzebruch and S. Jahnel from the Vieweg Verlag for encouragement and good advice.

Preface

There are already many good books on representation theory for all kinds of groups. Two of the best (in this author's opinion) are the one by A.W. Knapp: "Representation Theory for Semisimple Groups. An Overview based on Examples" [Kn1] and by G.W. Mackey: "Induced Representations in Physics, Probability and Number Theory" [Ma1]. The title of this text is a mixture of both these titles, and our text is meant as a very elementary introduction to these and, moreover, to the whole topic of group representations, even infinite-dimensional ones. As is evident from the work of Bargmann [Ba], Weyl [Wey] and Wigner [Wi], group representations are fundamental for the theory of atomic spectra and elementary physics. But representation theory has proven to be an inavoidable ingredient in other fields as well, particularly in number theory, as in the theory of theta functions, automorphic forms, Galois representations and, finally, the Langlands program. Hence, we present an approach as elementary as possible, having in particular these applications in mind.

This book is written as a summary of several courses given in Hamburg for students of Mathematics and Physics from the fifth semester on. Thus, some knowledge of linear and multilinear algebra, calculus and analysis in several variables is taken for granted. Assuming these prerequisites, several groups of particular interest for the applications in physics and number theory are presented and discussed, including the symmetric group $\mathfrak{S}_n$ as the leading example for a finite group, the groups $SO(2)$, $SO(3)$, $SU(2)$, and $SU(3)$ as examples of compact groups, the Heisenberg groups and $SL(2, \mathbf{R})$, $SL(2, \mathbf{C})$, resp. the Lorentz group $SO(3, 1)$ as examples for noncompact groups, and the Euclidean groups $E(n) = SO(n) \ltimes \mathbf{R}^n$ and the Poincaré group $\mathcal{P} = SO(3, 1)^+ \ltimes \mathbf{R}^4$ as examples for semidirect products.

This text would not have been possible without the assistance of my students and colleagues; it is a pleasure for me to thank them all. In particular, D. Bahns, S. Böcherer, O. v. Grudzinski, M. Hohmann, H. Knorr, J. Michaliček, H. Müller, B. Richter, R. Schmidt, and Chr. Schweigert helped in many ways, from giving valuable hints to indicating several mistakes. Part of the material was treated in a joint seminar with Peter Slodowy. I hope that a little bit of his way of thinking is still felt in this text and that it is apt to participate in keeping alive his memory. Finally, I am grateful to U. Schmickler-Hirzebruch and S. Jahnel from the Vieweg Verlag for encouragement and good advice.

Contents

Introduction

In this book, the groups enumerated in the Preface are introduced and treated as matrix groups to avoid as much as possible the machinery of manifolds, Lie groups and bundles (though some of it soon creeps in through the backdoor as the theory is further developed). Parallel to information about the structure of our groups we shall introduce and develop elements of the representation theory necessary to classify the unitary representations and to construct concrete models for these representations. As the main tool for the classification we use the *infinitesimal method* linearizing the representations of a group by studying those of the Lie algebra of the group. And as the main tools for the construction of models for the representations we use

– tensor products of the *natural* representation,

– representations given by smooth functions (in particular polynomials) living on a space provided with an action of the group,

and

– the machinery of *induced representations*.

Moreover, because of the growing importance in physics and the success in deriving branching relations, the procedure of *geometric quantization* and the *orbit method*, developed and propagated by Kirillov, Kostant, Duflo and many others shall be explained via its application to some of the examples above.

Besides the sources already mentioned, the author was largely influenced by the now classical book of Kirillov: "Elements of the Theory of Representations" [Ki] and the more recent "Introduction to the Orbit Method" [Ki1]. Other sources were the books by Barut and Raczka: "Theory of Group Representations and Applications" [BR], S. Lang: "SL(2, **R**)" [La], and, certainly, Serre: "Linear Representations of Finite Groups" [Se]. There is also the book by Hein: "Einführung in die Struktur- und Darstellungstheorie der klassischen Gruppen" [Hei], which follows the same principle as our text, namely to do as much as possible for matrix groups, but does not go into the infinite-dimensional representations necessary for important applications. Whoever is further interested in the history of the introduction of representation theory into the theory of automorphic forms and its development is referred to the classical book by Gelfand, Graev and Pyatetskii-Shapiro: "Representation Theory and Automorphic Forms" [GGP], Gelbart's: "Automorphic Forms on Adèle Groups" [Ge], and Bump's: "Automorphic Forms and Representations" [Bu]. More references will be given at the appropriate places in our text; as already said, we shall start using material only from linear algebra and analysis. But as we proceed more and more elements from topology, functional analysis, complex function theory, differential and symplectic geometry will be needed. We will try to introduce these as gently as possible but often will have to be very rudimentary and will have to cite the hard facts without the proofs, which the reader can find in the more refined sources.

To sum up, this text is prima facie about real and complex matrices and the nice and sometimes advanced things one can do with them by elementary means starting from a certain point of view: Representation theory associates to each matrix from a given group G another matrix or, in the infinite-dimensional case, an operator acting on a Hilbert space. One may want to ask, why study these representations by generally more complicated matrices or operators if the group is already given by possibly rather simple matrices? An answer to this question is a bit like the one for the pudding: the proof is in the eating. And we hope our text will give an answer. To the more impatient reader who wants an answer right away in order to decide whether to read on or not, we offer the following rough explanation. Certain groups G appear in nature as symmetry groups leaving invariant a physical or dynamical system. For example, the orthogonal group $O(3)$ is the symmetry group for the description of the motion of a particle in a central symmetric force field, and the Poincaré group $\mathcal{P}$ is the symmetry group for the motion of a free particle in Minkowski space. Then the irreducible unitary representations of G classify indivisible intrinsic descriptions of the system and, boldly spoken, can be viewed as "elementary particles" for the given situation. Following Wigner and his contemporaries, the parameters classifying the representations are interpreted as quantum numbers of these elementary particles...

The importance of representations for number theory is even more difficult to put into a nutshell. In the Galois theory of algebraic number fields (of finite degree) Galois groups appear as symmetry groups G. Important invariants of the fields are introduced via certain zeta- or L-functions, which are constructed using finite-dimensional representations of these Galois groups. Another aspect comes from considering smooth (holomorphic resp. meromorphic) functions in several variables which are periodic or have a more general covariant transformation property under the action of a given discrete subgroup of a continuous group G, like for instance $G = \mathrm{SL}(2, \mathbf{R})$ or G a Heisenberg or a symplectic group. Then these functions with preassigned types, e.g., theta functions or modular forms, generate representation spaces for (infinite-dimensional) representations of the respective group G.

Finally, we will give an overview over the contents of our text: In a *prologue* we will fix some notation concerning the groups and their actions that we later use as our first examples, namely, the general and special linear groups over the real and complex numbers and the orthogonal and unitary groups. Moreover, we present the symmetric group $\mathfrak{S}_n$ of permutations of n elements and some facts about its structure. We stay on the level of very elementary algebra and stick to the principle to introduce more general notions and details from group theory only when needed in our development of the representation theory. We follow this principle in the first chapter where we introduce the concept of linear representations using only tools from linear algebra. We define and discuss the fundamental notions of equivalence, irreducibility, unitarity, direct sums, tensor product, characters, and give some first examples.

The theory developed thus far is applied in the second chapter to the representations of finite groups, closely following Serre's exposition [Se]. We find out that all irreducible representations may be unitarized and are contained in the regular representation.

In the next step we move on to compact groups. To do this we have to leave the purely algebraic ground and take in topological considerations. Hence, in the third chapter, we define the notion of a topological and of a (real or complex) linear group, the central notion for our text. Following this, we refine the definition of a group representation by adding the usual continuity condition. Then we adequately modify the general concepts of the first chapter. We try to take over as much as possible from finite to compact groups. This requires the introduction of invariant measures on spaces with a (from now on) continuous group action, and a concept of integration with respect to these measures. In the forth chapter we concentrate on compact groups and prove that the irreducible representations are again unitarizable, finite-dimensional, fixed by their characters and contained in the regular representation. But their number is in general not finite, in contrast to the situation for finite groups. We state, but do not prove, the Peter-Weyl Theorem. But to get a (we hope) convincing picture, we illustrate it by reproducing Wigner's discussion of the representaions of SU(2) and SO(3). We use and prove that SU(2) is a double cover of SO(3). Angular momentum, magnetic and spin quantum numbers make an appearance, but for further application to the theory of atomic spectra we refer to [Wi] and the physics literature.

In a very short fifth chapter, we assemble some material about the representations of locally compact abelian groups. We easily get the result that every unitary irreducible representation is one-dimensional. But as can be seen from the example $G = \mathbf{R}$, their number need not be denumerable. More functional analysis than we can offer at this stage is needed to decompose a given reducible representation into a *direct integral* of irreducibles, a notion we not consider here.

Before starting the discussion of representations of other noncompact groups, we present in chapter 6 an important tool for the classification of representations, the *infinitesimal method*. Here, at first, we have to explain what a Lie algebra is and how to associate one to a given linear group. Our main ingredient is the matrix exponential function and its properties. We also reflect briefly on the notion of representations of Lie algebras. Here again we are on purely algebraic, at least in examples, easily accessible ground. We start giving examples by defining the derived representation $d\pi$ of a given group representation π. We do this for the Schrödinger representation of the Heisenberg group and the standard representation π_1 of SU(2). Then we concentrate on the classification of all unitary irreducible representations of SL(2, $\mathbf{R}$) via a description of all (integrable) representations of its Lie algebra. Having done this, we consider again the examples $\mathfrak{su}(2)$ and heis($\mathbf{R}$) (relating them to the theory of the *harmonic oscillator*) and give some hints concerning the general structure theory of *semisimple Lie algebras*. The way a general classification theory works is explained to some extent by considering Lie SU(3); we will see how *quarks* show up.

Chapters 7 and 8 are the core of our book. In the seventh chapter we introduce the concept of *induced representations*, which allows for the construction of (sometimes infinite-dimensional) representations of a given group G starting from a (possibly one-dimensional) representation of a subgroup H of G. To make this work we need again a bit more Hilbert space theory and have to introduce quasi-invariant measures on spaces with group action. We illustrate this by considering the examples of the Heisenberg group and $G = $ SU(2), where we rediscover the representations which we already know. Then we use the induction process to construct models for the unitary representations of

$\mathrm{SL}(2, \mathbf{R})$ and $\mathrm{SL}(2, \mathbf{C})$. In particular, we show how *holomorphic induction* arises in the discussion of the discrete series of $\mathrm{SL}(2, \mathbf{R})$ (here we touch complex function theory). We insert a brief discussion of the Lorentz group $G^L = \mathrm{SO}(3, 1)^0$ and prove that $\mathrm{SL}(2, \mathbf{C})$ is a double cover of G^L. To get a framework for the discussion of the representations of the Poincaré group G^P, which is a semidirect product of the Lorentz group with $\mathbf{R}^4$, we define semidirect products and treat Mackey's theory in a rudimentary form. We outline a recipe to classify and construct irreducible representations of semidirect products if one factor is abelian. We do not prove the general validity of this procedure as Mackey's *Imprimitivity Theorem* is beyond the scope of our book, but we apply it to determine the unitary irreducible representations of the Euclidean and the Poincaré group, which are fundamental for the classification of elementary particles.

Under the heading of *Geometric Quantization*, in the eighth chapter we take an alternative approach to some material from chapter 7 by constructing representations via the *orbit method*. Here we have to recall (or introduce) more concepts from higher analysis: manifolds and bundles, vector fields, differential forms, and in particular the notion of a symplectic form. We can again use the information and knowledge we already have of our examples $G = \mathrm{SL}(2, \mathbf{R})$, $\mathrm{SU}(2)$ and the Heisenberg group to get a feeling what should be done here. We identify certain spheres and hyperboloids as coadjoint orbits of the respective groups, and we construct line bundles on these orbits and representation spaces consisting of *polarized* sections of the bundles.

Finally, in the nineth and last chapter, we give a brief outlook on some examples where representations show up in number theory. We present the notion of an automorphic representation (in a rudimentary form) and explain its relation with theta functions and automorphic forms. We have a glimpse upon Hecke's and Artin's L-functions and mention the Artin conjecture.

We hope that some of the exercises and/or omitted proofs may give a starting point for a bachelor thesis, and also that this text motivates further studies in a master program in theoretical physics, algebra or number theory.

¡ Libro, afán
de estar en todas partes,
en soledad!

J. R. Jiménez

Chapter 0

Prologue: Some Groups and their Actions

This text is mainly on groups which some way or another come from physics and/or number theory and which can be described in form of a real or complex matrix group.

0.1 Several Matrix Groups

We will use the following notation:

The letter $\mathbf{K}$ indicates a field. The reader is invited to think of the field $\mathbf{R}$ of real or $\mathbf{C}$ of complex numbers. Most of the things we do at the beginning of our text are valid also for more general fields at least if they are algebraically closed and of characteristic zero, but as this is only an introduction for lack of space we will not go into this to a greater depth.

$M_{m,n}(\mathbf{K})$ denotes the $\mathbf{K}$-vector space of $m \times n$ matrices $A = (a_{ij})$ with $a_{ij} \in \mathbf{K}$ ($i = 1, \ldots, m, j = 1, \ldots, n$) and $M_n(\mathbf{K})$ stands for $M_{n,n}(\mathbf{K})$.

Our groups will be (for some n) subgroups of of the *general linear group* of invertible $n \times n$-matrices

$$\mathrm{GL}(n, \mathbf{K}) := \{A \in M_n(\mathbf{K}); \det A \neq 0\}.$$

As usual, we will denote the *special linear group* by

$$\mathrm{SL}(n, \mathbf{K}) := \{A \in M_n(\mathbf{K}); \det A = 1\},$$

the *orthogonal group* by

$$\mathrm{O}(n) := \{A \in M_n(\mathbf{R});\ ^tAA = E_n\},$$

resp. for $n = p + q$

$$\mathrm{O}(p, q) := \{A \in M_n(\mathbf{R});\ ^tAD_{p,q}A = D_{p,q}\},$$

where $D_{p,q}$ is the diagonal matrix having p times 1 and q times -1 in its diagonal, and the *unitary group*

$$\mathrm{U}(n) := \{A \in M_n(\mathbf{C});\ ^tA\bar{A} = E_n\},$$

resp. for $n = p + q$

$$U(p, q) := \{A \in M_n(\mathbf{R}); \, {}^tA D_{p,q} \bar{A} = D_{p,q}\}.$$

Again, addition of the letter S to the symbol of the group indicates that we take only matrices with determinant 1, e.g.

$$SO(n) := \{A \in O(n); \, \det A = 1\}.$$

These groups together with some other families of groups, in particular the *symplectic groups* showing up later, are known as *classical groups*.

Later on, we will often use subgroups consisting of certain types of block matrices, e.g. the *group of diagonal matrices*

$$A_n := \{D(a_1, \ldots, a_n); \, a_1, \ldots, a_n \in \mathbf{K}^*\},$$

where $D(a_1, \ldots, a_n)$ denotes the diagonal matrix with the elements $a_1, \ldots, a_n$ in the diagonal, or the *group of upper triangular matrices* (or *standard Borel group*) B_n consisting of matrices with zeros below the diagonal and the *standard unipotent group* N_n, the subgroup of B_n where all diagonal elements are 1.

In view of the importance for applications, moreover, we distinguish several types of *Heisenberg groups*: Thus the group N_3 we just defined, is mostly written as

$$\text{Heis}'(\mathbf{K}) := \left\{ g = \begin{pmatrix} 1 & x & z \\ & 1 & y \\ & & 1 \end{pmatrix}; \, x, y, z \in \mathbf{K} \right\}.$$

In the later application to theta functions it will become clear that, though it may seem more complicated, we shall better use the following description for the Heisenberg group.

$$\text{Heis}(\mathbf{K}) := \left\{ g = (\lambda, \mu, \kappa) := \begin{pmatrix} 1 & 0 & 0 & \mu \\ \lambda & 1 & \mu & \kappa \\ 0 & 0 & 1 & -\lambda \\ 0 & 0 & 0 & 1 \end{pmatrix}; \, \mu, \lambda, \kappa \in \mathbf{K} \right\},$$

and the "higher dimensional" groups, which for typographical reasons we here do not write as matrix groups

$$\text{Heis}(\mathbf{K}^n) := \{g = (x, y, z); \, x, y \in \mathbf{K}^n, z \in \mathbf{K}\}$$

with the multiplication law given by

$$gg' = (x + x', y + y', z + z' + {}^txy' - {}^tyx').$$

We suggest to write elements of $\mathbf{K}^n$ as columns.

Exercise 0.1: Write $\text{Heis}(\mathbf{K}^n)$ in matrix form. Show that $\text{Heis}(\mathbf{K}^n)$ for $n = 1$ and the other two Heisenberg groups above are isomorphic.

Exercise 0.2: Verify that the matrices

$$\begin{pmatrix} 0 & 1 \\ -1 & 0 \end{pmatrix}, \begin{pmatrix} z & 0 \\ 0 & \bar{z} \end{pmatrix}, \, z \in \mathbf{C}^* := \mathbf{C} \setminus \{0\}$$

generate (by taking all possible finite products) a non-abelian subgroup of $GL(2, \mathbf{C})$. This is called the *Weil group of* $\mathbf{R}$ and plays an important role in the epilogue at the end of our text.

0.2   Group Actions

Most groups "appear in nature" as *transformation groups* acting on some set or space. This motivates the introduction of the following concepts.

Let G be a group with neutral element e and let $\mathcal{X}$ be a set.

Definition 0.1: G *acts on $\mathcal{X}$ from the left* iff a map

$$G \times \mathcal{X} \longrightarrow \mathcal{X}, \quad (g, x) \longmapsto g \cdot x$$

is given, which satisfies the conditions

$$g \cdot (g' \cdot x) = (gg') \cdot x, \; e \cdot x = x$$

for all $g, g' \in G$ and $x \in \mathcal{X}$.

Remark 0.1: If $\operatorname{Aut} \mathcal{X}$ denotes the group of all bijections of $\mathcal{X}$ onto itself, the definition says that we have a group homomorphism $G \longrightarrow \operatorname{Aut} \mathcal{X}$ associating to every $g \in G$ the transformation $x \longmapsto g \cdot x$.

In this case the set $\mathcal{X}$ is also called a *left G-set*.

The group action is called *effective* iff no element except the neutral element e acts as the identity, i.e. the homomorphism $G \longrightarrow \operatorname{Aut} \mathcal{X}$ is faithful.

The group action is called *transitive* iff for every pair $x, x' \in \mathcal{X}$ there is a $g \in G$ with $x' = g \cdot x$.

For $x_0 \in \mathcal{X}$ we call the subset of $\mathcal{X}$

$$G \cdot x_0 := \{g \cdot x_0; \; g \in G\}$$

an *orbit of G* (through x_0) and the subgroup

$$G_{x_0} := \{g \in G; \; g \cdot x_0 = x_0\}$$

the *isotropy group* or the *stabilizing group* of x_0.

Example 0.1: $G = \operatorname{GL}(n, \mathbf{K})$ and its subgroups act on $\mathcal{X} = \mathbf{K}^n$ from the left by matrix multiplication

$$(A, x) \longmapsto Ax$$

for $x \in \mathbf{K}^n$ (viewed as a column).

Exercise 0.3: Assure yourself that $\operatorname{GL}(n, \mathbf{K})$ acts transitively on $\mathcal{X} = \mathbf{K}^n \setminus \{0\}$. Describe the orbits of $\operatorname{SO}(n)$.

Example 0.2: A group acts on itself from the left in three ways:

a) by left translation $(g, g_0) \longmapsto g \cdot g_0 = g g_0 =: \lambda_g g_0$,

b) by (the inverse of) right translation $(g, g_0) \longmapsto g \cdot g_0 = g_0 g^{-1} =: \rho_{g^{-1}} g_0$,

c) by conjugation $(g, g_0) \longmapsto g \cdot g_0 = g g_0 g^{-1} =: \kappa_g g_0$.

Left- and right translations are obviously transitive actions. For a given group, the determination of its *conjugacy classes*, i.e. the classification of the orbits under conjugation is a highly interesting question, as we shall see later.

Exercise 0.4: Determine a family of matrices containing exactly one representative for each conjugacy class of $G_1 = \mathrm{GL}(2, \mathbf{C})$ and $G_2 = \mathrm{GL}(2, \mathbf{R})$.

Definition 0.2: A set is called a *homogeneous space* iff there is a group G acting transitively on $\mathcal{X}$.

Remark 0.2: In this case one has $\mathcal{X} = G \cdot x$ for each $x \in \mathcal{X}$ and any two stabilizing groups G_x and $G_{x'}$ are conjugate. (Prove this as **Exercise 0.5.**)

Definition 0.3: For any G-set we shall denote by $\mathcal{X}/G$ or $\mathcal{X}_G$ the set of G-orbits in $\mathcal{X}$ and by $\mathcal{X}^G$ the set of *fixed points for* G, i.e. those points $x \in \mathcal{X}$ for which one has $g \cdot x = x$ for all $g \in G$.

If the set $\mathcal{X}$ has some structure, for instance, if $\mathcal{X}$ is a vector space, we will (tacidly) additionally require that a group action preserves this structure, i.e. in this case that $x \longmapsto g \cdot x$ is a linear map for each $g \in G$ (or later on, is continuous if $\mathcal{X}$ is a topological space). It is a fundamental question whether the orbit space $\mathcal{X}/G$ inherits the same properties as the space $\mathcal{X}$ may have. For instance, if $\mathcal{X}$ is a manifold, is this true also for $\mathcal{X}/G$? We will come back to this question later several times. Here let us only look at the following situation: Let $\mathcal{X}$ be a homogeneous G-space, $x_0 \in \mathcal{X}$, and $H = G_{x_0}$ the isotropy group. Then we denote by G/H the set of *left cosets* $gH := \{gh; h \in H\}$. These are also to be seen as H-orbits $H \cdot g$ where H acts on G by right translation. G/H is a G-set, G acting by $(g, g_0 H) \longmapsto g g_0 H$. Then we shall identify G/H and $\mathcal{X}$ via the map

$$G/H \longrightarrow \mathcal{X}, \quad gH \longmapsto g \cdot x_0.$$

This map is an example for the following general notion.

Definition 0.4: Let $\mathcal{X}$ and $\mathcal{X}'$ be G-sets and $f : \mathcal{X} \longrightarrow \mathcal{X}'$ be a map. The map f is called *G-equivariant* or a *G-morphism* iff one has for every $g \in G$ and $x \in \mathcal{X}$

$$g \cdot f(x) = f(g \cdot x).$$

Now we come back to the question raised above.

Exercise 0.6: Prove that G/H has the structure of a group iff H is not only a subgroup but a *normal* subgroup, i.e. one has $ghg^{-1} \in H$ for all $g \in G, h \in H$.

We mention here another useful fact: If H is a subgroup of G, one defines the *normalizer* of H in G as

$$N_G(H) := \{g \in G; \ gHg^{-1} = H\}.$$

It is clear that $N_G(H)$ is the maximal subgroup in G that has H as a normal subgroup. Then the group $\mathrm{Aut}\,(G/H)$ of G-equivariant bijections of G/H is isomorphic to $N_G(H)/H$. (Prove this as **Exercise 0.7.**)

Parallel to the normalizer, one defines the *centralizer*

$$C_G(H) := \{g \in G;\ ghg^{-1} = h \quad \text{for all } h \in H\}.$$

In particular for $H = G$, the centralizer specializes to the *center*

$$C_G(G) = \{g \in G; gh = hg \quad \text{for all } h \in G\} =: C(G).$$

As well as left actions one defines and studies *right actions* of a group G on a set or a space $\mathcal{X}$:

Definition 0.5: G acts from the right iff a map

$$G \times \mathcal{X} \longrightarrow \mathcal{X}, \quad (g, x) \longmapsto x \cdot g,$$

is given that satisfies the conditions

$$(x \cdot g) \cdot g' = x \cdot (gg'), \quad x \cdot e = x$$

for all $g, g' \in G$ and $x \in \mathcal{X}$.

Remark 0.3: Obviously it is easy to switch from right action to left action and vice versa using the antiautomorphism $g \longmapsto g^{-1}$ of G. Namely, on any right G-space $\mathcal{X}$ one can naturally define a left action by $g \cdot x := x \cdot g^{-1}$.

Right actions often show up in number theory and they are written there as $x^g := x \cdot g$ (for instance in the action of the Galois group on a number field).

0.3   The Symmetric Group

Though this text is meant to treat mainly *continuous* groups like the matrix groups set up in Section 0.1, some *discrete* groups as, for instance, $\mathbf{Z}$, $\mathrm{SL}(n, \mathbf{Z})$ (where the elements are matrices with determinant one and integers as entries), and in particular the symmetric group $\mathfrak{S}_n$, are unavoidable.

Definition 0.6: The *symmetric group* $\mathfrak{S}_n$ is the group of *permutations*, i.e. bijections, of a set of n elements.

We already introduced $\mathrm{Aut}\,\mathcal{X}$ as the group of bijections of a set $\mathcal{X}$, so we can take $\mathcal{X}_n := \{1, \ldots, n\}$ and have $\mathfrak{S}_n = \mathrm{Aut}\,\mathcal{X}_n$. We emphasize that (with the notions from Section 0.2) we treat permutations as left actions on the set $\mathcal{X}_n$.
There are several fundamental facts concerning this group, which the reader is supposed to know from Linear Algebra:

Remark 0.4: The number of elements of the symmetric group $\mathfrak{S}_n$ is $\#\,\mathfrak{S}_n = n!$

Remark 0.5: Each permutation $\sigma \in \mathfrak{S}_n$ can be written as a product of *transpositions*, i.e. permutations, which interchange exactly two elements.

For every $\sigma \in \mathfrak{S}_n$ one defines the *signum* of σ by

$$\mathrm{sgn}(\sigma) := \epsilon(\sigma) := \prod_{1 \leq i < j \leq n} \frac{\sigma(i) - \sigma(j)}{i - j}.$$

Remark 0.6: The map $\sigma \longmapsto \epsilon(\sigma)$ is a group homomorphism of $\mathfrak{S}_n$ to the subgroup $\{1, -1\}$ of the multiplicative group $\mathbf{R}^* = \mathbf{R} \setminus \{0\}$. Thus, the number of transpositions in a representation of σ as a product of transpositions is either even or odd.
The kernel of ϵ is usually written as $\mathfrak{A}_n$. It is a normal subgroup and called the *alternating group*.

Permutations are often written as products of *cycles*, i.e. permutations $(i_1, \ldots, i_r)$ with $i_j \longmapsto i_{j+1}$ for $j < r$ and $i_r \longmapsto i_1$ if $r > 1$ and the identity for $r = 1$. More precisely, one has

Remark 0.7: Each permutation is (up to order) a unique product of disjoint cycles.

From here it is not far to the fundamental fact:

Theorem 0.1: The number of conjugacy classes of $\mathfrak{S}_n$ is equal to the number $p(n)$ of *partitions* of n.

A *partition* of n is a sequence $(n_1, \ldots, n_r)$ of natural numbers $n_i \in \mathbf{N}$ with $n_i \geq n_j$ for $i < j$ and $\Sigma n_i = n$.

Example 0.3: The partitions of $n = 3$ are $(1,1,1)$, $(2,1)$, and (3).
The conjugacy classes of $\mathfrak{S}_3$ are

  $- \{id\}$

  $-$ the three two-element subgroups generated by transpositions, i.e.,
$\{(1, 2)\}, \{(2, 3)\}, \{(3, 1)\}$,

  $-$ the two three-element subgroups generated by the 3-cycles
$(1, 2, 3)$ resp. $(1, 3, 2)$.

Exercise 0.8: Determine the conjugacy classes of $\mathfrak{A}_4$.

$\mathfrak{S}_n$ can be realized as a matrix group: we associate to every $\sigma \in \mathfrak{S}_n$ the matrix $A = A(\sigma) = (a_{ij}) \in \mathrm{GL}(n, \mathbf{Z})$ with $a_{ij} = \delta_{i, \sigma^{-1} j}$.

A matrix like this is called a *permutation matrix*.

Exercise 0.9: Write down the matrices $A(\sigma)$ and verify that the map given by associating $\sigma \longmapsto A(\sigma)$ is in fact an isomorphism from $\mathfrak{S}_3$ to the subgroup of $\mathrm{GL}(3, \mathbf{R})$ consisting of the permutation matrices.

Chapter 1

Basic Algebraic Concepts for Group Representations

We first collect the basic definitions and concepts concerning representations of groups, using only algebraic methods. As is common, we introduce the concept of a *linear representation*, which is completely adequate and sufficient for the study of representations of finite groups. In a later chapter we will enrich the discussion with elements from topology to refine the notion of a representation to that of a *continuous representation*.

We start by stating the essential definitions and only afterwards illustrate these by some examples and constructions.

1.1 Linear Representations

Let G be a group and V a **K**-vector space. Here, in principle, we mean $\mathbf{K} = \mathbf{C}$, but most things go through as well for an algebraically closed field of characteristic zero. There are interesting phenomenae in the other cases, which are beyond the scope of this introduction. For an interested reader wanting some impression of this, we recommend Chapter 12 ff from Serre's book [Se], whose first Chapters together with §7-11 from Kirillov's book [Ki] are guidelines for this and our next section.

Definition 1.1: π is a *linear representation* of G in V iff π is a homomorphism of G into $\operatorname{Aut} V$ i.e. iff we have a map

$$\pi : G \longrightarrow \operatorname{Aut} V, \quad g \longmapsto \pi(g)$$

with

$$\pi(gg') = \pi(g)\pi(g') \quad \text{for all } g, g' \in G.$$

$\operatorname{Aut} V$ is equivalently denoted by $\operatorname{GL}(V)$ and stands for the group of all (linear) isomorphisms of V.

In the case of a finite-dimensional vector space V, say of dimension n, one says that π is *of degree n* or π is a *n-dimensional representation*.

Let $\mathcal{B} = (v_1, \ldots, v_n)$ be a basis of V. Then every $F \in \text{Aut } V$ is represented with respect to $\mathcal{B}$ by an invertible $n \times n$-matrix $A := M_{\mathcal{B}}(F)$, and we have an isomorphism of vector spaces $V \simeq \mathbf{K}^n$ and of groups $\text{Aut } V \simeq \text{GL}(n, \mathbf{K})$. Essentially equivalent to the definition above, but perhaps more down-to-earth for some readers is the definition:

An n-dimensional linear representation of a group G is a prescription π associating to each $g \in G$ a matrix $\pi(g) = A(g) \in \text{GL}(n, \mathbf{K})$ such that

$$A(gg') = A(g) \, A(g')$$

holds for all $g, g' \in G$.

As every group homomorphism transforms the neutral elements of the groups into one another, it is clear that this prescription associates the unit matrix E_n to the neutral element e of the group G, and that we have $\pi(e) = id_V$ in the situation of the general definition above.

If G is a matrix group, $G \subset \text{GL}(n, \mathbf{C})$, as in Section 0.1, obviously, we have the *natural representation* π_0 given by

$$\pi_0(A) = A$$

for each $A \in G$. Another even more ubiquous representation is the *trivial representation* $\pi = id$, associating the identity id_V to each $g \in G$.

Using the notion of group action introduced in Section 0.2, we can also say that a linear representation of G in V is the same thing as a left action of G on the vector space V.

At a first sight, it appears to be the main problem of representation theory to determine all representations of a given group. To make this a more sensible and accessible problem, one has to find some *building blocks* or *elementary particles* to construct general representations from, and, moreover, decide when representations are really different. We begin by searching for the building blocks.

Definition 1.2: Let π be a linear representation of G in V. π is called *irreducible*, iff there is no genuine π-invariant subspace V_0 in V.
A subspace $V_0 \subset V$ is π-*invariant* iff we have $\pi(g)v_0 \in V_0$ for all $g \in G$ and $v_0 \in V_0$.
If this is the case, $\pi_0 := \pi |_{V_0}$ is a representation of G in V_0 and this is called a *subrepresentation*.
So we can also say that π is irreducible, iff π has no genuine subrepresentation.

Next, we restrict our objects still more, in particular in view to the applications in physics.

We assume now that V is a *unitary* complex vector space, i.e. V is equipped with a scalar product

$$< ., . >: V \times V \longrightarrow \mathbf{C}, \quad (v, v') \longmapsto < v, v' >$$

which is

- linear in the second variable and antilinear in the first,
- *hermitian*: for every $v, v' \in V$ one has $< v, v' > = \overline{< v', v >}$, and
- *positive definite*: for every $v \in V$ one has $< v, v > \geq 0$ and $= 0$ exactly for $v = 0$.

For $V = \mathbf{C}^n$ we use the standard scalar product

$$< x, y >:= \sum_{i=1}^{n} \bar{x}_i y_i \quad \text{for all} \quad x, y \in \mathbf{C}^n.$$

Definition 1.3: A representation π of G in V is *unitary* iff each $\pi(g)$ is unitary, i.e. iff one has for every $v, v' \in V$ and every $g \in G$

$$< \pi(g)v, \pi(g)v' > = < v, v' > .$$

1.2 Equivalent Representations

If two representations π and π' of G in $\mathbf{K}$-vector spaces V resp. V' are given, it is obvious to look for G-equivariant maps $F : V \longrightarrow V'$:

Definition 1.4: A $\mathbf{K}$-linear map $F : V \longrightarrow V'$ is called an *intertwining operator* between π and π' iff one has for every $g \in G$

$$F\pi(g) = \pi'(g)F,$$

i.e. iff the following diagram commutes

$$
\begin{array}{ccc}
V & \xrightarrow{F} & V' \\
{\scriptstyle\pi(g)}\downarrow & & \downarrow{\scriptstyle\pi'(g)} \\
V & \xrightarrow{F} & V'.
\end{array}
$$

π and π' are called *equivalent* iff there is an isomorphism

$$F : V \longrightarrow V'$$

intertwining π and π'. In this case we write $\pi \sim \pi'$.

Remark 1.1: The space of intertwining operators between π and π' is again a vector space. It is denoted by $\mathrm{Hom}_G(V, V')$ or $\mathcal{C}(V, V')$. Moreover, we use the notation $\mathcal{C}(V) := \mathcal{C}(V, V)$ and $c(\pi, \pi') = c(V, V') = \dim \mathcal{C}(V, V')$. $c(\pi, \pi')$ is also called the *multiplicity* of π in π' and denoted by $\mathrm{mult}(\pi, \pi')$.

Representations π and π' with $c(\pi, \pi') = c(\pi', \pi) = 0$ are called *disjoint*.

At this point it is possible to state as our principal task:

Determine the *unitary* (linear) *dual* $\hat{G}$ of a given group G, i.e. the set of equivalence classes of unitary irreducible representations of G.

1.3   First Examples

We hope the reader gets the feeling that the following special cases prepare the ground
for the later more general considerations.

Example 1.1: In Section 1.1 we already mentioned that every matrix group G has its
natural representation π_0, i.e. for each real or complex matrix group $G \subset \mathrm{GL}(n, \mathbf{K})$ one
has the representation in $V = \mathbf{C}^n$ associating to every $A \in G$ the matrix A itself. These
natural representations are obviously unitary for $G = \mathrm{SO}(n)$ or $\mathrm{SU}(n)$ but in general
neither unitary nor irreducible, as the following example shows:

Example 1.2: Let g be an element of the group $G = \mathrm{SO}(2)$. Then we usually write

$$g = r(\vartheta) := \begin{pmatrix} \alpha & \beta \\ -\beta & \alpha \end{pmatrix}, \quad \alpha = \cos\vartheta, \beta = \sin\vartheta, \ \vartheta \in \mathbf{R}.$$

For each $k \in \mathbf{Z}$ we put

$$\chi_k(r(\vartheta)) := \exp(ik\vartheta).$$

Hence $\pi = \chi_k$ is a one-dimensional unitary representation in $V = \mathbf{C}$, as, by the addition
theorems for the trigonometric functions, we have

$$\chi_k(r(\vartheta)r(\vartheta')) = \chi_k(r(\vartheta + \vartheta')) = \exp(ik(\vartheta + \vartheta')) = \chi_k(r(\vartheta))\chi_k(r(\vartheta)).$$

For $U = (1/\sqrt{2})\begin{pmatrix} 1 & i \\ i & 1 \end{pmatrix}$ we have the relation

$$Ur(\vartheta)U^{-1} = \begin{pmatrix} \alpha - i\beta & 0 \\ 0 & \alpha + i\beta \end{pmatrix} = \begin{pmatrix} e^{-i\vartheta} & 0 \\ 0 & e^{i\vartheta} \end{pmatrix}.$$

This can be interpreted as follows. The matrix U is the matrix for an intertwining
operator F intertwining the natural representation π_0 with the representation π_1 on
$V = \mathbf{C}^2$ given by

$$\pi_1(r(\vartheta)) = \begin{pmatrix} e^{-i\vartheta} & 0 \\ 0 & e^{i\vartheta} \end{pmatrix}.$$

This representation is reducible and we see that the natural representation is equivalent
to a reducible one. Or we realize that the vectors

$$e_1' := Ue_1 = (1/\sqrt{2})\begin{pmatrix} 1 \\ i \end{pmatrix}, \quad e_2' := Ue_2 = (1/\sqrt{2})\begin{pmatrix} i \\ 1 \end{pmatrix}$$

span nontrivial π_0-invariant subspaces of $V = \mathbf{C}^2$. But if we look at π_0 as a real representation, i.e. in the $\mathbf{R}$-vector space $V_0 = \mathbf{R}^2$, π_0 is in fact irreducible.

Example 1.3: In Section 0.3 we introduced the symmetric group $\mathfrak{S}_n$. $G = \mathfrak{S}_3$ consists
of the elements

$$id = (1), (1,2) =: \sigma, (1,3), (2,3), (1,2,3) =: \tau, (1,3,2).$$

As there are relations like $\sigma^2 = id, \tau^3 = id$ and

$$\sigma\tau\sigma = \tau^2 = (1,3,2), \ \sigma\tau = (2,3), \tau\sigma = (1,3),$$

we see that each element $g \in \mathfrak{S}_3$ can be written as a product of (in general) several factors σ and τ. We indicate this here (and later on in a similar fashion in other cases) by the notation

$$\mathfrak{S}_3 = <\sigma, \tau>.$$

We easily find one-dimensional representations in $V = \mathbf{C}$, namely the trivial representation

$$\pi_1(g) = 1 \quad \text{for all } g \in \mathfrak{S}_3$$

and the signum representation

$$\pi_2(g) = \text{sgn } g \in \{\pm 1\} \quad \text{for all } g \in \mathfrak{S}_3.$$

One also has a natural three-dimensional representation π_0 given in $V = \mathbf{C}^3$ by the permutation matrices from Exercise 0.9. For instance, one has

$$\pi_0(\tau) = A(\tau) = \begin{pmatrix} 0 & 0 & 1 \\ 1 & 0 & 0 \\ 0 & 1 & 0 \end{pmatrix}.$$

This representation is also called the *permutation representation*. It can equivalently be described as follows.
We write

$$V = \mathbf{C}^3 = \sum_{i=1}^{3} e_i \mathbf{C} \ni w = e_1 z_1 + e_2 z_2 + e_3 z_3$$

with the canonical basis vectors $e_1 = {}^t(1, 0, 0), \ldots, e_3 = {}^t(0, 0, 1)$ and $z_1, z_2, z_3 \in \mathbf{C}$. Then π_0 is given by

$$\pi_0(g)w = \sum_i e_{g(i)} z_i = \sum_i e_i z_{g^{-1}(i)}.$$

Exercise 1.1: Verify that this is true.

π_0 is unitary (verify this also) but not irreducible: $V_1 := (e_1 + e_2 + e_3)\mathbf{C}$ is an invariant subspace, and as we have

$$\pi_0(g)(e_1 + e_2 + e_3) = e_{g(1)} + e_{g(2)} + e_{g(3)} = e_1 + e_2 + e_3,$$

$\pi_0 |_{V_1}$ is the trivial representation π_1. As one has for $w = \sum z_i e_i$

$$<e_1 + e_2 + e_3, w> = \sum z_i,$$

$V_3 := \{w, \sum z_i = 0\}$ is the orthogonal complement to V_1 and as $\sum_i z_i = \sum_i z_{g^{-1}(i)}$ holds for all $g \in \mathfrak{S}_3$, V_3 is an invariant subspace.

Exercise 1.2: Show that $a := e_1\zeta + e_2 + e_3\zeta^2$ and $b := e_1 + e_2\zeta + e_3\zeta^2$ is a basis for V_3, if $\zeta := e^{2\pi i/3}$. Determine the matrices of $\pi_2 := \pi_0 |_{V_3}$ with respect to this basis and show that π_2 is irreducible.

Remark 1.2: By the general methods to be developed later, it will be clear that all irreducible representations of $\mathfrak{S}_3$ are equivalent to π_1, π_2, or π_3, i.e. the equivalence classes of π_i ($i=1,2,3$) constitute the unitary dual of $\mathfrak{S}_3$.

Exercise 1.3: Prove this in a direct elementary way (see [FH] §1.3).

Example 1.4: A rather general procedure to construct representations is hidden in the following simple example.

We take $G \subset \mathrm{GL}(2, \mathbf{C})$ and as vector space V the space $V_m = \mathbf{C}\,[u, v]_m$ of homogeneous polynomials P of degree m in two variables u, v. By counting, we see dim $V = m + 1$. There are two essentially equivalent ways to define a representation of G on V_m:
For $P \in V_m$ and $g = \begin{pmatrix} a & b \\ c & d \end{pmatrix} \in G$ resp. $g^{-1} = \begin{pmatrix} a' & b' \\ c' & d' \end{pmatrix}$ we put

$$(\lambda_m(g)P)(u, v) := P(a'u + b'v, c'u + d'v)$$

or

$$(\rho_m(g)P)(u, v) := P(au + cv, bu + dv).$$

Obviously $\lambda(g)P$ and $\rho(g)P$ are both again homogeneous polynomials of degree m. The functional equations $\lambda(gg') = \lambda(g)\lambda(g')$ and $\rho(gg') = \rho(g)\rho(g')$ can be verified directly. It becomes also clear from the general discussion in the following Example 1.5.
We remark that λ_m and ρ_m are irreducible unitary representations of $G = \mathrm{SU}(2)$, if V_m is provided with a suitable scalar product. This will come out later in section 4.2.

Example 1.5: Let $\mathcal{X}$ be a G-set with a left G-action $x \longmapsto g \cdot x$ and $V = \mathcal{F}(\mathcal{X})$ a vector space of complex functions $f : \mathcal{X} \longrightarrow \mathbf{C}$ such that for $f \in V$ we also have $f_g \in V$ where f_g is defined by

$$f_g(x) = f(g^{-1}x).$$

Remark 1.3: The prescription

$$(\lambda(g)f)(x) := f(g^{-1}x)$$

defines a representation λ of G on V.

Proof: One has $(gg') \cdot x = g \cdot g' \cdot x$ and hence

$$\lambda(gg')f(x) = f((gg')^{-1} \cdot x) = f(g'^{-1}g^{-1} \cdot x) = f(g'^{-1} \cdot g^{-1} \cdot x)$$

and

$$\lambda(g)\lambda(g')f(x) = \lambda(g)f_{g'}(x) = f_{g'}(g^{-1} \cdot x) = f(g'^{-1} \cdot g^{-1} \cdot x).$$

An important special case is given by $\mathcal{X} = \mathbf{C}^n$ and $G \subset \mathrm{GL}(n, \mathbf{C})$ acting by matrix multiplication $g \cdot x = gx$ (as usual x is here a column) on V, a space of continuous functions in n variables or, say, a subspace of homogeneous polynomials of fixed degree (for $n = 2$ we then have the last example).

Analogously, one treats a set $\mathcal{X}$ with a right G-action $x \longmapsto x \cdot g$ and a vector space $V = \mathcal{F}(\mathcal{X})$ of complex functions f on $\mathcal{X}$ closed under $f \longmapsto f^g$ where f^g is defined by $f^g = f(x \cdot g)$.

Remark 1.4: The prescription

$$(\rho(g)f)(x) := f(x \cdot g)$$

defines a representation ρ of G on V.

Proof: One has $x \cdot (gg') = x \cdot g \cdot g'$ and hence

$$\rho(gg')f(x) = f(x \cdot (gg')) = f(x \cdot g \cdot g')$$

and

$$\rho(g)\rho(g')f(x) = \rho(g)f^{g'}(x) = f(x \cdot g \cdot g').$$

For $\mathcal{X} = \mathbf{C}^n$ and V again a space of functions in n variables, this time viewed as a row, and $G \subset \mathrm{GL}(n, \mathbf{C})$, we have again the action $x \longmapsto x \cdot g = xg$ by matrix multiplication. For $n = 2$ this leads naturally to the second version of Example 1.4.

Another (as, later to be seen, crucial) application of this procedure is given by making G act on itself, i.e. $\mathcal{X} = G$, by left or right translation and $V = L^2(G)$, the space of square integrable functions with respect to a left resp. right invariant Haar measure (these notions will be explained later). We then get the *left (right) regular representation* of G.

Example 1.6: Let $G = \mathrm{Heis}(\mathbf{R})$ be the Heisenberg group from Section 0.1, in the form of the group of triples $g = (\lambda, \mu, \kappa), \lambda, \mu, \kappa \in \mathbf{R}$ with the multiplication law

$$gg' = (\lambda + \lambda', \mu + \mu', \kappa + \kappa' + \lambda\mu' - \lambda'\mu),$$

and $V = L^2(\mathbf{R})$ the space of quadratic Lebesgue integrable functions f on $\mathcal{X} = \mathbf{R}$ (with norm $\| f \|_2 = (\int_{\mathbf{R}} \overline{f(x)}f(x)dx)^{1/2}$). Then for $m \in \mathbf{R}^*$, $\pi = \pi_m$ given by

$$\pi_m(g)f(x) := e^{2\pi i m(\kappa + (\lambda + 2x)\mu)} f(x + \lambda)$$

is a unitary irreducible representation of $\mathrm{Heis}(\mathbf{R})$. We call it the *Schrödinger representation*.

Exercise 1.4: Prove that π_m is a unitary representation. (The proof of the irreducibility will become easier later when we have prepared some more tools.)

Example 1.7: Prove (as **Exercise 1.5**) that the Weil group of $\mathbf{R}$, $W_{\mathbf{R}}$ from Exercise 0.2 in Section 0.1, has the irreducible representations $\pi = \pi_0$ and $\pi = \pi_m, m \in \mathbf{Z} \setminus \{0\}$ given by

$$\pi_0\left(\begin{pmatrix} 0 & 1 \\ -1 & 0 \end{pmatrix}\right) := \det \begin{pmatrix} 0 & 1 \\ -1 & 0 \end{pmatrix} = 1, \quad \pi_0\left(\begin{pmatrix} z & 0 \\ 0 & \bar{z} \end{pmatrix}\right) := \det \begin{pmatrix} z & 0 \\ 0 & \bar{z} \end{pmatrix} = z\bar{z},$$

and for $m \neq 0$

$$\pi_m\left(\begin{pmatrix} z & 0 \\ 0 & \bar{z} \end{pmatrix}\right) := \begin{pmatrix} z^m & 0 \\ 0 & \bar{z}^m \end{pmatrix}, \quad \pi_m\left(\begin{pmatrix} 0 & 1 \\ -1 & 0 \end{pmatrix}\right) := \begin{pmatrix} 0 & 1 \\ (-1)^m & 0 \end{pmatrix}.$$

1.4 Basic Construction Principles

Proceeding from a given representation (π, V) of a group G one can construct several new representations using standard notions from (multi-)linear algebra. We suppose that the reader has some knowledge of direct sums, tensor products and dual spaces. If this is not the case, we hope that one can follow in spite of it as we stay on a very elementary level and do enough paraphrasing to get some understanding. But sooner or later the reader should consult any book where these notions appear, for instance take Fischer's "Lineare Algebra" [Fi], Lang's "Algebra" [La], or the appendices to [Ki1] or [Be].

1.4.1 Sum of Representations

Let (π, V) and (π', V') be (linear) representations of the same group G. Then we define the *direct sum* $\pi \oplus \pi'$ of π and π' by

$$(\pi \oplus \pi')(g)(v \oplus v') := \pi(g)v \oplus \pi'(g)v' \quad \text{for all } v \oplus v' \in V \oplus V'.$$

For $V = \mathbf{K}^n$, $V' = \mathbf{K}^m$ and $\pi(g) = A(g) \in \mathrm{GL}(n, \mathbf{K})$, $\pi'(g) = A'(g) \in \mathrm{GL}(m, \mathbf{K})$ in matrix form, the sum is given by the block diagonal matrix

$$(\pi \oplus \pi')(g) = \begin{pmatrix} A(g) & 0 \\ 0 & A'(g) \end{pmatrix} \in \mathrm{GL}(n + m, \mathbf{K}).$$

1.4.2 Tensor Product of Representations

Let (π, V) and (π', V') be as above and $V \otimes V'$ the tensor product of V and V'. Then we define the *tensor product* $\pi \otimes \pi'$ of π and π' by

$$(\pi \otimes \pi')(g)(v \otimes v') := \pi(g)v \otimes \pi'(g)v' \quad \text{for all } v \otimes v' \in V \otimes V'.$$

For $V = \mathbf{K}^n$, $V' = \mathbf{K}^m$ and $\pi(g) = A(g) \in \mathrm{GL}(n, \mathbf{K})$, $\pi'(g) = A'(g) \in \mathrm{GL}(m, \mathbf{K})$, the tensor product is given by the *Kronecker product* of the matrices $A(g)$ and $A'(g)$

$$(\pi \otimes \pi')(g) = \begin{pmatrix} a_{1,1}A'(g) & \cdots & a_{1,n}A'(g) \\ & \cdots & \\ a_{n,1}A'(g) & \cdots & a_{n,n}A'(g) \end{pmatrix} \in \mathrm{GL}(nm, \mathbf{K}).$$

We will not go into the definition of the tensor product (as to be found in the afore mentioned sources) but simply recall the following. If V has a basis $(e_i)_{i \in I}$ and V' a basis $(f_j)_{j \in J}$, then $V \otimes V'$ has a basis $(e_i \otimes f_j)_{(i,j) \in I \times J}$.

In $V \otimes V'$ we have the rules

$$\begin{aligned} (v_1 + v_2) \otimes v' &= v_1 \otimes v' + v_2 \otimes v', \\ v \otimes (v_1' + v_2') &= v \otimes v_1' + v \otimes v_2', \\ \lambda(v \otimes v') &= (\lambda v) \otimes v' = v \otimes (\lambda v') \quad \text{for all } v, v_1, v_2 \in V, v', v_1', v_2' \in V', \lambda \in \mathbf{K}. \end{aligned}$$

It is clear that one can also define tensor products of more than two factors, and in particular tensor powers like $\otimes^3 V = V \otimes V \otimes V$ (It is not difficult to see that these products are associative.)

Moreover, there are subspaces of tensor powers with a preassigned symmetry property. In particular we are interested in p-fold symmetric tensors S^pV and p-fold antisymmetric tensors $\wedge^pV$.

For instance, if V is three-dimensional with basis (e_1, e_2, e_3),

- $\otimes^2V$ has dimension 9 and a basis

$$(e_1 \otimes e_1, e_1 \otimes e_2, e_1 \otimes e_3, e_2 \otimes e_1, \ldots, e_3 \otimes e_3),$$

- S^2V has dimension 6 and a basis

$$e_1 \otimes e_1, e_1 \otimes e_2 + e_2 \otimes e_1, e_1 \otimes e_3, e_2 \otimes e_2, e_2 \otimes e_3 + e_3 \otimes e_2, e_3 \otimes e_3.$$

- $\wedge^2V$ has dimension 3 and a basis

$$e_1 \wedge e_2 := e_1 \otimes e_2 - e_2 \otimes e_1, e_1 \wedge e_3 := e_1 \otimes e_3 - e_3 \otimes e_1, e_2 \wedge e_3 := e_2 \otimes e_3 - e_3 \otimes e_2.$$

More generally, one has symmetric and antisymmetric subspaces in $\otimes^pV$ with bases

$$e_{i_1} \cdot \ldots \cdot e_{i_p} := \sum_{g \in \mathfrak{S}_p} e_{i_{g(1)}} \otimes \cdots \otimes e_{i_{g(p)}}, \quad i_1 \leq \cdots \leq i_p,$$

resp.

$$e_{i_1} \wedge \cdots \wedge e_{i_p} := \sum_{g \in \mathfrak{S}_p} \text{sgn} g \; e_{i_{g(1)}} \otimes \cdots \otimes e_{i_{g(p)}}, \quad i_1 < \cdots < i_p.$$

For instance, if V is three-dimensional, S^pV can be identified with the space $\mathbf{K}[u, v, w]_p$ of homogeneous polynomials of degree p in three variables. And if π is a representation of G on V, the prescription $e_i \longmapsto \pi(g)e_i$ defines by linear continuation representations $S^p\pi$ and $\wedge^p\pi$ on S^pV resp. $\wedge^pV$.

It is one of the most important construction principles for finite-dimensional representations to take the natural representation π_0 and to get (other) irreducible representations by taking tensor products and *reduce* these to sums of irreducibles. This will be analysed later.

1.4.3 The Contragredient Representation

We use the notation V^* to denote the *dual space* of a given $\mathbf{K}$-vector space. This space consists of $\mathbf{K}$-linear maps $\varphi : V \longrightarrow \mathbf{K}$, is again a $\mathbf{K}$-vector space (of the same dimension as V if the dimension is finite), and is written as

$$V^* = \text{Hom}(V, \mathbf{K}) = \{\varphi : V \longrightarrow K, \varphi \; \mathbf{K} - \text{linear}\}.$$

We shall also use the notation

$$\varphi(v) =: \; <\varphi, v> \qquad \text{for all } \varphi \in V^*, v \in V$$

and, if $\dim V = n$ with basis $(e_1, \ldots, e_n)$, we denote $(e_1^*, \ldots, e_n^*)$ for the *dual basis* for V^*, i.e. with $<e_i^*, e_j> = \delta_{i,j} (= 1$ for $i = j$ and $= 0$ for $i \neq j)$.

If π is a representation of G on V, we define the *contragredient representation* π^* on V^* by

$$(\pi^*(g)\varphi)(v) := \varphi(\pi(g^{-1})v) \quad \text{for all } \varphi \in V^*, v \in V,$$

or equivalently by the condition

$$< \pi(g)^*\varphi, \pi(g)v >=< \varphi, v > \quad \text{for all } \varphi \in V^*, v \in V.$$

The proof that we really get a representation is already contained in the example 1.5 in section 1.3.

It is clear from linear algebra that in matrix form we have $A^*(g) = {}^t A(g^{-1})$.

Exercise 1.6: Verify this.

1.4.4   The Factor Representation

If (π_1, V_1) is a subrepresentation of the representation (π, V) of G, one has also an associated representation on the factor space V/V_1, called the *factor representation*.

There is still another essential construction principle, namely the concept of *induced representations*. It goes back to Frobenius and is also helpful in our present algebraic context. But we leave it aside here and postpone it for a more general treatment in Chapter 7.

1.5   Decompositions

We now have at hand procedures to construct representations and we have the notion of irreducible representations which we want to look at as the building blocks for general representations. To be able to do this, we still have to clarify some notions.

We defined in Section 1.1 that (π, V) is irreducible, iff there is no nontrivial subrepresentation. Now we add:

Definition 1.5: (π, V) is *decomposable*, iff there exists an invariant subspace $V_1 \subset V$ which admits an invariant complementary subspace V_2, i.e. with $V = V_1 \oplus V_2$. We then write $\pi = \pi_1 + \pi_2$.

Representations for which every nontrivial subrepresentation admits an invariant complement are called *completely reducible*.

Complete reducibility is desirable as it leads (at least in the case of finite-dimensional representations) immediately to the possibility to decompose a given representation into irreducible representations. Therefore facts like the following are very important.

Theorem 1.1: Let (π, V) be the representation of a finite group G and (π_1, V_1) a subrepresentation. Then there exists an invariant complementary subspace V_2 .

Proof: Let $<,>'$ be any scalar product for V (such a thing exists as is obvious at least for finite-dimensional $V \simeq \mathbf{K}^n$). We can find a G-invariant scalar product $<,>$, namely the one defined by

$$< v, v' >:= \sum_{g \in G} < \pi(g)v, \pi(g)v' >' \quad \text{for all } v, v' \in V.$$

Then $V_2 := \{v \in V, < v, v_1 > = 0 \quad \text{for all } v_1 \in V_1\}$ is a complementary subspace to V_1 which is π-invariant:
For $v \in V_2, g \in G$ one has, as $<,>$ is in particular $\pi(g^{-1})$-invariant,

$$< \pi(g)v, v_1 > = < \pi(g^{-1})\pi(g)v, \pi(g^{-1})v_1 > = < v, \pi(g^{-1})v_1 > = 0$$

if $v_1 \in V_1$ and hence also $\pi(g^{-1})v \in V_2$.

Remark 1.5: If G is any group and (π, V) a unitary representation of G, again every subrepresentation (π_1, V_1) has an invariant complement (π_2, V_2) given as in the proof above by $V_2 := V_1^{\perp} = \{v \in V, < v, v_1 > = 0 \quad \text{for all } v_1 \in V_1\}$ because V is already provided with a π-invariant scalar product.

Remark 1.6: If G is compact, the theorem holds as well, as the sum in the definition of the invariant scalar product can be replaced by an integral. This will be done later in Chapter 4.

Remark 1.7: As an immediate consequence we have that finite-dimensional unitary representations of an arbitrary group are completely reducible.

Remark 1.8: Every finite-dimensional representation of a finite (or compact) group is *unitarizable* as the representation space can be given a π-invariant scalar product by the Theorem 1.1 above.

Example 1.7: We have already seen an example for a decomposition of a representation in Example 1.3 in Section 1.3, namely the permutation representation of $G = \mathfrak{S}_3$. The standard example that **not** every representation is completely reducible and not even decomposable is the following.
Let $G = \mathbf{R}$ be the additive group of the real numbers and π a two-dimensional representation of G in $V = \mathbf{C}^2$ given by

$$\mathbf{R} \ni b \longmapsto \begin{pmatrix} 1 & b \\ 0 & 1 \end{pmatrix} =: A(b).$$

Here $V_1 := e_1 \mathbf{C}$ is an invariant subspace: we have $A(b)e_1 = e_1$ for all $b \in \mathbf{R}$. If

$$V_2 := v\mathbf{C} = \begin{pmatrix} x \\ y \end{pmatrix} \mathbf{C}$$

is an invariant complementary subspace, we have

$$A(b)v = \begin{pmatrix} x + by \\ y \end{pmatrix} = \lambda \begin{pmatrix} x \\ y \end{pmatrix}$$

for a certain $\lambda \in \mathbf{C}$, i.e. $x + by + \lambda x$ and $y = \lambda y$. As y must be nonzero, we have $\lambda = 1$. This leads to $by = 0$ for all b and hence a contradiction.

The following considerations are important in a general representation theory. But as we will not use them here so much they are only mentioned (for proofs see for instance [Ki] p.115 f).

Every representation (π, V) is either indecomposable or the sum of two representations $\pi = \pi_1 + \pi_2$. The same alternative applies to the representations π_1 and π_2 and so on. In the general case this process can continue infinitely. We say that π is *finite*, if every family of π-invariant subspaces V_i of V that is strictly monotone with respect to inclusion is finite. In this case there exists a strictly monotone finite collection of invariant subspaces

$$V = V_0 \supset V_1 \supset \cdots \supset V_n = \{0\}$$

such that that the representations π_i appearing on V_i/V_{i+1} are irreducible. One has the *Jordan - Hölder Theorem*: The length of such a chain of subspaces and the equivalence classes of the representations π_i (up to order) are uniquely defined by the equivalence class of π.

If we want to discuss the question whether the decomposition of a given representation (π, V) is the direct sum of irreducibles, we need criteria for irreducibility. A central criterium is a consequence of the following famous statement.

Theorem 1.2: (*Schur's Lemma*): If two linear representations (π, V) and (π', V') are irreducible, then every intertwining operator $F \in \mathcal{C}(\pi, \pi')$ is either zero or invertible. If $V = V', \pi = \pi'$ and $\dim V = n$, then F is a *homothety*, i.e. $F = \lambda \, id, \lambda \in \mathbf{C}$.

Proof: i) As defined in Section 1.2 an intertwining operator for π and π' is a linear map $F : V \longrightarrow V'$ with

(1.1) $\pi'(g)F(v) = F(\pi(g)v)$ for all $g \in G, v \in V$.

Then $\operatorname{Ker} F = \{v \in V, F(v) = 0\}$ is a π-invariant subspace of V because for $v \in \operatorname{Ker} F$ we have $F(v) = 0$ and by (1.1) $F(\pi(g)v) = \pi'(g)F(v) = 0$, i.e. $\pi(g)v \in \operatorname{Ker} F$. By a similar reasoning it is clear that $\operatorname{Im} F = \{F(v), v \in V\}$ is an invariant subspace of V'. The irreducibility of π implies $\operatorname{Ker} F = \{0\}$ or $\operatorname{Ker} F = V$ and by the irreducibility of π' we have $\operatorname{Im} F = \{0\}$ or $= V'$. Hence F is the zero map or an isomorphism.

ii) If $V = V'$, the endomorphism $F : V \longrightarrow V$ has an eigenvalue $\lambda \in \mathbf{C}$ (as the characteristical polynomial $P_F(t) := \det(F - tE) \in \mathbf{C}[t]$ has a zero $\lambda \in \mathbf{C}$ by the fundamental theorem of algebra). Hence $F' := F - \lambda E$ is a linear map with $\operatorname{Ker} F' \neq \{0\}$. From part i) we then deduce $F' = 0$, i.e. $F = \lambda E$.

Remark 1.9: The first part of the Theorem is also expressed as follows: Two irreducible representations of the same group are either equivalent or *disjoint*.
The second part can be written as $\mathcal{C}(\pi) = \mathbf{C}$.

There are several more refined versions of Schur's Lemma which come up if one sharpens the notion of a linear representation by adding a continuity requirement as we shall do later on. For the moment, though it is somewhat redundant, we add another version which we take over directly from Wigner ([Wi] p. 75) because it contains an irreducibility criterium used in many physics papers and its proof is a very instructive example for matrix calculus.

Theorem 1.3 (*the finite-dimensional unitary Schur*): Let $(\pi, \mathbf{C}^n)$ be a unitary matrix representation of a group G, i.e. with $\pi(g) = A(g) \in U(n)$. Let $M \in GL(n, \mathbf{C}^n)$ be a matrix commuting with all $A(g)$, i.e.

(1.2) $$MA(g) = A(g)M \quad \text{for all } g \in G.$$

Then M is a scalar multiple of the unit matrix, $M = \lambda E_n, \lambda \in \mathbf{C}$.

Proof: i) First, we show that we may assume M is a Hermitian matrix: We take the adjoint (with $A \longmapsto A^* := {}^t\bar{A}$) of the commutation relation (1.2) and get

$$A(g)^* M^* = M^* A(g)^*,$$

multiply this from both sides by $A(g)$ and, using the unitarity $A(g)A(g)^* = E_n$, get

$$M^* A(g) = A(g) M^*.$$

Hence, if M commutes with all $A(g)$, not only M^* does but also the matrices

$$H_1 := M + M^*, \ H_2 := i(M - M^*),$$

which are Hermitian. It is therefore sufficient to show that every Hermitian matrix, which commutes with all $A(g)$, is a scalar matrix, since if H_1 and H_2 are multiples of E_n, so must be $2M = H_1 - iH_2$.

ii) If M is Hermitian, one of the fundamental theorems from linear algebra (see for instance [Ko], p.256 or [Fi] Kapitel 4) tells that M can be diagonalized, i.e. conjugated into a diagonal matrix D by a matrix $U \in U(n)$

$$UMU^{-1} = D.$$

Then we put $\tilde{A}(g) := UA(g)U^{-1}$. $\tilde{A}(g)$ is unitary as can be easily verified. From the commutation relation (1.2), we have $D\tilde{A}(g) = \tilde{A}(g)D$. With $D = (d_{i,j})$, $d_{i,j} = 0$ for $i \neq j$, $\tilde{A}(g) = (\tilde{a}_{i,j})$, this leads to

$$d_{i,i}\tilde{a}_{i,j} = \tilde{a}_{i,j}d_{j,j}.$$

Hence for $d_{i,i} \neq d_{j,j}$ we have $\tilde{a}_{i,j} = \tilde{a}_{j,i} = 0$, that is, all $\tilde{A}(g)$ must have zeros at all intersections of rows and columns, where the diagonal elements are different. This would mean that $g \longmapsto \tilde{A}(g)$ resp. $g \longmapsto A(g)$ are reducible representations. Hence, as this is not the case, we have that all diagonal elements are equal and D is a scalar matrix $\lambda E_n, \lambda \in \mathbf{C}$.

The proof shows that we can sharpen the statement from the Theorem to the irreducibiliy criterium:

Corollary: If there exists a nonscalar matrix which commutes with all matrices of a finite-dimensional unitary representation π, then the representation is reducible. If there exists none, it is irreducible.

This Corollary will help to prove the irreducibility in many cases. Here we show the usefulness of Schur's Lemma by proving another fundamental fact (later we shall generalize this to statements valid also in the infinite-dimensional cases):

Theorem 1.4: Any finite-dimensional irreducible representation of an abelian group is one-dimensional.

Proof: Let (π, V) be a representation of G. Then

$$F : V \longrightarrow V, \quad v \longmapsto \pi(g_0)v$$

is an intertwining operator for each $g_0 \in G$, as we have

$$
\begin{aligned}
\pi(g)F(v) &= \pi(g)\pi(g_0)v \\
&= \pi(gg_0)v, && \text{as } \pi \text{ is a representation,} \\
&= \pi(g_0 g)v, && \text{as } G \text{ is abelian,} \\
&= \pi(g_0)\pi(g)v, \\
&= F(\pi(g)v).
\end{aligned}
$$

The second part of Schur's Lemma says that F is a homothety, i.e. we have a $\lambda \in \mathbf{C}$ with

$$F(v) = \pi(g_0)v = \lambda v \quad \text{for all } v \in V.$$

As $g_0 \in G$ was chosen arbitrarily, we see that $V_0 = v\mathbf{C}$ is an invariant subspace of V. For an irreducible representation π this implies $V = V_0$, i.e. V is one-dimensional.

To finish our statements on decompositions in this section, we state here without proofs (for these see [Ki] p.121ff) some standard facts about the kind of uniqueness one can expect for a decomposition of a completely reducible representation. By the way, this theorem is a reflex of a central theorem from Algebra about the structure of modules over a ring (a representation (π, V) will be seen also as module over the *group ring* $\mathbf{C}[G]$).

Reminding the notions of disjoint representations from Section 1.2 and of finite representations from above, we add the following notion.

Definition 1.6: A representation π is said to be *primary*, iff it cannot be represented as a sum of two disjoint representations.

As an example one may think here of a representation of the form π^k, $k \in \mathbf{N}$, the sum of k representations (π, V) given on the k-th sum $V^k = V \oplus \cdots \oplus V$.
The central fact can be stated like this:

Theorem 1.5: Let π be a completely reducible finite representation in the space V. Then there exists a unique decomposition of V into a sum of invariant subspaces W_j, for which the representations $\pi_j = \pi \mid_{W_j}$ are primary and pairwise disjoint. Every invariant subspace $V' \subset V$ has the property $V' = \oplus_j(V \cap W_j)$.

Among other things, the proof uses the notion of *projection operators*, known from linear algebra. As we have to use these here later on, we recall the following concept.

Definition 1.7: Suppose that the space V of a representation π is decomposed into a direct sum of invariant subspaces V_i, i.e., $V = \oplus_{i=1}^{m} V_i$. Then the *projection operator* P_i *onto* V_i parallel to the remaining V_j is defined by $P_i v = v_i$ if $v = (v_1, \ldots, v_i, \ldots, v_m)$, $v_i \in V_i$, $i = 1, \ldots, m$.

P_i has the following properties

- a) $P_i^2 = P_i$,
- b) $P_i P_j = P_j P_i = 0$ for $i \neq j$,
- c) $\sum_{i=1}^{m} P_i = 1$,
- d) $P_i \in \mathcal{C}(\pi)$.

One can prove that every collection of operators (i.e. linear maps) in V that enjoys these properties produces a decomposition of V into a sum of invariant subpaces $V_i = P_i V$.

Later on, we shall treat explicit examples of decompositions.

1.6 Characters of Finite-dimensional Representations

The following notions need the finite-dimensionality of a representation. For the infinite-dimensional cases there are generalizations using distributions and/or infinitesimal methods which we only touch briefly later on because they are beyond the scope of this book. So in this Section (π, V) is always a finite-dimensional complex representation of the group G with $\dim V = n$.

Definition 1.8: We call *character* of π the complex function χ_π (or equivalently χ_V) on G given by
$$\chi_\pi(g) := \operatorname{Tr} \pi(g) \quad \text{for all } g \in G.$$

As usual Tr denotes the *trace*. If π is given by the matrices $A(g) = (a_{ij}(g)) \in \mathrm{GL}(n, \mathbf{C})$, one has
$$\chi_\pi(g) = \sum_{i=1}^{n} a_{ii}.$$

Conjugate matrices $A \sim A'$, i.e. with $A' = TAT^{-1}, T \in \mathrm{GL}(n, \mathbf{C})$ have the same trace. Therefore the trace of a representation is well defined independently of the choice of the matrix representation. And for conjugate $g, g' \in G$ we have
$$\chi_\pi(g) = \chi_\pi(g').$$

This is also expressed by saying the character is a *class function*.
Obviously one has for the neutral element e of G
$$\chi_\pi(e) = n = \dim V.$$

We assume moreover that π is unitary (we shall see soon that this is not really a restriction if the group is finite or compact). Then $\pi(g)$ is in matrix form $A(g)$ conjugate to a diagonal matrix $D = D(\lambda_1, \ldots, \lambda_n)$, where $\lambda_1, \ldots, \lambda_n$ are the eigenvalues of $A(g)$. Hence:

Remark 1.10: $\chi_\pi(g) = \sum_{i=1}^{n} \lambda_i$.

If G is finite, each element g has finite order and so has $\pi(g)$. Thus the eigenvalues λ_i must have $|\lambda_i| = 1$ and we have $\lambda_i^{-1} = \overline{\lambda_i}$.

Using this, from Remark 1.10 we deduce an important fact.

Remark 1.11: $\chi_\pi(g^{-1}) = \overline{\chi_\pi(g)}$.

Finally we state two very useful formulae.

Theorem 1.6: Let (π, V) and (π', V) be two finite-dimensional representations of a group G. Then we have

(1.3)
$$\chi_{\pi \oplus \pi'} = \chi_\pi + \chi_{\pi'}$$
(1.4)
$$\chi_{\pi \otimes \pi'} = \chi_\pi \times \chi_{\pi'}.$$

Proof: The first equation is simply the expression of the fact that the trace of a block matrix $\begin{pmatrix} A & \\ & A' \end{pmatrix}$ is the sum of the traces of the two matrices A and A'. The second relation can be seen by inspection of the trace of the Kronecker product in 1.4.2.

Exercise 1.7: Let (π, V) be a finite-dimensional representation of G. Show that for every $g \in G$ one has the equations

$$\chi_{\wedge^2 \pi}(g) = (1/2)(\chi_\pi(g)^2 - \chi_\pi(g^2)),$$

$$\chi_{S^2 \pi}(g) = (1/2)(\chi_\pi(g)^2 + \chi_\pi(g^2)).$$

As we shall find out in the next chapters, the irreducible representations of finite and compact groups (where in the last case we have to add a continuity condition) are all finite-dimensional. Then the characters of these representations already contain enough information to fix these representations in a way which we will discuss more precisely now.

Chapter 2

Representations of Finite Groups

As we are mainly interested in continuous groups like SO(3) etc., this topic plays no central rôle in our text. But here we can learn a lot about how representation theory works. Thus we present the basic facts. We will not do it by using the abstract notions offered by modern algebra (as for instance in [FH]), where eventually the whole theory can be put into the nutshell of one page. Instead we follow very closely the classic presentation from the first pages of Serre's book [Se].

In this chapter G is always a finite group of order $\#G = m$.

2.1 Characters as Orthonormal Systems

We introduce the *group ring* (or *group algebra*): Let $\mathbf{K}$ be a field and G a group with $\#G = m$. Then we denote the space of maps from G to $\mathbf{K}$ by

$$\mathbf{K}G := \{u : G \longrightarrow \mathbf{K}\}.$$

$\mathbf{K}G$ is a $\mathbf{K}$-vector space as we have an addition $u + u'$ defined by

$$(u + u')(g) := u(g) + u'(g) \quad \text{for all } g \in G$$

and scalar multiplication λu by $(\lambda u)(g) := \lambda u(g)$ for $u \in \mathbf{K}G$ and $\lambda \in \mathbf{K}$. But we also have a *multiplication uu'* defined by

$$(uu')(g) := \sum_{a,b \in G, ab=g} u(a)u'(b).$$

Remark 2.1: One can show that $\mathbf{K}G$ is an *associative $\mathbf{K}$-algebra* (for this notion see Section 6.1).

There is an alternative way to describe $\mathbf{K}G$, namely as a set of formal sums $\sum_{g \in G} u(g)g$ built from the values $u(g)$ of u in $g \in G$, i.e.

$$\mathbf{K}G := \{\sum_{g \in G} u(g)g, u(g) \in \mathbf{K}\}.$$

Here we have addition

$$\sum u(g)g + \sum u'(g)g := \sum (u(g) + u'(g))g$$

and multiplication given by

$$\sum u(g)g \cdot \sum u'(g)g := \sum v(g)g, \quad \text{with } v(g) := \sum_{ab=g} u(a)u'(b).$$

As we restrict our interest in this text to complex representations, in the sequel we write $\mathcal{H} = \mathbb{C}G$ and $\mathcal{H}_0 = \mathbb{C}_{cl}G$ for the algebra of *class functions*, i.e. with $u(g) = u(g')$ for $g \sim g'$. If (π, V) is a representation of G, by Remark 1.8 in 1.5 we may assume that it is unitary. Conjugate elements of G have the same character. Hence the character $u = \chi_\pi$ is not only an element in $\mathcal{H}$ but also in $\mathcal{H}_0$.

Our first aim is to prove that, for a complete system $\pi_1, \ldots, \pi_h$ of irreducible representations, the characters $\chi_{\pi_1}, \ldots, \chi_{\pi_h}$ are an orthonormal system (=: ON-system) in $\mathcal{H}$. Here the scalar product $<,>$ is given by

$$< u, v > := (1/m) \sum_{t \in G} \overline{u(t)} v(t) \quad \text{for all } u, v \in \mathcal{H}.$$

(It is clear that this is a scalar product.)
We will also use the following bilinear form

$$(u, v) := (1/m) \sum_{t \in G} u(t^{-1}) v(t).$$

If $u = \chi_\pi$ is a character, by the Remark 1.11 in 1.6 we have $\overline{u(t)} = u(t^{-1})$ and hence $(\chi_\pi, v) = < \chi_\pi, v >$ for $v \in \mathcal{H}$.

Theorem 2.1: Let π and π' be irreducible representations of G and $\chi = \chi_\pi$ as well as $\chi' = \chi_{\pi'}$ their characters. Then we have

$$< \chi, \chi' > = 0, \text{ if } \pi \not\sim \pi',$$
$$< \chi, \chi > = 1.$$

This is also written as $< \chi_\pi, \chi'_\pi > = \delta_{\pi, \pi'}$.

The proof will be a consequence of two propositions.

Proposition 2.1: Let π, π' be irreducible, $F : V \longrightarrow V'$ a linear map, and

$$F^0 := (1/m) \sum_{g \in G} \pi'(g^{-1}) F \pi(g).$$

Then one has

1) $F^0 = 0$, if $\pi \not\sim \pi'$ ("case 1"),
2) $F^0 = \lambda \, id$, if $V = V', \pi = \pi'$, where $\lambda = (1/\dim V) \text{Tr} \, F$ ("case 2").

Proof: F^0 is an intertwining operator for π and π': We have

$$\pi'(g)^{-1}F^0\pi(g) \;=\; (1/m)\sum_{t\in G}\pi'(g)^{-1}\pi'(t)^{-1}F\pi(t)\pi(g)$$

$$=\; (1/m)\sum_{t\in G}\pi'(tg)^{-1}F\pi(tg) = F^0.$$

By application of Schur's Lemma, we have $F^0 = 0$ or $F^0 = \lambda\,id$ in the respective cases, and because of

$$\mathrm{Tr}\,F^0 = (1/m)\sum_{t\in G}\mathrm{Tr}(\pi(t^{-1})F\pi(t)) = \mathrm{Tr}\,F$$

and $\mathrm{Tr}\,id = n = \dim V$, we get $\lambda = (1/n)\mathrm{Tr}\,F$.

Proposition 2.2: For π and π' as in Proposition 2.1 given in matrix form by $\pi(g) = A(g)$ and $\pi'(g) = B(g)$, we have

$$(1/m)\sum_{t\in G}B_{ij}(t^{-1})A_{kl}(t) \;=\; 0 \quad\text{for all } i,j,k,l \text{ in case } 1,$$

$$(1/m)\sum_{t\in G}A_{ij}(t^{-1})A_{kl}(t) \;=\; (1/n)\delta_{il}\delta_{jk} \quad\text{for all } i,j,k,l \text{ in case } 2.$$

Proof: With $F = (F_{ij})$ Proposition 2.1 expressed in matrix form

$$(1/m)\sum_{t\in G}\sum_{jk}B_{ij}(t^{-1})F_{jk}A_{kl}(t) = F_{il}^0$$

states

$$(1/m)\sum_{t\in G}\sum_{jk}B_{ij}(t^{-1})F_{jk}A_{kl}(t) = 0 \quad\text{for all } i,l \text{ in case } 1,$$

and this requires that all the coefficients of all F_{jk} are zero. Hence, we have the first relation. In the second case we have

$$(1/m)\sum_{t\in G}\sum_{jk}A_{ij}(t^{-1})F_{jk}A_{kl}(t) = \lambda\delta_{il} = (1/n)\sum_{jk}F_{jk}\delta_{jk}\delta_{il}.$$

Comparing the coefficients of F_{jk} on both sides leads to the second relation.

Proof of the Theorem: We have $\chi'(t) = \sum B_{ii}(t)$ and $\chi(t) = \sum A_{ii}(t)$ and hence

$$<\chi,\chi> \;=\; (1/m)\sum_{t\in G}\chi(t^{-1})\chi(t) = (1/m)\sum_{t\in G}\sum_{ij}A_{ii}(t^{-1})A_{jj}(t)$$

$$=\; (1/n)\sum_{ij}\delta_{ij}\delta_{ij} = 1$$

by the second relation in Proposition 2.2, and

$$<\chi,\chi'> = (1/m)\sum_{t\in G}\chi'(t^{-1})\chi(t) = (1/m)\sum_{t\in G}\sum_{ij}B_{ii}(t^{-1})A_{jj}(t)$$

by the first relation in Proposition 2.2.

Corollary 1: Let (π, V) be a finite-dimensional representation of G with character $\chi := \chi_\pi$ and (π_i, V_i) irreducible representations with characters $\chi_i := \chi_{\pi_i}$ and

$$V = V_1 \oplus \cdots \oplus V_k.$$

Let moreover be (π', V') an irreducible representation with character $\chi' = \chi_{\pi'}$. Then we have

(2.1) $$\# \{V_i, \ V_i \simeq V'\} = \ <\chi', \chi> .$$

From 1.2 we know that this number is also recognized as $\text{mult}(\pi', \pi)$.

Proof: By Theorem 1.6 in 1.6 we have $\chi = \sum_i \chi_i$. Theorem 2.1 in 2.1 says that the summands in

$$<\chi', \chi> = \sum_i <\chi', \chi_i>$$

are either 0 or 1 if $\chi' \not\sim \chi_i$ or $\chi' \sim \chi_i$. Hence $<\chi', \chi>$ counts the number of irreducible components π_i contained in π which are equivalent to π'.

Corollary 2: $\text{mult}(\pi', \pi)$ is independent of the decomposition as in Corollary 1.

Proof: $<\chi', \chi>$ does not depend on the decomposition.

Corollary 3: If two representations π and π' have the same character $\chi_\pi = \chi_{\pi'}$, they are equivalent.

Proof: By Corollary 1 the multiplicities are equal for all irreducible components of π and π' and so both representations are equivalent.

Corollary 4: If π decomposes into the irreducible representations $\pi_1, \ldots, \pi_h$ with respective multiplicities m_i, i.e. one has

$$V = m_1 V_1 \oplus \cdots \oplus m_h V_h,$$

or, as we also write

$$V = V_1^{m_1} \oplus \cdots \oplus V_h^{m_h},$$

one has

$$<\chi, \chi> = \sum_{i=1}^{h} m_i^2.$$

Proof: This follows immediately from Theorem 2.1 in 2.1, as $\chi = \sum m_i \chi_i$ and the χ_i are an ON-system.

This Corollary specializes to the following important irreducibility criterium.

Corollary 5: π is irreducible exactly if $<\chi, \chi> = 1$.

2.2   Regular Representation and its Decomposition

The regular representation λ of the finite group G with $\#G = m$ is defined on the vector space V spanned by a basis e_t indexed by the elements t of G, i.e. $V = <e_t>_{t \in G}$ by the prescription

$$\lambda(g)e_t := e_{gt},$$

i.e. we have $\lambda(g)v = \sum_{t \in G} z_{(g^{-1}t)}e_t$ for $v = \sum_{t \in G} z_t e_t, z_t \in \mathbf{C}$.
V can be identified with $\mathcal{H}$ concerning the vector space structure.
Then we have

$$\lambda(g)u = \sum_{t \in G} u(g^{-1}t)t \quad \text{for all} \ \ u = \sum_{t \in G} u(t)t \in \mathcal{H}.$$

If $g \neq e$, we have $gt \neq t$ for all $t \in G$, which shows that the diagonal elements of $\lambda(g)$ are zero. In particular we have $\operatorname{Tr} \lambda(g) = 0$ and $\operatorname{Tr} \lambda(e) = \operatorname{Tr} E_m = m$. So we have proven:

Proposition 2.3: The character χ_λ of the regular representation λ of G is given by

$$\begin{aligned} \chi_\lambda(e) &= \#G = m, \\ \chi_\lambda(g) &= 0 \quad \text{for all} \ g \neq e. \end{aligned}$$

Proposition 2.4: Every irreducible representation π_i is contained in the regular representation with multiplicity equal to its dimension n_i, i.e. $\operatorname{mult}(\pi_i, \lambda) = n_i = \dim V_i$.

Proof: According to Corollary 1 in 2.1 we have

$$\operatorname{mult}(\pi_i, \lambda) = <\chi_{\pi_i}, \chi_\lambda>$$

and by the definition of the scalar product also

$$\operatorname{mult}(\pi_i, \lambda) = (1/m) \sum_{t \in G} \chi_\lambda(t^{-1})\chi_{\pi_i}(t).$$

By Proposition 2.3 and the general fact that the character of the neutral element equals the dimension, we finally get

$$\operatorname{mult}(\pi_i, \lambda) = (1/m) \cdot m \cdot \chi_{\pi_i}(e) = n_i.$$

Proposition 2.5: Let $\pi_1, \ldots, \pi_h$ be a complete system representing the equivalence classes of irreducible representations of G. Then the dimensions $n_i = \dim V_i$ satisfy the relations

a) $\sum_{i=1}^{h} n_i^2 = m$,

and, if $t \in G, t \neq e$,

b) $\sum_{i=1}^{h} n_i \chi_{\pi_i}(t) = 0.$

Proof: By Proposition 2.4 we have $\chi_\lambda(t) = \Sigma n_i \chi_i(t)$ for all $t \in G$. Taking $t = e$ we obtain a) and for $t \neq e$ we obtain b).

This result is very useful for the determination of the irreducible representations of a finite group: suppose one has constructed some mutually nonequivalent irreducible representations of dimensions $n_1, \ldots, n_h$. In order that they be **all** irreducible representations (up to equivalence) it is necessary and sufficient that one has $n_1^2 + \cdots + n_h^2 = m$.

In 1.3 we already discussed the example $G = \mathfrak{S}_3$. Here we have $m = 6$. And we singled out the two one-dimensional representations $\pi_1 = 1$ and $\pi_2 = sgn$ and the two-dimensional representation π_3. As we have here $1^2 + 1^2 + 2^2 = 6 = m$, we see that we got all irreducible representations. Moreover, we have a simple way to check the irreducibility of π_3 by using our Corollary 5. in 2.1

As an easy **Exercise 2.1**, we recommend to do this and to determine the characters of all the representations π_1, π_2, π_3, and π_0 introduced there. One can see that that one has $\pi_0 = \pi_1 + \pi_3$ consistent with our Corollary 4 in 2.1.

One can prove more facts in this context, for instance the dimensions n_i all divide the order m of G (see [Se] p.53]).

2.3 Characters as Orthonormal Bases and Number of Irreducible Representations

We sharpen Theorem 2.1 from 2.1 to the central fact that the characters $\chi_1, \ldots, \chi_h$ belonging to a complete set of equivalence classes of irreducible representations of G form an orthonormal basis of the space of class functions $\mathcal{H}_0$. We recall that a function f defined on G is a *class function* if one has $f(g) = f(tgt^{-1})$ for all $g, t \in G$.

Proposition 2.6: Let f be a class function on G and (π, V) a representation of G. Let π_f be the endomorphism of V defined by

$$\pi_f := \sum_{t \in G} f(t)\pi(t).$$

If π is irreducible with $\dim V = n$ and character χ, then π_f is a scalar multiple of the identity, i.e. $\pi_f = \lambda \, id_V$, with

$$\lambda = (1/n) \sum_{t \in G} \chi(t) f(t) = (m/n) < \bar{\chi}, f > .$$

Proof: π_f is an intertwining operator for π: We have

$$\pi(s)^{-1}\pi_f\pi(s) = \sum_{t \in G} f(t)\pi(s)^{-1}\pi(t)\pi(s) = \sum_{t \in G} f(t)\pi(s^{-1}ts).$$

As f is a class function, we get by $t \longmapsto s^{-1}ts = t'$

$$\pi(s)^{-1}\pi_f\pi(s) = \sum_{t' \in G} f(t')\pi(t') = \pi_f.$$

By the second part of Schur's Lemma, $\pi_f = \lambda id$ is a scalar multiple of the identity. We have $\mathrm{Tr}\, \lambda \, id = n\lambda$ and

$$\mathrm{Tr}\, \pi_f = \Sigma f(t)\mathrm{Tr}\, \pi(t) = \Sigma f(t)\chi(t).$$

Hence

$$\lambda = (1/n)\Sigma f(t)\chi(t) = (m/n) < \bar{\chi}, f > .$$

As the characters are class functions, they are elements of $\mathcal{H}_0$.

Theorem 2.2: The characters $\chi_1, \ldots, \chi_h$ form an ON-basis of $\mathcal{H}_0$.

Proof: Theorem 2.1 from Section 2.1 says that the (χ_i) form an ON-system in $\mathcal{H}_0$. We have to prove that they generate $\mathcal{H}_0$. That is, we have to show that every $f \in \mathcal{H}_0$, orthogonal to all $\bar{\chi}_i$, is zero. Let f be such an element. For each representation π of G, we put $\pi_f = \Sigma f(t)\pi(t)$ as above. Proposition 2.6 shows that π_f is zero if π is irreducible. But as each π may be decomposed into irreducible representations, π_f is always zero. We apply this to the regular representation , i.e. to $\pi = \lambda$, and compute for the first basis vector of the representation space for λ

$$\pi_f e_1 = \sum_t f(t)\lambda(t)e_1 = \sum_t f(t)e_t.$$

Since π_f is zero, we have $\pi_f e_1 = 0$ and this requires $f(t) = 0$ for all $t \in G$. Hence f is zero.

As each element of $\mathcal{H}_0$ is fixed by associating a complex number to each conjugacy class $g_\sim$ of G, the dimension of $\mathcal{H}_0$ as a **C**-vector space equals the number of conjugacy classes. Hence we have:

Corollary 1: The number of irreducible representations of G (up to equivalence) is equal to the number of conjugacy classes.

Corollary 2: Let g be an element of G and $c(g) := \#\{g'; g' \sim g\}$ the number of elements in the conjugacy class of g. Then we have

$$
\begin{array}{llll}
a) & \sum_{i=1}^{h} \overline{\chi_i(g)}\chi_i(g') & = & m/c(g) \quad \text{if} \quad g' = g, \\
b) & & = & 0 \qquad\quad \text{if} \quad g' \not\sim g.
\end{array}
$$

Proof: Let f_g be the characteristic function of the class of g, i.e. $f_g(g') = 1$ if $g' \sim g$ and $= 0$ else. Since we have $f_g \in \mathcal{H}_0$, by the Theorem, we can write

$$f_g = \sum_{i=1}^{h} \lambda_i \chi_i \text{ with } \lambda_i = <\chi_i, f_g> = (c(g)/m)\overline{\chi_i(g)}.$$

For each $t \in G$ we then have

$$f_g(t) = (c(g)/m)\sum_{i=1}^{h} \overline{\chi_i(g)}\chi_i(t).$$

And because of $f_g(t) = 1$ for $t \sim g$ and $= 0$ for $t \not\sim g$, we get a) and b).

Remark 2.2: We recall the statement from 0.3 that the number of conjugacy classes of the symmetric group $\mathfrak{S}_n$ equals the number $p(n)$ of partitions of n.
It is a nice excercise to redo the example of $G = \mathfrak{S}_3$ in the light of the general facts now at hand: We have $p(3) = 3$ and thus know of three irreducible representations π_1, π_2, π_3 with respective characters χ_1, χ_2, χ_3. Moreover we know that $G = \mathfrak{S}_3$ can be generated by the transposition $\sigma = (1, 2)$ and the cycle $\tau = (1, 2, 3)$. The three conjugacy classes are represented by $e = id, \sigma$, and τ. We have

$$\sigma^2 = e, \ \tau^3 = e, \sigma\tau = \tau^2\sigma.$$

Hence, for each representation π the equation $\pi(\sigma)^2 = id$ holds. So there are two one-dimensional representations, namely $\pi_1 = 1$ and $\pi_2 = sgn$ with

$$\chi_1(e) = 1, \quad \chi_1(\sigma) = 1, \quad \chi_1(\tau) = 1,$$
$$\chi_2(e) = 1, \quad \chi_2(\sigma) = -1, \quad \chi_2(\tau) = 1.$$

The third representation π_3 must have $\dim \pi_3 = n$ with $1^2 + 1^2 + n^2 = 6$, i.e. $n = 2$, as we already concluded at the end of 2.2. The value of χ_3 can be deduced (Proposition 2.4 in 2.2) from the relation with the character χ_λ of the regular representation λ

$$\chi_1 + \chi_2 + 2\chi_3 = \chi_\lambda.$$

As we have $\chi_\lambda(e) = 6$ and $\chi_\lambda(g) = 0$ for $g \neq e$, we conclude

$$\chi_3(e) = 2, \chi_3(\sigma) = 0, \chi_3(\tau) = -1.$$

These formulae should also be read off from the results of Exercise 2.1.

There is a lot more to be said about the representations of finite groups in general and of the symmetric group in particular. Here, we only mention that the irreducible representations of $\mathfrak{S}_n$ are classified by the *Young tableaus* realizing a permutation of n. And the following theorem about the canonical decomposition of a representation of a finite group, which is a special case of our Theorem 1.5 in 1.5 and can be proved rather easily (see [Se] p. 21):

Let $\pi_1, \dots, \pi_h$ be as before irreducible representations representing each exactly one equivalence class and let (π, V) be any representation. Let $V = U_1 \oplus \cdots \oplus U_k$ be a direct sum decomposition of V into spaces belonging to irreducible representations. For $i = 1, \dots, h$ denote by W_i the direct sum of those of the $U_1, \dots, U_k$ which are isomorphic to V_i (the W_i belong to the primary representations introduced at the end of 1.5). Then we have

$$V = W_1 \oplus \cdots \oplus W_h,$$

which is called the *canonical decomposition*. Its properties are as follows:

Theorem 2.3: i) The decomposition $V = W_1 \oplus \cdots \oplus W_h$ does not depend on the initially chosen decomposition of V into irreducible subspaces.
ii) The projection P_i of V onto W_i associated to the decomposition is given by

$$P_i = (n_i/m) \sum_{t \in G} \overline{\chi_i(t)} \pi(t), \quad n_i = \dim V_i.$$

We stop here our treatment of finite groups and propose to the reader to treat the following examples.

Exercise 2.2: Determine the irreducible representations of the cyclic group C_n and of the dihedral group D_n (This is the group of rotations and reflections of the plane which preserve a regular polygon with n vertices, it is generated by a rotation r and a reflection s fulfilling the relations $r^n = e, s^2 = e, srs = r^{-1}$).

Exercise 2.3: Determine the characters of $\mathfrak{S}_4$.

Chapter 3

Continuous Representations

Among the topological groups, compact groups are the most easy ones to handle. Since we would like to treat their representations next, it is quite natural that the appearance of topology leads to an additional requirement for an appropriate definition, namely a continuity requirement. We will describe this first and indicate the necessary changes for the general concepts from Sections 1.1 to 1.5 (to be used in the whole text later). In the next chapter we specialize to compact groups. We will find that their representation theory has a lot in common with that of the finite groups, in particular the fact that all irreducible representations are finite-dimensional and contained in the regular representation (the famous Theorem of Peter and Weyl). But there is an important difference: the number of equivalence classes of irreducible representations may be infinite.

We do not prove all the modifications in the general theorems when linear representations are specialized to continuos representations, but concentrate in the next chapters on an explicit description of the representation theory for SU(2) resp. SO(3) and the other examples mentioned in the introduction.

3.1   Topological and Linear Groups

We follow [Ki] p.22:

Definition 3.1: A *topological group* is a set that is simultaneously a group and a topological space in which the group and topological structures are connected by the following condition:

The map

$$G \times G \longrightarrow G, \quad (a, b) \longmapsto ab^{-1}$$

is continuous.

Remark 3.1: This requirement is equivalent to the following three conditions which are more convenient for checking

1.) $(a, b) \longmapsto ab$ is continuous in a and in b,
2.) $a \longmapsto a^{-1}$ is continuous at the point $a = e$,
3.) $(a, b) \longmapsto ab$ is continuous in both variables together at the point (e, e).

Every abstract group can be regarded as a topological group if one gives it the discrete topology where every subset is open.

All the matrix groups presented in Section 0.1 and later on are topological groups: As they are subsets of $M_n(\mathbf{R}) \simeq \mathbf{R}^{n^2}$, they inherit the (standard) topology from $\mathbf{R}^{n^2}$. Open balls in $M_n(\mathbf{R})$ (and analogously in $M_n(\mathbf{C})$) are

$$K_\rho(A) := \{B \in M_n(\mathbf{R}), \| B - A \| < \rho\}, \rho \in \mathbf{R}_{>0}, A \in M_n(\mathbf{R}),$$

where

(3.1) $$\| A \| := \sqrt{< A, A >}, \; < A, B > := \operatorname{Re} \operatorname{Tr} {}^t\bar{A}B.$$

This topology is consistent with the group structure: multiplication of two matrices gives a matrix where the coefficients of the product matrix are polynomials in the coefficients of the factors and thus the two continuity requirements 1.) and 3.) in the remark above are fulfilled. Similarly with condition 2.), as the inverse of a matrix has coefficients which are polynomials in the coefficients of the matrix divided by the determinant, another polynomial here not equal to zero, i.e. we have again a continuous function.

For $A \in \mathrm{U}(n)$ we have ${}^t\bar{A}A = E$, i.e. $\| A \|^2 = n$, so we have an example of a compact group. $\mathrm{Heis}(\mathbf{R}), \mathrm{SL}(2, \mathbf{R})$ and $\mathrm{SO}(3, 1)$ are not compact but these groups together with all others in this text are *locally compact*, i.e. each element admits a neighbourhood with compact closure (as $\mathbf{R}^n$ does). Moreover, they are closed subgroups of some $\mathrm{GL}(N, \mathbf{R})$. This, in principle, has to be verified in each case. A practical tool is the following standard criterium.

Remark 3.2: G is closed if for every convergent sequence (A_j), $A_j \in G$ with $\lim A_j \in \mathrm{GL}(n, \mathbf{R})$, we have $\lim A_j \in G$.

An example for a different procedure is the following: $\mathrm{SL}(n, \mathbf{R})$ is a closed subgroup of $\mathrm{GL}(n, \mathbf{R})$ as $\det: \mathrm{GL}(n, \mathbf{R}) \longrightarrow \mathbf{R}^*$ is a continuous map and $\mathrm{SL}(n, \mathbf{R})$ is the inverse image of the closed subset $\{1\} \subset \mathbf{R}^*$.

Now we have the possibility to define the class of groups we will restrict to in this text:

Definition 3.2: A group is called a *linear group* if there is an $n \in \mathbf{N}$ such that G is isomorphic (as abstract group) to a closed subgroup of $\mathrm{GL}(n, \mathbf{R})$ or $\mathrm{GL}(n, \mathbf{C})$.

Remark 3.3: More generally, one calls also *linear* those groups, which are isomorphic to a closed subgroup of $\mathrm{GL}(n, \mathbf{H})$, $\mathbf{H}$ the quaternion skewfield, which in this text will only appear in a later application.

Remark 3.4: In another kind of generality one treats *Lie groups*, which are topological groups and at the same time differentiable manifolds such that

$$G \times G \longrightarrow G, \quad (a, b) \longmapsto ab \text{ and } G \longrightarrow G, \quad a \longmapsto a^{-1},$$

are differentiable maps. As we try to work without the (important) notion of a manifold as long as possible, it is quite adequate to stay with the concept of linear groups as defined above (which comprises all the groups we are interested in here).

While treating topological notions, let us remind that a group G is called *(path-)connected* if for any two $a, b \in G$ there is a continuous map of a real interval mapping one endpoint of the interval to a and the other to b.

Exercise 3.1: Show that $SU(2)$ and $SL(2, \mathbf{R})$ are connected and $GL(2, \mathbf{R})$ is not.

Before we come to the definition of continuous representations, let us adapt the notion of a G-set from Section 0.2 to the case that G is a topological group.

Definition 3.3: A set $\mathcal{X}$ is called a *(left-)topological G-space* iff it is a left G-set as in 0.2 and $\mathcal{X}$ is also provided with a topology such that the map

$$G \times \mathcal{X} \longrightarrow \mathcal{X}, \quad (g, x) \longmapsto gx,$$

is continuous.

In the sequel, G-space will abbreviate the content of this definition.

If the space $\mathcal{X}$ is Hausdorff (as all our examples will be), then the stabilizing subgroup $G_x \subset G$ of each point $x \in \mathcal{X}$ is a closed subgroup of G. Conversely, if H is a closed subgroup of a Hausdorff group G, then the space G/H of left cosets gH, $g \in G$, is a homogeneous Hausdorff G-space when given the usual quotient space topology. If $\mathcal{X}$ is any other topological G-space with the same stabilizing group $G_x = H$ for $x \in \mathcal{X}$, then the natural map

$$\Phi : G/H \longrightarrow \mathcal{X}, \quad gH \longmapsto gx$$

is 1-1 and continuos. If G and $\mathcal{X}$ are locally compact, then Φ is a homeomorphism and allows an identification of both spaces.

3.2 The Continuity Condition

From now on a *group* will mean a locally compact topological group. Then an appropriate representation space V should be a linear topological space. To make things easier for us, we will suppose that $V = \mathcal{H}$ is a separable complex Hilbert space (i.e. having denumerable Hilbert space basis $(e_j)_{j \in I}, I \simeq \mathbf{N}$), as for instance $L^2(S^1)$ with basis given by the functions $e_j(t) := \exp(2\pi i j t), j \in \mathbf{Z}$. For the finite-dimensional case this is nothing new: we have as before $V = \mathbf{C}^n$.

If we write again $GL(V)$, in the finite-dimensional case, we have the same set as before provided with the topology given by an isomorphism $GL(V) \simeq GL(n, \mathbf{C})$. In the infinite-dimensional case, $V = \mathcal{H}$, $GL(V) = GL(\mathcal{H})$ is meant as the group $L(\mathcal{H})$ of linear bounded operators in $\mathcal{H}$. We recall here the following fundamental notions relevant for the infinite-dimensional cases (from e.g. [BR] p.641f) :

- An operator T with domain $D(T) \subset \mathcal{H}$ is said to be *continuous* at a point $v_0 \in D(T)$ iff for every $\epsilon > 0$ there exists a $\delta = \delta(\epsilon) > 0$ such that from $\| v - v_0 \| < \delta$ with $v \in D(T)$ we can conclude $\| Tv - Tv_0 \| < \epsilon$.

- An operator T is said to be *bounded* iff there exists a constant C such that one has $\| Tv \| \leq C \| v \|$ for all $v \in D(T)$.

- The *norm* $\| T \|$ of a bounded operator T is defined as the smallest C such that

$$\| Tv \| \leq C \| v \| \quad \text{for all} \ \| v \| \leq 1$$

Remark 3.5: A bounded linear operator is uniformly continuous. Conversely, if a linear operator T is continuous at a point v_0 (e.g. $v_0 = 0$), then T is bounded.

Exercise 3.2: Prove this.

- Let A be a linear operator in $\mathcal{H}$ with dense domain $D(A) \subset \mathcal{H}$. Then one can prove that there are $v, v' \in \mathcal{H}$ such that $< Au, v > = < u, v' >$ holds for all $u \in \mathcal{H}$. One sets $v' =: A^* v$, verifies that A^* is a linear operator, and calls A^* the *adjoint* of A.

- For operators A and B in $\mathcal{H}$ one writes $B \supset A$ iff one has $D(B) \supset D(A)$ and $Bu = Au$ for all $u \in D(A)$. Then B is called an *extension* of A.

- A linear operator A with dense domain $D(A)$ is called *symmetric* iff one has $A^* \supset A$ and

$$< Au, v > = < u, Av > \quad \text{for all} \ u, v \in D(A).$$

- A linear operator A with dense domain is called *self-adjoint* iff one has $A^* = A$. One can prove that a linear symmetric operator A with $D(A) = \mathcal{H}$ is bounded and self-adjoint.

If not mentioned otherwise, $L(\mathcal{H})$ is provided with the *strong operator topology* ([BR]), i.e. we have as basis of open neighbourhoods

$$B_{\rho,u}(T_0) := \{T \in L(\mathcal{H}); \| Tu - T_0 u \| < \rho\}, \ \rho \in \mathbf{R}_{>0}, \ u \in \mathcal{H}, \ T_0 \in L(\mathcal{H}).$$

In the sequel a *representation* of G shall mean the following.

Definition 3.4: A linear representation $(\pi, \mathcal{H})$ is said to be *continuous* iff π is *strongly continuous*, i.e. the map

$$G \longrightarrow \mathcal{H}, \quad g \longmapsto \pi(g)v$$

is continuous for each $v \in V$.

Remark 3.6: It is quite obvious that one could also ask for the following continuity requirements:
    a) The map $G \times \mathcal{H} \longrightarrow \mathcal{H}$, $(g, v) \longmapsto \pi(g)v$ is continuous.
    b) The map $G \longrightarrow GL(\mathcal{H})$, $g \longmapsto \pi(g)$ is continuous.
and
    c) For each $v \in V$, the map $G \longrightarrow \mathcal{H}$, $g \longmapsto \pi(g)v$ is continuous in $g = e$ and there is a uniform bound for $\| \pi(g) \|$ in some neighbourhood of e ([Kn] p.10).

Definition 3.5: A representation (π, V) is bounded iff we have

$$\sup_{g \in G} \| \pi(g) \| < \infty.$$

In all examples in this section and in most cases of the following sections all these conditions are fulfilled. We will not discuss the hypotheses necessary so that all these conditions are equivalent and refer to this to [Ki] p.111, [Kn] p.10 and [BR] p.134. But we will show for some examples how the continuity of a representation comes out. Later we will not examine continuity and leave this to the reader's diligence. (To check this and the question whether a representation from the examples to be treated is bounded or not may even be (part of) the subject of a bachelor thesis.)

Example 3.1: The linear representation $\pi = \chi_k$, $k \in \mathbf{Z}$ of $G = SO(2)$ in $V = \mathbf{C}$ given by

$$\chi_k(r(\theta)) := e^{ik\vartheta} \quad \text{for all } g = r(\vartheta) = \begin{pmatrix} \cos\vartheta & \sin\vartheta \\ -\sin\vartheta & \cos\vartheta \end{pmatrix}, \vartheta \in \mathbf{R},$$

is obviously continuous: The map

$$(g, z) \longmapsto e^{ik\vartheta} z$$

is continuous in $g = r(\vartheta)$ for each $z \in \mathbf{C}$ or even contiuous in both variables g and z, because the exponential function is (in particular) continuous and an open neighborhood of $e \in SO(2)$ is homeomorphic to an open interval in $\mathbf{R}$ containing 0.

Later we will modify the notions of reducibility and irreducibility for continuous representations. But here we can already prepare the way how to do this and prove the following:

Remark 3.7: Each irreducible continuous representation π of $G = SO(2)$ is equivalent to some $\chi_k, k \in \mathbf{Z}$.

Proof: We look at (π, V) only as a linear representation and apply Schur's Lemma from 1.5. As G is abelian, we can conclude by Theorem 1.4 following from Schur's Lemma in 1.5 that $V = \mathbf{C}$. Hence we can assume $\pi : SO(2) \longrightarrow \mathbf{C}^*$. As $SO(2)$ is homeomorphic to $\mathbf{R}/\mathbf{Z}$, we have a continuous map $r : \mathbf{R} \longrightarrow \mathbf{R}/\mathbf{Z}$. Now we use the assumption that π is continuous and we have a sequence of continuous maps

$$\mathbf{R} \longrightarrow SO(2) \longrightarrow \mathbf{C}^*, \quad t \longmapsto r(t) \longmapsto \pi(r(t)),$$

and in consequence a continuous map

$$\mathbf{R} \ni t \longmapsto \pi(r(t)) =: \chi(t) \in \mathbf{C}^*.$$

As π is a linear representation, χ has the property

$$\chi(t_1 + t_2) = \chi(t_1)\chi(t_2), \ \chi(t) = 1 \quad \text{for all } t \in 2\pi\mathbf{Z}.$$

Here we have to recall a fact from standard calculus (see for instance [Fo] p.75): Every continuous function $\chi : \mathbf{R} \longrightarrow \mathbf{C}^*$ obeying the functional equation $\chi(t_1 + t_2) = \chi(t_1)\chi(t_2)$ is an exponential function, i.e. $\chi(t) = c\exp(t\zeta)$ for $c \in \mathbf{C}^*, \zeta \in \mathbf{C}$. The condition $\chi(2\pi n) = 1$ for all $n \in \mathbf{Z}$ fixes $c = 1$ and $\zeta = ki$ for $k \in \mathbf{Z}$.

By the way, already here, we can can anticipate a generalization of Proposition 2.4 in 2.2 from finite to compact groups: $L^2(SO(2))$ can be identified with a space of periodic functions. And from the theory of Fourier series one knows that this space has as a Hilbert space basis the trigonometric functions $e_j, j \in \mathbf{Z}$, with

$$e_j(t) := e^{2\pi ijt}, \ t \in \mathbf{R}.$$

This can be translated into the statement that each irreducible continuous representation of SO(2) is contained (with multiplicity one) in the left- (or right-)regular representation λ resp. ρ of SO(2) and λ and ρ are direct sums of the $\chi_j, j \in \mathbf{Z}$.

Example 3.2: Let G be SO(3) and, for fixed $m \in \mathbf{N}$, let π_m be the representation of G given on the space V_m of all homogeneous polynomials of degree m in 3 variables, i.e.

$$V_m := \{P \in \mathbf{C}[x_1, x_2, x_3]; P \text{ homogeneous, } \deg P = m\},$$

by

$$\pi(g)P(x) := P(g^{-1}x), \ x = {}^t(x_1, x_2, x_3) \text{ (a column).}$$

π_m is a linear representation as is already clear from the general consideration in 1.3. The continuity condition is fulfilled since SO(3) inherits the topology from $\mathrm{GL}(n, \mathbf{R})$ and matrix multiplication $x \longmapsto g^{-1}x$ is a continuous map and polynomials P are continuous functions in their arguments.

This example is similarly constructed as the example $\pi = \pi_j$ for $G = \mathrm{SU}(2)$ introduced in 1.3, which is continuous by the same reasoning as are also the other examples in 1.3 and later on. So we will now take the continuity of our representations for granted.

Moreover, let us announce here that we shall have to discuss the relationship of the two examples $G = \mathrm{SO}(3)$ and $G = \mathrm{SU}(2)$ very thoroughly later. The representations π_j of SU(2) will come out as irreducible, but here we see already that for instance for $m = 2$ (π_m, V_m) has an invariant subspace $V_0 := (x_1^2 + x_2^2 + x_3^2)\mathbf{C}$ since each element $g \in \mathrm{SO}(3)$ acts on $\mathbf{R}^3$ leaving invariant the spheres given by $\sum x_i^2 = \rho^2$.

The addition of the continuity condition to the definition of the representation has natural consequences for the notions related to the definition of the linear representation in Sections 1.1 and 1.2: Here only *closed* subspaces are relevant. Hence in the future, we use definitions modified as follows.

- Suppose that there is a *closed* subspace V_1 of the space V of a representation (π, V) of G, invariant under all $\pi(g), g \in G$. Then $\pi_1 := \pi|_V$ is called a (topological) *subrepresentation* of π.

- The representation π_2 on the factor (or quotient) space V/V_1 is called a (topological) *factor representation*.

The representation (π, V) is called

- (topologically) *irreducible* iff it admits no nontrivial subrepresentation,

- (topologically) *decomposable* iff there exist closed subspaces V_1 and V_2 in V such that $V = V_1 \oplus V_2$. In this case we write $\pi = \pi_1 + \pi_2$, where $\pi_i := \pi|_{V_i}$,

- *completely reducible* (or *discretely decomposable*) iff it can be expressed as a direct sum of irreducible subrepresentations,

- *unitary* iff V is a Hilbert space $\mathcal{H}$ and every $\pi(g)$ respects the scalar product in $\mathcal{H}$.

- If (π, V) and (π', V') are continuous representations of G, then a bounded linear map (or *operator*) $F : V \longrightarrow V'$ is called an *intertwining operator* for π and π' iff one has

$$F\pi(g) = \pi'(g)F \qquad \text{for all } g \in G.$$

Then Schur's Lemma here appears in the following form.

Theorem 3.1: If π and π' are unitary irreducible representations in the Hilbert spaces $\mathcal{H}$ resp. $\mathcal{H}'$ and $F : \mathcal{H} \longrightarrow \mathcal{H}'$ is an intertwining operator, then F is either an isomorphism of Hilbert spaces or $F = 0$.

This theorem has as a consequence the following irreducibility criterium, which is a generalization to the infinite-dimensional case of the similar one we proved for the finite-dimensional linear representations without the continuity condition in 1.5. We will call it *the unitary Schur*:

Theorem 3.2: A unitary representation $(\pi, \mathcal{H})$ is irreducible iff the only operators commuting with all the $\pi(g)$ are multiples of the identity.

The proofs of these theorems (see e.g. [BR] p. 143/4 or [Kn] p.12) use some tools from functional analysis (the *spectral theorem*) which we do not touch at this stage.

We indicate another very useful notion (in particular concerning unitary representations).

Definition 3.6: A representation π of G in V is said to be *cyclic* iff there is a $v \in V$ (called a *cyclic vector* for V) such that the closure of the linear span of all $\pi(g)v$ is V itself.

Theorem 3.3: Every unitary representation $(\pi, \mathcal{H})$ of G is the direct sum of cyclic representations.

As the **proof** is rather elementary and gives some indications how to work in the infinite-dimensional case and how the continuity condition works, we give it here (following [BR] p.146):
Let $\mathcal{H}_{v_1}$ be the closure of the linear span of all $\pi(g)v_1, g \in G$ for any $0 \neq v_1 \in \mathcal{H}$. Then $\mathcal{H}_{v_1}$ is π-invariant: Indeed, let $\mathcal{H}'_{v_1}$ be the linear span of all $\pi(g)v_1$, then for each $u \in \mathcal{H}_{v_1}$ we have a sequence $(u_n), u_n \in \mathcal{H}'_{v_1}$ which converges to u. Obviously, one has $\pi(g)u_n \in \mathcal{H}'_{v_1}$. The continuity of each $\pi(g)$ implies $\pi(g)u_n \longrightarrow \pi(g)u$ and hence we have $\pi(g)u \in \mathcal{H}_{v_1}$ and $\mathcal{H}_{v_1}$ is invariant. Thus the subrepresentation $\pi_1 = \pi \mid_{\mathcal{H}_{v_1}}$ is cyclic with cyclic vector v_1.
If $\mathcal{H}_{v_1} \neq \mathcal{H}$, choose $0 \neq v_2 \in \mathcal{H}^{\perp}_{v_1} = \mathcal{H} \setminus \mathcal{H}_{v_1}$, consider the closed linear span $\mathcal{H}_{v_2}$ which is π-invariant and orthogonal to $\mathcal{H}_{v_1}$, and continue like this if $\mathcal{H} \neq \mathcal{H}_{v_1} \oplus \mathcal{H}_{v_2}$.
Let ξ denote the family of all collections $\{\mathcal{H}_{v_i}\}$, each composed of a sequence of mutually orthogonal, invariant and cyclic subspaces. We order the family by means of the

inclusion relation. Then ξ is an ordered set to which *Zorn's Lemma* applies: It assures the existence of a maximal collection $\{\mathcal{H}_{v_i}\}_{max}$. By the separability of $\mathcal{H}$, there can be at most a countable number of subspaces in $\{\mathcal{H}_{v_i}\}_{max}$ and their direct sum, by the maximality of $\{\mathcal{H}_{v_i}\}_{max}$ must coincide with $\mathcal{H}$.

As a consequence of this theorem, we have a very convenient irreducibility criterium for unitary representations :

Corollary: A unitary representation $(\pi, \mathcal{H})$ of G is irreducible iff every nonzero $u \in \mathcal{H}$ is cyclic for π.

Proof: i) If π is irreducible, every $0 \neq u \in \mathcal{H}$ is cyclic as to be seen from the first part of the proof above.

ii) Suppose that $\mathcal{H}_1 \subset \mathcal{H}$ is a nontrivial invariant subspace of $\mathcal{H}$ and $0 \neq u \in \mathcal{H}_1$. Due to the invariance of $\mathcal{H}_1$ we have $\pi(g)u \in \mathcal{H}_1$. Moreover, the closure of the linear span of all $\pi(g)u$ is contained in $\mathcal{H}_1$ but by assumption is also equal to $\mathcal{H}$. Hence, we have a contradiction and π is irreducible.

To close these general considerations let us recall the following. If $\mathcal{H}$ and $\mathcal{H}'$ are Hilbert spaces with scalar products $<,>$ resp. $<,>'$, then we can define a scalar product in $\mathcal{H} \otimes \mathcal{H}'$ by the formula

$$< v \otimes v', w \otimes w' >:=< v, w >< v', w' > .$$

If either $\mathcal{H}$ or $\mathcal{H}'$ is finite-dimensional, then the space $\mathcal{H} \otimes \mathcal{H}'$ equipped with this scalar product is complete. If both $\mathcal{H}$ and $\mathcal{H}'$ are infinite-dimensional, we complete $\mathcal{H} \otimes \mathcal{H}'$ by the norm defined by the scalar product above and denote this by $\mathcal{H} \hat{\otimes} \mathcal{H}'$.

3.3   Invariant Measures

We want to extend the results from finite groups to compact groups. The tool to do this is given by replacing the finite sum $\sum_{g \in G}$ by an integration $\int_G$. To be a bit more precise, one has to recall some notions from measure and integration theory, which moreover are inevitable in all the following sections. For the whole topic we recommend the classical books *Measure Theory* by Halmos ([Ha]) and *Abstract Harmonic Analysis* by Hewitt-Ross ([HR]) or, for summaries, the appropriate chapters in [Ki] p.129ff and [BR] p.67ff.

If $\mathcal{X}$ is a topological space with the family $\mathfrak{V} = \{O_i\}_{i \in I}$ of open sets, it is also a *Borel space* with a family $\mathfrak{B}$ of *Borel sets* B, where these sets are obtained from open and closed sets by the operation of countable unions, countable intersections and complementation. A little more generally, a *Borel space* is a set with a family $\mathfrak{B}$ of subsets B such that $\mathfrak{B}$ is a σ-*algebra*, i.e. one has

$$\begin{aligned} \mathcal{X} \setminus B \in \mathfrak{B} \quad &\text{for} \quad B \in \mathfrak{B}, \\ \bigcap_{i \in I} B_i \in \mathfrak{B} \quad &\text{for} \quad B_i \in \mathfrak{B}, i \in I \simeq \mathbf{N}, \\ \bigcup_{i \in I} B_i \in \mathfrak{B} \quad &\text{for} \quad B_i \in \mathfrak{B}, i \in I \simeq \mathbf{N}. \end{aligned}$$

Let $(\mathcal{X}, \mathfrak{B})$ and $(\mathcal{X}', \mathfrak{B}')$ be two Borel spaces. A map $F : \mathcal{X} \longrightarrow \mathcal{X}'$ is called a *Borel map* iff one has $F^{-1}(B') \in \mathfrak{B}$ for all $B' \in \mathfrak{B}'$.

Definition 3.7: We call μ a *measure* on the Borel space $(\mathcal{X}, \mathfrak{B})$ if μ is a map

$$\mu : \mathfrak{B} \longrightarrow [0, \infty] := \mathbf{R}_{\geq 0} \cup \{\infty\},$$

where μ is σ-additive, i.e.

i) $\mu(\cup B_i) = \sum \mu(B_i)$ if the B_i are pairwise disjoint $(i \in I \simeq \mathbf{N})$, and

ii) there is a covering of $\mathcal{X}$ by sets B_i of finite measure, i.e. we have $B_i \in \mathfrak{B}, i \in I \simeq \mathbf{N}$ with $\mu(B_i) < \infty$ and $\cup B_i = \mathcal{X}$.

The most obvious example is $\mathcal{X} = \mathbf{R}$ provided with the Lebesgue-measure $\mu =: \lambda$, which is defined by

$$\lambda([a, b)) := b - a \quad \text{for } a, b \in \mathbf{R}, \, a \leq b.$$

As we know from elementary integration theory (or see immediately), λ is *translation invariant*. This is a special incidence of the following general machinery.

Let $\mathcal{X}$ be a space with a family $\mathfrak{V}$ of open sets and a family $\mathfrak{B}$ of Borel sets B and let μ be a measure on $\mathcal{X}$. We then say $(\mathcal{X}, \mathfrak{B}, \mu)$ is a *measure space*. If we have a right G-action

$$(G \times \mathcal{X}) \longrightarrow \mathcal{X}, \quad (g, x) \longmapsto xg,$$

we can define a measure μg by

$$\mu g(B) = \mu(Bg) \quad \text{for all } B \in \mathfrak{B}.$$

Then the measure μ is called *right-invariant* iff one has $\mu g = \mu$ for all $g \in G$. Similarly μ is called *left-invariant* iff we have a left action and $g\mu = \mu$ for all $g \in G$ where $g\mu(B) = \mu(gB)$ for $B \in \mathfrak{B}$.

Now we take $\mathcal{X} = G$, a locally compact group. Then a right- or left-invariant measure on G is called a *right-* or *left-Haar measure*. And we have as central result ([Ki] p.130):

Theorem 3.4 (Haar): On every locally compact group with countable basis for the topology, there exists a nonzero left-invariant (and similarly a right-invariant) measure. It is defined uniquely up to a numerical factor.

We will not go into the proof of the Theorem (see for instance [HR] Ch.IV, 15) but give later on several concrete examples for its validity. The most agreable case is the following:

Definition 3.8: G is called *unimodular* iff it has a measure which is simultaneously left- and right-invariant. Such a measure is also called *biinvariant* or simply *invariant*.

To make these up to now only abstract concepts work in our context, we have to recall a more practical aspect due to the general integration theory:

Let $(\mathcal{X}, \mathfrak{B}, \mu)$ be a measure space and $f : \mathcal{X} \longrightarrow \mathbf{C}$ a *measurable function*, i.e. f is the sum $f = f_1 - f_2 + i(f_3 - f_4)$ of four real nonnegative functions f_j $(j = 1, .., 4)$ having the property

$$\{x \in \mathcal{X}; f_j(x) > a\} \in \mathfrak{B} \quad \text{for all } a > 0.$$

For measurable nonnegative functions f general integration theory provides a linear map

$$f \longmapsto \mu(f) = \int_{\mathcal{X}} f(x)d\mu(x) \in \mathbf{C} \cup \{\infty\}.$$

We call a measurable function f *integrable* iff $\int |f(x)| \, d\mu(x) < \infty$.
If $\mathcal{X} = G$ as above, one usually writes

$$
\begin{aligned}
d\mu(g) \quad =: \quad & d_r g \quad \text{if } \mu \text{ is right} - \text{invariant}, \\
=: \quad & d_l g \quad \text{if } \mu \text{ is left} - \text{invariant}, \\
=: \quad & dg \quad \text{if } \mu \text{ is bi} - \text{invariant}.
\end{aligned}
$$

The following fact is central for applications to representation theory.

Proposition 3.1: Let G be a locally compact group. Then there is a continuous homomorphism Δ_G, called the *modular function* of the group G, into the multiplicative group of positive real numbers for which the following equalities hold

$$d_r(gg') = \Delta_G(g)d_r g', \quad d_l(gg') = \Delta_G(g')^{-1}d_l g,$$

$$d_r g = const.\Delta_G(g)d_l g, \quad d_l g = const.\Delta_G(g)^{-1}d_r x,$$

$$d_r(g^{-1}) = \Delta_G(g)^{-1}d_r g = const.d_l g, \quad d_l(g^{-1}) = \Delta_G(g)d_l g.$$

As Kirillov does in [Ki] p.130, we propose the proof as an exercise based on the application of Haar's Theorem. There is a proof in [BR] p.68/9, but the modular function there is the inverse to our function, which here is defined following Kirillov.

Proposition 3.2: If G is compact, then the Haar measure is finite and biinvariant.

Proof: The image of a compact group under the continuous homomorphism Δ_G is again a compact group. The multiplicative group $\mathbf{R}_{>0}$ has only $\{1\}$ as a compact subgroup. So by Proposition 3.1, we have a biinvariant measure which can be normalized to $\int_G dg = 1$.

3.4 Examples

Most of the noncompact groups to be treated later on are also unimodular, e.g. the group $SL(2,\mathbf{R})$ or the Heisenberg group $\text{Heis}(\mathbf{R})$. We will show here how this can be proved.

Example 3.3: $G = \text{Heis}(\mathbf{R}) = \{g = (\lambda, \mu, \kappa); \; \lambda, \mu, \kappa \in \mathbf{R}\}$ with

$$gg' = (\lambda + \lambda', \mu + \mu', \kappa + \kappa' + \lambda\mu' - \lambda'\mu).$$

As $\text{Heis}(\mathbf{R})$ is isomorphic to $\mathbf{R}^3$ as a vector space, we try to use the usual Lebesgue measure for $\mathbf{R}^3$. So we take

$$dg := d\lambda d\mu d\kappa.$$

The left translation $\lambda_{g_0} : g \longmapsto g_0 g =: g'$ acts as

$$(\lambda, \mu, \kappa) \longmapsto (\lambda' := \lambda_0 + \lambda, \mu' := \mu_0 + \mu, \kappa' := \kappa_0 + \kappa + \lambda_0\mu - \lambda\mu_0).$$

The Jacobian of this map is

$$J_{g_0}(g) = \begin{pmatrix} 1 & & -\mu_0 \\ & 1 & \lambda_0 \\ & & 1 \end{pmatrix}$$

with $\det J_{g_0}(g) = 1$. Hence we see that dg is left-invariant. Similarly we prove the right-invariance.

Remark 3.8: Without going into the general theory of integration on manifolds, we indicate that, for calculations like this, it is very convenient to associate the infinitesimal measure element with an appropriate *exterior* or *alternating differential form* and then check if this form stays invariant. In the example above we take

$$\omega' := d\lambda' \wedge d\mu' \wedge d\kappa'$$

and calculate using the the general rules for alternating differential forms (we shall say a bit more about this later in section 8.1), in particular

$$d\mu \wedge d\lambda = -d\lambda \wedge d\mu, \quad d\mu \wedge d\mu = 0,$$

as follows

$$\omega' \circ \lambda_{g_0} = d\lambda \wedge d\mu \wedge (d\kappa + \lambda_0 d\mu - \mu_0 d\lambda) = d\lambda \wedge d\mu \wedge d\kappa = \omega.$$

Example 3.4: $G = SO(2)$

We introduced as standard notation

$$SO(2) \ni r(\vartheta) = \begin{pmatrix} \cos\vartheta & \sin\vartheta \\ -\sin\vartheta & \cos\vartheta \end{pmatrix}.$$

Here we have $dg = d\vartheta$ as biinvariant measure with

$$\int_G dg = \int_0^{2\pi} d\vartheta = 2\pi.$$

Example 3.5: $G = SU(2)$

This group will be our main example for the representation theory of compact groups. As standard parametrization we use the same as Wigner does in [Wi] p.158: The general form of a two–dimensional unitary matrix

$$g = \begin{pmatrix} a & b \\ c & d \end{pmatrix}.$$

of determinant one is

$$g = \begin{pmatrix} a & b \\ -\bar{b} & \bar{a} \end{pmatrix}, \quad a, b \in \mathbf{C} \text{ with } |a^2| + |b^2| = 1.$$

With

$$s(\alpha) := \begin{pmatrix} e^{i\alpha} \\ & e^{-i\alpha} \end{pmatrix} \text{ and } r(\vartheta) = \begin{pmatrix} \cos\vartheta & \sin\vartheta \\ -\sin\vartheta & \cos\vartheta \end{pmatrix}$$

one has a decomposition

$$\mathrm{SU}(2) \ni g = \begin{pmatrix} a & b \\ -\bar{b} & \bar{a} \end{pmatrix} = s(-\alpha/2)r(-\beta/2)s(-\gamma/2), \quad \alpha, \gamma \in [0, 2\pi], \beta \in [0, \pi].$$

In particular, one can see that $\mathrm{SU}(2)$ is generated by the matrices $s(\alpha), r(\beta)$.

Exercise 3.3: Prove this.

Later on we shall see later that the angles α, β, γ correspond to the Euler angles for the three-dimensional rotation introduced by Wigner in [Wi] p.152. In Hein [He] one finds all these formulas, but unfortunately in another normalization, namely one has $z_{Hein} = a, u_{Hein} = -b, \alpha_{Hein} = \beta, \beta_{Hein} = -\alpha, \gamma_{Hein} = -\gamma$. Our parametrization is one-to-one for $\alpha, \gamma \in [0, 2\pi], \beta \in [0, \pi]$ up to a set of measure zero. An invariant normalized measure is given by

$$dg := (1/16\pi^2) \sin\beta \, d\alpha d\beta d\gamma.$$

Exercise 3.4: Verify this.

Example 3.6: An example for a group G which is **not** unimodular is the following

$$G = \{g = \begin{pmatrix} a & b \\ & 1 \end{pmatrix}; \ a \in \mathbf{R}^*, b \in \mathbf{R}\}$$

We have $gg' = \begin{pmatrix} aa' & ab' + b \\ & 1 \end{pmatrix} =: g''$, hence

$$da'' \wedge db'' = a' da \wedge (b' da + db) = a' da \wedge db$$

so that $d_r g = a^{-1} dadb$ and, from $g'g = \begin{pmatrix} a'a & a'b + b' \\ & 1 \end{pmatrix} =: g^*$

$$da^* \wedge db^* = a'^2 da \wedge db,$$

we see $d_l g = a^{-2} dadb$. Hence, we have here $\Delta_G(g) = a^{-1}$.

Exercise 3.5: Determine Δ_G for

$$G = \{g = \begin{pmatrix} 1 & x \\ & 1 \end{pmatrix}\begin{pmatrix} y^{1/2} \\ & y^{-1/2} \end{pmatrix}; \ y > 0, x \in \mathbf{R}\}.$$

Chapter 4

Representations of Compact Groups

As already mentioned, the representation theory of compact groups generalizes the theory of finite groups. We shall closely follow the presentation in [BR], and cite from [BR] p.166: "The representation theory of compact groups forms a bridge between the relatively simple representation theory of finite groups and that of noncompact groups. Most of the theorems for the representations of finite groups have direct analogues for compact groups and these results in turn serve as the starting point for the representation theory of noncompact groups."

4.1 Basic Facts

Let G be a compact group provided with an invariant measure μ normalized such that $\int_G dg = 1$ and let π be a representation of G in a Hilbert space $\mathcal{H}$. We recall that all our representations are meant to be linear und continuous. The following statements are easy to prove for the finite-dimensional case $\dim \mathcal{H} < \infty$. Most times this is quite sufficient because it will come out that all irreducible representations of compact groups are finite-dimensional. To prepare a feeling for the noncompact cases where infinite-dimensionality is essential, we state the results here for general $\mathcal{H}$. But we will report on the proofs only when not too much general Hilbert space theory is needed.

Proposition 4.1: Let $(\pi, \mathcal{H})$ be a representation of G in the Hilbert space $\mathcal{H}$ with scalar product $<,>'$. Then there exists a new scalar product $<,>$ defining a norm equivalent to the initial one, relative to which the map $g \longmapsto \pi(g)$ defines a unitary representation of G.

Proof: i) As to be expected from the case of finite groups, we define $<,>$ by

$$< u, v > := \int_G < \pi(g)u, \pi(g)v >' \, dg.$$

Sesquilinearity and hermiticity of $<,>$ are obvious. Let us check that we can conclude $u = 0$ from

$$< u, u > = \int_G < \pi(g)u, \pi(g)u >' \, dg = 0 :$$

From $\int_G < \pi(g)u, \pi(g)u >' dg = 0$ one deduces that $< \pi(g)u, \pi(g)u >' = 0$ almost everywhere. If $g \in G$ is such that $\pi(g)u = 0$, then $\pi(g)^{-1}\pi(g)u = u = 0$.

ii) $<,>$ is π-invariant:

$$< \pi(g')u, \pi(g')v > \; = \; \int_G < \pi(g)\pi(g')u, \pi(g)\pi(g')v >' dg$$

$$= \; \int_G < \pi(gg')u, \pi(gg')v >' dg = < u, v >,$$

because dg is left-invariant.

iii) The norms $\| \cdot \|$ and $\| \cdot \|'$ induced by $<,>$ resp. $<,>'$ are equivalent:
We have

$$\| u \|^2 \; = \; \int_G < \pi(g)u, \pi(g)u >' dg$$

$$\leq \; \sup_{g \in G} \| \pi(g) \|^2 \int_G < u, u >' dg = N^2 \| u \|'^2$$

with $N := \sup_{g \in G} \| \pi(g) \|$. And from

$$\| u \|'^2 \; = \; < \pi(g^{-1})\pi(g)u, \pi(g^{-1})\pi(g)u >'$$

$$\leq \; \sup_{g \in G} \| \pi(g^{-1}) \|^2 < \pi(g)u, \pi(g)u >' = N^2 \| \pi(g)u \|'^2$$

it follows that

$$\| u \|'^2 \; = \; \int_G < u, u >' dg$$

$$\leq \; N^2 \int_G < \pi(g)u, \pi(g)u >' dg = N^2 < u, u >= N^2 \| u \|^2 .$$

Hence we have $N^{-1} \| u \|' \leq \| u \| \leq N \| u \|'$ and $\| \cdot \|$ and $\| \cdot \|'$ are equivalent.

iv) Equivalent norms define the same families of open balls and thus equivalent topologies, i.e. if $g \longmapsto \pi(g)v, v \in \mathcal{H}$ is continuous for the topology belonging to $\| \cdot \|'$, it is also continuous for the topology deduced from $\| \cdot \|$.

We now turn to the already announced result that every unitary irreducible representation is finite-dimensional. In its proof we use an interesting tool, namely the *Weyl operator* K_u defined for elements u, v from a Hilbert space $\mathcal{H}$ with π-invariant scalar product by

$$K_u v := \int_G < \pi(g)u, v > \pi(g)u \, dg.$$

Remark 4.1: The Weyl operator has the following properties
   i) K_u is bounded,
   ii) K_u commutes with every $\pi(g)$, i.e. $K_u \pi(g) = \pi(g) K_u$.

Proof: We perform the standard calculations using Cauchy–Schwarz inequality.

i) $$\begin{aligned} \parallel K_u v \parallel &\leq \int_G |< \pi(g)u, v >| \parallel \pi(g)u \parallel dg \\ &\leq \int_G \parallel \pi(g)u \parallel^2 \parallel v \parallel dg = \parallel u \parallel^2 \parallel v \parallel, \end{aligned}$$

ii) $$\begin{aligned} \pi(g')K_u v &= \int_G < \pi(g)u, v > \pi(g'g)u dg \\ &= \int_G < \pi(g'g)u, \pi(g')v > \pi(g'g)u dg \\ &= \int_G < \pi(g)u, \pi(g')v > \pi(g)u dg = K_u \pi(g')v \end{aligned}$$

for all $g' \in G$ and $v \in \mathcal{H}$.

Theorem 4.1: Every irreducible unitary representation π of G in a Hilbert space $\mathcal{H}$ is finite-dimensional.

Proof: By the last remark, K_u is an intertwining operator for π. Hence by Schur's Lemma, K_u is a homothety, i.e. $K_u = \lambda(u)id_\mathcal{H}$ and, in consequence,

$$(4.1) \qquad \begin{aligned} < K_u v, v > &= \int_G < v, \pi(g)u >< \pi(g)u, v > dg \\ &= \int_G |< \pi(g)u, v >|^2 dg = \lambda(u) \parallel v \parallel^2 . \end{aligned}$$

By interchanging the rôles of u and v and using the equality (which is equivalent to the unimodularity of G)

$$\int f(g^{-1})dg = \int f(g)dg,$$

we get

$$\begin{aligned} \lambda(v) \parallel u \parallel^2 &= \int |< \pi(g)v, u >|^2 dg = \int |< u, \pi(g)v >|^2 dg \\ &= \int |< \pi(g^{-1})v, u >|^2 dg = \int |< \pi(g)u, v >|^2 dg \\ &= \lambda(u) \parallel v \parallel^2 . \end{aligned}$$

Hence we have $\lambda(u) = c \parallel u \parallel^2$ for all $u \in \mathcal{H}$ with a constant $c \in \mathbf{C}$. In (4.1) we put $u = v, \parallel v \parallel = 1$ and get

$$\int |< \pi(g)u, u >|^2 dg = \lambda(u) = c.$$

As the nonnegative continuous function $g \longmapsto |< \pi(g)u, u >|$ assumes the value $\parallel u \parallel = 1$ at $g = e$, we must have $c > 0$.

Let $\{e_i\}_{i=1,\ldots,n}$ be a set of ON-vectors in $\mathcal{H}$. In (4.1) we now put $u = e_k$ and $v = e_1$ to obtain

$$\int |< \pi(g)e_k, e_1 >|^2 dg = \lambda(e_k) \parallel e_1 \parallel^2 = c.$$

and

$$\begin{aligned} nc &= \sum_{k=1}^n \int_G |< \pi(g)e_k, e_1 >|^2 dg = \int_G \sum_{k=1}^n |< \pi(g)e_k, e_1 >|^2 dg \\ &\leq \int_G \parallel e_1 \parallel^2 dg = 1. \end{aligned}$$

The last inequality is a special case of Parseval's inequality saying that for a unitary matrix representation $g \longmapsto A(g)$ one has $\sum_k |A_{k1}(g)|^2 \leq 1$. So finally, one has $n \leq 1/c$, i.e. the dimension of $\mathcal{H}$ must stay finite.

We already proved that a finite-dimensional unitary representation is completely reducible. This result can be sharpened for compact groups G:

Theorem 4.2: Every unitary representation π of G is a direct sum of irreducible finite-dimensional unitary subrepresentations.

The proof uses some more tools from functional analysis and so we skip it here (see [BR] p.169/170). But we mention the appearance of the important notion of a *Hilbert–Schmidt operator*, which is an operator A in $\mathcal{H}$, s.t. for an arbitrary basis $\{e_i\}_{i \in I}$ of $\mathcal{H}$ we have

$$\sum_{i \in I} \| A e_i \|^2 < \infty.$$

Again, as in the theory of finite groups, we have ON-relations for the matrix elements of an irreducible representation.

Theorem 4.3: Let π and π' be two irreducible unitary representations of G and $A(g)$ resp. $A'(g)$ their matrices with respect to a basis $\{e_i\}$ of $\mathcal{H}$ and $\{e'_k\}$ of $\mathcal{H}'$. Then one has relations

$$
\begin{aligned}
\int \overline{A_{ij}(g)} A'_{kl}(g) dg &= 0 && \text{if } \pi \nsim \pi' \\
&= (1/n)\delta_{ik}\delta_{jl} && \text{if } \pi \sim \pi' \text{ and } n := \dim \mathcal{H}.
\end{aligned}
$$

Proof: As to be expected, one applies Schur's Lemma (Theorem 3.1): We introduce a matrix f_{ij} with entries $(f_{ij})_{kl} = \delta_{ik}\delta_{jl}$ and an operator

$$F_{ij} := \int_G \pi(g) f_{ij} \pi'(g^{-1}) dg.$$

This is an operator intertwining the representations π' and π since we have

$$
\begin{aligned}
\pi(g') F_{ij} &= \int_G \pi(g'g) f_{ij} \pi'(g^{-1}) dg \\
&= \int_G \pi(g) f_{ij} \pi'(g^{-1}g') dg = F_{ij} \pi'(g').
\end{aligned}
$$

Hence, if $\pi \nsim \pi'$, we have $F_{ij} = 0$ or in matrix form for all (r,t)

$$
\begin{aligned}
(F_{ij})_{rt} &= \int A_{ri}(g) A'_{jt}(g^{-1}) dg \\
&= \int A_{ri}(g) \overline{A'_{tj}(g)} dg = 0.
\end{aligned}
$$

For $\pi = \pi'$, Schur's Lemma requires $F_{ij} = \lambda_{ij} \mathrm{id}$. Hence for $(r,t) \neq (i,j)$, the orthogonality relations just obtained are still satisfied. For $(r,t) = (i,j)$, we have

$$
\begin{aligned}
(F_{ij})_{ij} &= \int A_{ii}(g)\overline{A_{jj}(g)} dg &&= \lambda_{ij} = 0 && \text{if } i \neq j, \\
&= \int |A_{ii}(g)|^2 dg &&= \lambda_{ii} && \text{if } i = j.
\end{aligned}
$$

We have $\operatorname{Tr} F_{ii} id = n\lambda_{ii}$ and from the definition of F_{ij},

$$\operatorname{Tr} F_{ii} = \int_G \operatorname{Tr} \left(A(g) f_{ii} A(g^{-1}) \right) dg = \operatorname{Tr} f_{ii} = 1$$

and hence $\lambda_{ii} = 1/n$.

Similar to the case of finite groups we have a nice fact:

Theorem 4.4: Every irreducible unitary representation π of G is equivalent to a sub-representation of the right regular representation.

Proof: Let $A(g) = (A_{jk}(g))_{j,k=1,\ldots,n}$ be a matrix form of $\pi(g)$ and let $\mathcal{H}_\pi$ be the subspace of $L^2(G)$ spanned by the vectors e_k with $e_k(g) := \sqrt{n} A_{1k}(g), k = 1, \ldots, n$ (these are an ON-system by the preceding theorem). We have

$$
\begin{aligned}
\rho(g_0) e_k(g) &= e_k(g g_0) = \sqrt{n} A(g g_0) \\
&= \sqrt{n} \Sigma_j A_{1j}(g) A_{jk}(g_0) = \Sigma_j \sqrt{n} A_{1j}(g) A_{jk}(g_0) \\
&= \Sigma_j e_j(g) A_{jk}(g_0).
\end{aligned}
$$

Hence ρ restricted to $\mathcal{H}_\pi$ is a subrepresentation equivalent to π.

Exercise 4.1: Show the same fact for the left regular representation.

Remark 4.2: With the same reasoning as in the case of finite groups, the characters χ_π of finite-dimensional irreducible unitary representations π, defined by $\chi_\pi(g) := \operatorname{Tr} \pi(g)$ resp. $\chi_\pi(g) := \Sigma_i A_{ii}(g) = \Sigma_i < A(g) e_i, e_i >$ for a matrix form with respect to a basis $e_1, \ldots, e_n$, have the following properties.

1) The characters are class functions, i.e. $\chi(g_0 g g_0^{-1}) = \chi(g)$ for all $g, g_0 \in G$.
2) We have $\chi(g^{-1}) = \overline{\chi(g)}$.
3) If $\pi \sim \pi'$, then $\chi_\pi = \chi'_\pi$.
4) $\int \overline{\chi_\pi(g)} \chi'_\pi(g) dg = 0$ if $\pi \not\sim \pi'$ and $= 1$ if $\pi \sim \pi'$.

If π is any finite-dimensional representation of G, then we can decompose it into irre-ducible representations π_i appearing with multiplicities m_i and we have

$$\chi_\pi = \sum_{i=1}^h m_i \chi_{\pi_i}.$$

Due to the ON-relation 4) in Remark 4.2 above, we obtain

$$m_i = \int_G \overline{\chi_\pi(g)} \chi_{\pi_i}(g) dg$$

and

$$\sum_{i=1}^h m_i^2 = \int_G \overline{\chi_\pi(g)} \chi_\pi(g) dg.$$

This formula specializes to the very useful irreducibility criterium:

Corollary: The finite-dimensional representation π is irreducible iff

$$\int_G \overline{\chi_\pi(g)}\chi_\pi(g)dg = 1.$$

The big difference to the theory of finite groups is that for compact groups the number of equivalence classes of irreducible representations no longer necessarily is a finite number. We already know this from the example $G = \mathrm{SO}(2)$ in section 3.2. The central fact is now the famous *Peter–Weyl Theorem*:

Theorem 4.5: Let $\hat{G} = \{\pi_i\}_{i\in I}$ be the unitary dual of G, i.e. a complete set of representatives of the equivalence classes of irreducible unitary representations of G. For every $i \in I$ let $A^i(g) = (A^i_{jk}(g))_{j,k=1,\ldots,n_i}$ be a matrix form of π_i and

$$Y^i_{jk}(g) := \sqrt{n_i}\, A^i_{jk}(g).$$

Then we have
  1) The functions $Y^i_{jk}, i \in I, j, k = 1, \ldots, n_i$ form a complete ON-system in $L^2(G)$.
  2) Every irreducible unitary representation π of G occurs in the decomposition of the right regular representation with a multiplicity equal to the dimension of π, i.e.

$$\mathrm{mult}\,(\pi_i, \rho) = n_i.$$

  3) Every **C**-valued continuous function f on G can be uniformly approximated by a linear combination of the (Y^i_{jk}).
  4) The characters $(\chi_{\pi_i})_{i\in I}$ generate a dense subspace in the space of continuous class functions on G.

The proof of these facts use standard techniques from higher analysis. We refer to [BR] p. 173 - 176 or [BtD] p.134.

4.2   The Example $G = \mathrm{SU}(2)$

Representation theory owes a lot to the work of H. Weyl and E. Wigner. As we have the feeling that we can not do better, we will discuss the example $G = \mathrm{SU}(2)$ following Wigner's presentation in [Wi] p.163 - 166:

We put

$$G = \mathrm{SU}(2) \ni g = \begin{pmatrix} a & b \\ -\bar{b} & \bar{a} \end{pmatrix}, |a|^2 + |b|^2 = 1$$

and for j a half integral nonnegative number, we let $V^{(j)}$ be the space $\mathbf{C}[x,y]_{2j}$ of homogeneous polynomials of degree $2j$ in two variables. Thus we have $\dim V^{(j)} = 2j+1$. We choose as a basis of $V^{(j)}$ monomials f_p, which are conveniently normalized by

$$f_p(x,y) := \frac{x^{j+p}}{\sqrt{(j+p)!}} \frac{y^{j-p}}{\sqrt{(j-p)!}}, \quad p = -j, -j+1, \ldots, j.$$

We define the representation π_j by the action

$$\left(g, \begin{pmatrix} x \\ y \end{pmatrix}\right) \longmapsto g^{-1}\begin{pmatrix} x \\ y \end{pmatrix},$$

i.e. we put

$$\pi_j(g)f_p(x,y) := f_p(\bar{a}x - by, \bar{b}x + ay),$$

and get by a straightforward computation

$$\pi_j(g)f_p(x,y) = \Sigma_{p'} f_{p'}(x,y) A^j_{p'p}(g)$$

with

$$A^j_{p'p}(g) = \sum_{k=0}^{j+p}(-1)^k \frac{\sqrt{(j+p)!(j-p)!(j+p')!(j-p')!}}{(j-p'-k)!(j+p-k)!k!(k+p'-p)!} a^{j-p'-k}\bar{a}^{j+p-k}b^k\bar{b}^{k+p'-p}$$

in particular

(4.2) $$A^j_{jp}(g) = \sqrt{\frac{(2j)!}{(j+p)!(j-p)!}}.$$

Exercise 4.2: Verify these formulae (at least) for $j = 1$.

Theorem 4.6: These representations $\pi_j, j \in (1/2)\,\mathbf{N}_0$, are unitary, irreducible, and - up to equivalence - there are no other such representations of SU(2).

Proof: i) **Unitarity**

Wigner's proof relies on the fact that the polynomials f_p are normalized so that we have

$$\sum_{p=-j}^{j} \bar{f}_p f_p = \sum_{p=-j}^{j} \frac{1}{(j+p)!(j-p)!} \mid x^2 \mid^{j+p} \mid y^2 \mid^{j-p} = \frac{(\mid x^2 \mid + \mid y^2 \mid)^{2j}}{(2j)!}.$$

By a similar computation, we conclude from the definition of π_j

$$\sum_{p=-j}^{j} \mid \pi_j(g)f_p(x,y) \mid^2 = \sum_{p=-j}^{j} \frac{\mid \bar{a}x - by \mid^{2(j+p)} \mid \bar{b}x - ay \mid^{2(j-p)}}{(j+p)!(j-p)!}$$

$$= \frac{(\mid \bar{a}x - by \mid^2 + \mid \bar{b}x - ay \mid^2)^{2j}}{(2j)!} = \frac{(\mid x \mid^2 + \mid y \mid^2)^{2j}}{(2j)!},$$

where the last equality follows from the unitarity of g. So we have the invariance

$$\Sigma_p \mid \pi(g)f_p \mid^2 = \Sigma_p \mid f_p \mid^2.$$

Substituting here $\pi(g)f_p = \Sigma_{p'} f_{p'} A^j_{p'p}(g)$, we get

$$\Sigma_p(\Sigma_{p'} \bar{f}_{p'} \bar{A}^j_{p'p}(g) \Sigma_{p''} f_{p''} A^j_{p''p}(g)) = \Sigma_p \bar{f}_p f_p.$$

If we know that the $(2j+1)^2$ functions $\bar{f}_{p'} f_{p''}$ are linearly independent, we see that we have

$$\Sigma_p \bar{A}^j_{p'p}(g) A^j_{p''p}(g) = \Sigma_p \bar{A}^j_{p'p}(g)^t A^j_{pp''}(g) = \delta_{p'p''},$$

i.e. the unitarity of $A^j(g)$. Hence we still have to show the linear independence of the functions $\bar{f}_{p'} f_{p''}$: We look at a relation

$$\sum_{p',p''} c_{p'p''} \bar{x}^{j+p'} \bar{y}^{j-p'} x^{j+p''} y^{j-p''} = 0, \ c_{p'p''} \in \mathbf{C}.$$

Then in particular for $x \in \mathbf{R}$ with $q = 2j + p' + p''$, the coefficient of x^q has to vanish. After division by $\bar{y}^j y^{3j-q}$ this requires for

$$\sum_{p'} c_{p',q-2j-p'}(y/\bar{y})^{p'} = 0.$$

But this condition implies that $c_{p',q-2j-p'} = 0$ (and thus the linear independence of the $\bar{f}_{p'} f_{p''}$), because we can write $y/\bar{y} = e^{it}, t \in \mathbf{R}$, and a relation

$$\Sigma c_{p',q-2j-p'} e^{itp} = 0 \quad \text{for all } t \in \mathbf{R}$$

requires that all the coefficients c must vanish.

ii) Irreducibility

Wigner's proof of the irreducibility of π_j is by direct application of the criterium which here appears as Corollary to the *finite-dimensional unitary Schur* (Theorem 1.3) in Section 1.5, namely by showing that any matrix M, which commutes with all $A^j(g)$, must necessarily be a constant matrix:
We take

$$g = s(-\alpha/2) = \begin{pmatrix} e^{-i\alpha/2} & \\ & e^{i\alpha/2} \end{pmatrix}.$$

Then the matrix $A^j(g)$ specializes to the diagonal matrix

(4.3) $$A^j_{p'p}(s(-\alpha/2)) = \delta_{p'p} e^{ip\alpha}.$$

It is a standard fact that each matrix M commuting with all such matrices must be diagonal. Just after the introduction of the matrices $A^j(g)$ we remarked (in (9.9)) that no element in the last row of $A^j(g)$ vanishes identically. Then by equating the elements of the j-th row of $A^j(g)M$ and $MA^j(g)$ we conclude that

$$A^j_{jk}(g)M_{kk} = M_{jj}A^j_{jk}(g), \text{ i.e. } M_{jj} = M_{kk},$$

and M is a scalar multiple of the unit matrix.

One can also find a proof by application of the criterium given in the Corollary at the end of Section 4.1 using the character of π_j. We will see this as a byproduct of the following third part.

iii) Completeness

Wigner's proof that there are - up to equivalence - no other irreducible representations of SU(2) than the π_j can be seen as a direct proof of (part of) the Peter–Weyl Theorem from 4.1 in this special case:
Characters are class functions. Since each unitary matrix can be diagonalized by conjugation with a unitary matrix, in each conjugacy class of SU(2) we find a diagonal matrix of the type

$$s(\alpha) = \begin{pmatrix} e^{i\alpha} & \\ & e^{-i\alpha} \end{pmatrix}, \alpha \in [0, 2\pi).$$

Conjugation by $w = \begin{pmatrix} & 1 \\ -1 & \end{pmatrix}$ transforms

$$\begin{pmatrix} & 1 \\ -1 & \end{pmatrix} \begin{pmatrix} \zeta^{-1} & \\ & \zeta \end{pmatrix} \begin{pmatrix} & -1 \\ 1 & \end{pmatrix} = \begin{pmatrix} \zeta & \\ & \zeta^{-1} \end{pmatrix}.$$

Hence every conjugacy class is exactly represented by a matrix

$$s(\alpha) := \begin{pmatrix} e^{i\alpha} & \\ & e^{-i\alpha} \end{pmatrix}, \; \alpha \in [0, \pi), \; \text{resp.} \; s(\alpha/2), \; \alpha \in [0, 2\pi).$$

As we have (see (4.3) in ii))

$$A^{j}_{p'p}(s(-\alpha/2)) = \delta_{p'p} e^{ip\alpha}$$

its trace is

$$\chi_{\pi_j}(s(-\alpha/2)) = \sum_{p=-j}^{j} e^{ip\alpha} =: \xi_j(\alpha).$$

So we have

$$
\begin{aligned}
\xi_0(\alpha) &= 1, \\
\xi_{1/2}(\alpha) &= e^{-i\alpha/2} + e^{i\alpha/2} = 2\cos(\alpha/2), \\
\xi_1(\alpha) &= e^{-i\alpha} + 1 + e^{i\alpha} = 2\cos\alpha + 1, \\
\xi_{3/2}(\alpha) &- \xi_{1/2}(\alpha) = 2\cos(3\alpha/2), \quad \text{etc.}
\end{aligned}
$$

It is now evident that $\mathrm{SU}(2)$ can have no other representations than the $\pi_j, j \in (1/2)\mathbf{N}_0$: For the character of such a representation must, after multiplication by a weighting function, be orthogonal to all ξ_j, and therefore to $\xi_0, \xi_{1/2}, \xi_1 - \xi_0, \xi_{3/2} - \xi_{1/2}, \ldots$. But a function, which is orthogonal to $1, 2\cos(\alpha/2), 2\cos\alpha, 2\cos(3\alpha/2), \ldots$ in $[0, 2\pi)$ must vanish according to Fourier's Theorem.

Exercise 4.3: Verify the irreducibilty of π_j using the character criterium (Corollary to Remark 4.2 in Section 4.1).

The explicit determination of the characters $\chi_{\pi_j} = \xi_j$ of the irreducible representations π_j is very useful if one wants to decompose a given representation π.

Example 4.1: For $\pi = \pi_1 \otimes \pi_1$, by Theorem 1.6 in Section 1.6, we have

$$
\begin{aligned}
\chi_\pi(s(\alpha/2)) &= \chi_{\pi_1}(s(\alpha/2))^2 = (e^{i\alpha} + 1 + e^{-i\alpha})^2 \\
&= \xi_2(\alpha) + \xi_1(\alpha) + \xi_0(\alpha),
\end{aligned}
$$

i.e.

$$\pi_1 \otimes \pi_1 = \pi_0 + \pi_1 + \pi_2.$$

Exercise 4.4: Verify the decomposition ([Wi] p.185)

$$\pi_j \otimes \pi_{j'} = \sum_{\ell=|j-j'|}^{j+j'} \pi_\ell.$$

4.3   The Example $G = \mathrm{SO}(3)$

One can construct representations of $\mathrm{SO}(3)$ by a procedure similar to the one used in the case of $\mathrm{SU}(2)$, namely by using homogeneous polynomials in three variables. But as we already remarked in the Example 3.2 in Section 3.2, we do not get irreducible representations without some further considerations. So let us discuss this a bit later and first follow a method suggested by H. Weyl deducing the representation theory of $\mathrm{SO}(3)$ from the result just obtained in 4.2 and the fact that $\mathrm{SU}(2)$ is a double cover of $\mathrm{SO}(3)$:

The elements g of $\mathrm{SO}(3)$ are real 3×3-matrices consisting of three ON-rows (or columns) (a_1, a_2, a_3), i.e. we have six conditions $< a_j, a_k >= \delta_{jk}$ for the nine entries in g. Hence we expect three free parameters. It is a classical fact that the elements of $\mathrm{SO}(3)$ are parametrized by three *Euler angles* α, β, γ. Again we follow Wigner's notation ([Wi] p.152):

Proposition 4.2: For every $g \in \mathrm{SO}(3)$ there exist angles $\alpha, \beta, \gamma \in \mathbf{R}$, such that

$$g = S_3(\alpha) S_2(\beta) S_3(\gamma).$$

Here $S_3(\alpha)$ and $S_3(\gamma)$ are rotations about the z-axis (or $e_3 \in \mathbf{R}^3$) through α resp. γ and $S_2(\beta)$ is a rotation about the y-axis (or $e_2 \in \mathbf{R}^3$), i.e. we put

$$S_3(\alpha) := \begin{pmatrix} \cos\alpha & -\sin\alpha & \\ \sin\alpha & \cos\alpha & \\ & & 1 \end{pmatrix},$$

$$S_2(\alpha) := \begin{pmatrix} \cos\alpha & & -\sin\alpha \\ & 1 & \\ \sin\alpha & & \cos\alpha \end{pmatrix},$$

$$S_1(\alpha) := \begin{pmatrix} 1 & & \\ & \cos\alpha & -\sin\alpha \\ & \sin\alpha & \cos\alpha \end{pmatrix}.$$

Exercise 4.5: Prove this.

Remark 4.3: i) There are several notational conventions for the Euler angles in common use. In Hein [He] p.56 we find a statement (with proof) analogous to our Proposition 4.2 with $S_2(\beta)$ replaced by $S_1(\beta)$. For a parametrization of $\mathrm{SO}(3)$ one needs only $\alpha, \gamma \in [0, 2\pi)$ and $\beta \in [0, \pi]$. This parametrization is not everywhere injective: for $\beta = 0$ α and γ are only fixed up to their sum. Goldstein [Go] p.145-148 gives an overview of other descriptions.

ii) $\mathrm{SO}(3)$ is generated by rotations $S_3(t)$ and $S_2(t)$ or by rotations of type $S_3(t)$ and $S_1(t)$.

The key fact for our discussion here is given by the following theorem (we shall later see that this is a special case of an analogous statement about $\mathrm{SL}(2, \mathbf{C})$ covering the (restricted) Lorentz group $\mathrm{SO}(3, 1)^+$):

Theorem 4.7: There is a surjective homomorphism

$$\varphi : \mathrm{SU}(2) \longrightarrow \mathrm{SO}(3)$$

with $\ker \varphi = \{\pm E_2\}$.
This homomorphism can be chosen such that

$$\varphi(s(\alpha/2)) = S_3(\alpha) \text{ and } \varphi(r(-\beta/2)) = S_2(\beta).$$

Proof: As in [Wi] p.158 we consider the *Pauli matrices*

$$s_x := (\begin{smallmatrix} & 1 \\ 1 & \end{smallmatrix}), \; s_y := (\begin{smallmatrix} & i \\ -i & \end{smallmatrix}), \; s_z := (\begin{smallmatrix} -1 & \\ & 1 \end{smallmatrix}).$$

The three-dimensional **R**-vector space V of hermitian 2×2-marices H with trace zero is the linear span of the Pauli matrices

$$
\begin{aligned}
V \; &:= \; \{H \in M_2(\mathbf{C}); \, {}^tH = \bar{H}, \mathrm{Tr}\, H = 0\} \\
&= \; \{H = (\begin{smallmatrix} -z & x+iy \\ x-iy & z \end{smallmatrix}) = xs_x + ys_y + zs_z; \; x, y, z \in \mathbf{R}\}.
\end{aligned}
$$

$G = \mathrm{SU}(2)$ acts on V by conjugation, i.e. we have a map

$$G \times V \longrightarrow V, \quad (g, H) \longmapsto gHg^{-1} = gH^t\bar{g} =: H',$$

which for H as above and $g = (\begin{smallmatrix} a & b \\ -\bar{b} & \bar{a} \end{smallmatrix})$ gives by a straightforward computation

$$H' = (\begin{smallmatrix} -z' & x'+iy' \\ x'-iy' & z' \end{smallmatrix})$$

with

$$
\begin{aligned}
x' &= (1/2)(a^2 + \bar{a}^2 - b^2 - \bar{b}^2)x + (i/2)(a^2 - \bar{a}^2 + b^2 - \bar{b}^2)y + (\bar{a}\bar{b} + ab)z \\
(4.4) \quad y' &= (i/2)(\bar{a}^2 - a^2 + b^2 - \bar{b}^2)x + (1/2)(a^2 + \bar{a}^2 + b^2 + \bar{b}^2)y + i(\bar{a}\bar{b} - ab)z \\
z' &= -(\bar{a}b + a\bar{b})x + i(\bar{a}b - a\bar{b})y + (a\bar{a} - b\bar{b})z.
\end{aligned}
$$

E.g. from the fact that the map det is multiplicative, we deduce

$$\det H = x^2 + y^2 + z^2 = \det H' = x'^2 + y'^2 + z'^2$$

and if we write (4.4) as a matrix relation with column ${}^t(x, y, z) = \mathbf{x}$

$$\mathbf{x}' = A(g)\mathbf{x},$$

we see that multiplication with $A(g)$ respects the euclidean quadratic form, i.e. we have $A(g) \in \mathrm{O}(3)$. Moreover, by a continuity argument, we have $A(g) \in \mathrm{SO}(3)$: as every $g \in \mathrm{SU}(2)$ can be deformed contiuously to $g = E_2$ and det is a continuous function, we cannot have $\det A(g) = -1$.

From the fact that conjugation is a group action, we see that

$$\varphi : g \longmapsto A(g) \in SO(3)$$

is a group homomorphism. As $\ker \varphi = \{g \in SU(2); gH^t\bar{g} = H \quad \text{for all } H \in V\}$, we can deduce easily that $\ker \varphi = \{\pm E_2\}$ (by choosing appropriate matrices H).
The fact that φ is surjective is more delicate: If $g = s(\alpha/2)$, i.e. $a = e^{i\alpha/2}, b = 0$, we see from (4.4) that we have

$$A(s(\alpha/2)) = S_3(\alpha).$$

If $g = r(\beta/2)$, we see from (4.4) that we have

$$A(r(\beta/2)) = S_2(-\beta).$$

We already know that $SO(3)$ is generated by matrices of type S_2 and S_3 and so we are done.

Now we use this result to describe representations of $SO(3)$:

If a representation $\pi : SU(2) \longrightarrow GL(V)$ *factorizes through* $SO(3)$, i.e. one has a homomorphism $\pi' : SO(3) \longrightarrow GL(V)$ with $\pi = \pi' \cdot \varphi$ resp. a commutative diagram

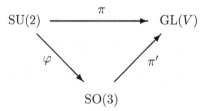

then, this way, π induces the representation π' of $SO(3)$. Obviously this is the case for a representation π with $\pi(-E_2) = id$, where we get a unique prescription by putting

$$\pi'(A) := \pi(g) \quad \text{if } A = \varphi(g).$$

For $\pi(-E_2) = -id$ one gets a *double valued representation* by putting

$$\pi'(A) := \pm\pi(g) \text{ if } A = \varphi(g) = \varphi(-g).$$

(This is a special case of a *projective representation,* which will be discussed below.)

In consequence, we know the unitary dual of $SO(3)$: The representations π_j of $SU(2)$ with $j \in \mathbf{N}_0$ are exactly those with $\pi(-E_2) = \pi(E_2) = id$ and thus induce (up to equivalence) all irreducible unitary representations π' of $SO(3)$: If there were any π' not in this family, it would composed with φ produce a representation of $SU(2)$, which by our completeness result for $SU(2)$ would be equivalent to a π_j.
For $g \in SU(2)$ and $\varphi(g) = g'$, Wigner ([Wi] p.167) conjugates the $\pi'(g') := \pi_j(g) = A^j(g)$ from 4.2 by the diagonal matrix $M = ((i)^{-2k}\delta_{kl})$ and calls the result $D^j(\{\alpha, \beta, \gamma\})$ with $\{\alpha, \beta, \gamma\}$ representing the Euler angles of $g' \in SO(3)$ in the parametrization chosen above. Hence we have from the explicit form of the matrices $A^j(g)$ obtained in 4.2 the explicit form

$$D^j(\{\alpha, \beta, \gamma\})_{p'p} = \sum_k (-1)^k \frac{\sqrt{(j+p)!(j-p)!(j+p')!(j-p')!}}{(j+p'-k)!(j+p-k)!k!(k+p'-p)!}$$
$$\times \quad e^{ip'\alpha}(\cos(\beta/2))^{2j+p+p'-2k}(\sin(\beta/2))^{2k+p'-p}.$$

The characters of these (projective) representations are

$$\chi_{\pi'_j}(S_3(\alpha)) = \sum_{p=-j}^{j} e^{ip\alpha} = 1 + 2\cos\alpha + \cdots + 2\cos(j\alpha), \; j \text{ integral}$$

$$= 2\cos(\alpha/2) + 2\cos(3\alpha/2) + \cdots + 2\cos(j\alpha), \; j \text{ half} - \text{integral}.$$

The double valued representations π'_j, j half-integral, led to H. Weyl's explanation of the *spin* of an electron. As said above, they are special examples of projective representations, which are very natural for reasons from physics: In quantum theory the state of a system is not so much described by an element v of a Hilbert space $\mathcal{H}$ but by an element $v_\sim$ of the associated projective space $\mathbf{P}(\mathcal{H})$ consisting of the one-dimensional subspaces of $\mathcal{H}$. In general, one writes for a **K**-vector space $\mathbf{P}(V) := \{v_\sim; \, v \in V, v \neq 0\}$ where $v_\sim$ is the class of v for the equivalence $v \sim v'$ if there is a $\lambda \in \mathbf{K}^*$ such that $v' = \lambda v$. Having this in mind, it is appropriate to introduce the following.

Definition 4.1: A *projective representation* $\tilde{\pi}$ of G in the **K**-vector space V is a map

$$\tilde{\pi} : G \longrightarrow GL(V),$$

which enjoys the property

$$\tilde{\pi}(gg') = c(g, g')\tilde{\pi}(g)\tilde{\pi}(g'),$$

where

$$c : G \times G \longrightarrow \mathbf{K}^*$$

with

$$c(g, g')c(gg', g'') = c(g, g'g'')c(g', g'') \quad \text{for all } g, g', g'' \in G.$$

c is also called a *system of multipliers.*

Remark 4.4: The functional equation for c is equivalent to the associativity

$$\tilde{\pi}(g(g'g'')) = \tilde{\pi}((gg')g'').$$

Exercise 4.6: Prove this.

Every linear representation is also a projective representation. And analogously to the case of linear representations, one also adds a continuity condition and discusses in general these representations.

In the second part of his book [Wi], Wigner describes the application of these results to the theory of *spin* and atomic spectra. Though we are tempted to do more, from this we only take over that, roughly, an irreducible unitary (projective) representation π'_j of SO(3) corresponds to a particle having three-dimensional rotational symmetry. The number j specifying the representation then is its angular momentum quantum number or its spin and the indices numbering a basis of the representation space are interpreted as magnetic quantum numbers. We strongly recommend to the reader interested in this topic to look into Wigner's book or any text from physics.

To finish this chapter, we briefly indicate another approach to the determination of the representations of $G = \mathrm{SO}(3)$, which is a first example for the large theory of *eigenspace representations* promoted in particular by Helgason.

From analysis it is well known that the three-dimensional euclidean Laplace operator

$$\Delta = \partial_x^2 + \partial_y^2 + \partial_z^2$$

is $\mathrm{SO}(3)$ invariant, i.e. for $f \in C^\infty(\mathbf{R}^3)$ we have

$$(\Delta f)(g\mathbf{x}) = \Delta f(\mathbf{x}), \quad \text{for all } g \in G, \mathbf{x} = {}^t(x, y, z) \in \mathbf{R}^3.$$

Hence the space V_ℓ of homogeneous polynomials of degree ℓ in three variables, which satisfy the Laplace equation

$$\Delta f = 0,$$

is invariant under the action $f \longmapsto f^g$ with $f^g(\mathbf{x}) = f(g^{-1}\mathbf{x})$ of $G = \mathrm{SO}(3)$. We denote by $D^\ell(g)$ the matrix form of the representation π_ℓ of G on V_ℓ given by $\pi_\ell(g)f = f^g$. To solve the Laplace equation, one usually introduces spherical coordinates (r, φ, ϑ) with

$$
\begin{aligned}
x &= r \sin \vartheta \cos \varphi, \\
y &= r \sin \vartheta \sin \varphi, \\
z &= r \cos \vartheta.
\end{aligned}
$$

Homogeneous polynomials f of degree ℓ in x, y, z have the form $f = r^\ell Y(\vartheta, \varphi)$. If this f is introduced into the Laplace equation written in the spherical coordinates

$$(\partial_r^2 + (2/r)\partial_r + (1/r^2)(\partial_\vartheta^2 + \cot \vartheta \, \partial_\vartheta + (1/(\sin \vartheta)^2)\partial_\varphi^2))f = 0$$

r drops out and the differential equation

$$(\ell(\ell + 1) + (\partial_\vartheta^2 + \cot \vartheta \, \partial_\vartheta + (1/(\sin \vartheta)^2)\partial_\varphi^2))Y = 0$$

in the variables (ϑ, φ) results. The $(2\ell+1)$ linearly independent solutions $Y = Y_{\ell m}$ (with $m = -\ell, -\ell+1, \ldots, \ell$) are known as *spherical harmonics* of the ℓ-th degree. They have the form

$$Y_{\ell m}(\vartheta, \varphi) = \Phi_m(\varphi)\Theta_{\ell m}(\vartheta)$$

where

$$
\begin{aligned}
\Phi_m(\varphi) &:= \frac{1}{\sqrt{2\pi}} e^{im\varphi}, \\
\Theta_{\ell m}(\vartheta) &:= (-1)^m \left(\frac{2\ell + 1}{2} \frac{(\ell - m)!}{(\ell + m)!}\right)^{1/2} (\sin \vartheta)^m \frac{d^m}{d(\cos \vartheta)^m} P_l(\cos \vartheta), \\
\Theta_{\ell, -m}(\vartheta) &:= \left(\frac{2\ell + 1}{2} \frac{(\ell - m)!}{(\ell + m)!}\right)^{1/2} (\sin \vartheta)^m \frac{d^m}{d(\cos \vartheta)^m} P_l(\cos \vartheta),
\end{aligned}
$$

with $m \geq 0$ in the last two equations. The P_ℓ are the *Legendre polynomials* defined by

$$P_\ell(x) := \frac{1}{2^\ell \ell!} \frac{d^\ell}{dx^\ell}(x^2 - 1)^\ell.$$

This leads to the matrix form D^ℓ for π'_ℓ with

$$\pi'_\ell(g) r^\ell Y_{\ell m}(\vartheta, \varphi) = \sum_{m'=-\ell}^{\ell} Y_{\ell m'}(\vartheta, \varphi) D^\ell_{m'm}(g).$$

As one easily deduces, we have in particular

$$D^\ell_{m'm}(S_3(\alpha)) = e^{im\alpha} \delta_{m'm}.$$

By applying Schur's commutation criterium, one can prove here, like we did already above, that D^ℓ is irreducible for all $\ell = 0, 1, 2, \ldots$.

The conjugacy classes of $\mathrm{SO}(3)$ are parametrized by the angle α, $\alpha \in [0, \pi)$. For the characters we obtain the formula

$$\chi^{(\ell)}(S_3(\alpha)) = \sum_{m=-\ell}^{\ell} e^{im\alpha} = 1 + 2\cos\alpha + \cdots + 2\cos(\ell\alpha).$$

Again we can see that we got **all** irreducible representations as $\{\cos n\varphi\}_{n \in \mathbf{N}_0}$ is a complete system in $L^2([0, \pi))$ by Fourier's Theorem.

Chapter 5

Representations of Abelian Groups

We have seen that
– finite groups have (up to equivalence) finitely many irreducible representations and these are all finite-dimensional, and
– compact groups have (up to equivalence) denumerably many irreducible representations and these are also all finite-dimensional.
Abelian topological groups G (with prototype $G = \mathbf{R}$ or $= \mathbf{C}$) are something of an extreme in the opposite direction. Here we know that unitary irreducible representations are all one-dimensional (we proved this as a consequence of the "Unitary Schur" in 1.5). But in general the unitary dual is not denumerable. Since in this text we are mainly interested in groups like $\mathrm{SL}(2, \mathbf{R})$ and $\mathrm{Heis}(\mathbf{R})$, where unitary irreducible representations will turn out to be infinite-dimensional, we discuss here only briefly some notions of general interest, following [BR] p.159 - 165 and [Ki] p.167 - 173.

5.1   Characters and the Pontrjagin Dual

Definition 5.1: A *character* of an abelian locally compact group G is a continuous function $\chi : G \longrightarrow \mathbf{C}$, which satisfies

$$|\chi(g)| = 1 \text{ and } \chi(gg') = \chi(g)\chi(g') \quad \text{for all } g, g' \in G,$$

i.e. a character is a (one-dimensional) continuous irreducible unitary representation of G.

Remark 5.1: We already defined the *unitary dual* $\hat{G}$ of any group G as the set of equivalence classes of unitary irreducible representations of G. For G abelian, $\hat{G}$ consists just of all characters of G. It is also called the *Pontrjagin dual* of G.

Remark 5.2: For G abelian, $\hat{G}$ is also a group: We define the composition of two characters χ and χ' by

$$(\chi\chi')(g) := \chi(g)\chi'(g) \quad \text{for all } g \in G.$$

If G is a topological group, $\hat{G}$ is also. More precisely, if G is locally compact (resp. compact, resp. discrete), $\hat{G}$ is locally compact (resp. discrete, resp. compact). One has $G \simeq \hat{\hat{G}}$.

Example 5.1: For $G = \mathbf{R}^n$ every character χ has the form

$$\chi(x) = \exp(i(\hat{x}_1 x_1 + \cdots + \hat{x}_n x_n)) = \exp(i < \hat{x}, x >), \text{ with } \hat{x} \in \mathbf{R}^n, \text{ for } x \in \mathbf{R}^n,$$

i.e. we have $\hat{G} = G$.

Example 5.2: For $G = S^1 = \{\zeta \in \mathbf{C}; |\zeta| = 1\}$ every character has the form

$$\chi(\zeta) = \zeta^k, \ k \in \mathbf{Z},$$

as we discussed already in 3.2. I.e., here we have $\hat{G} = \mathbf{Z}$.

Example 5.3: For $G = \mathbf{Z}$, every character has the form

$$\chi(m) = e^{im\varphi}, \ \varphi \in [0, 2\pi),$$

i.e. we have $\hat{G} = S^1$ (and, hence, the self-duality of $\mathbf{Z}$).

5.2   Continuous Decompositions

While studying representation theory for compact groups, we generalized the Fourier series

$$f(t) = \sum_{k \in \mathbf{Z}} c(k)e^{ikt}, \ c(k) := (1/(2\pi)) \int_0^{2\pi} e^{-ikt} f(t)dt$$

for $G = \mathrm{SO}(2) \simeq S^1$ to other compact groups. Now, the analogon of Fourier series for $G = \mathbf{R}$ is the *Fourier transform* associating to a sufficiently nice function f its Fourier transform

(5.1) $$\mathcal{F}f(k) = \hat{f}(k) := (1/\sqrt{2\pi}) \int_{-\infty}^{\infty} \overline{\chi}_k(t) f(t)dt, \ \chi_k(t) := e^{ikt}.$$

This may be interpreted in the following way: In the case of compact groups we had a (denumerable) direct sum decomposition of the regular representation ρ into irreducible subrepresentations π_j

$$\rho = \sum_{j \in J} \pi_j.$$

For $G = \mathbf{R}$ we analogously have the regular representation ρ given on $\mathcal{H} = L^2(\mathbf{R}, dx)$ by

$$\rho(t)f(x) = f(x + t) \quad \text{for all } x, t \in \mathbf{R}.$$

Suppose $\mathcal{H}_1$ would be the space of a one-dimensional subrepresentation. Then we have for every $f_1 \in \mathcal{H}_1$

$$\rho(t)f_1(x) = f_1(x + t) = \lambda(t)f_1(x) \quad \text{for all } x, t \in \mathbf{R}.$$

Hence f_1 has to be an exponential function and we get a contradiction, because $L^2(\mathbf{R})$ contains no exponential function $\neq 0$. But (5.1) indicates that ρ can be decomposed into a *direct integral* of the irreducible one-dimensional representations $\chi_k, k \in \mathbf{R}$. For a proper generalization one here needs more tools from functional analysis, in particular replacement of the usual measure (as recalled in 3.3) by the notion of a *spectral measure* $E(.)$, i.e. a measure which takes operators as values and allows to decompose (linear) operators into a direct integral. Following [BR] p.649 resp. [Ki] p.57, we present them briefly as they are also essential for subsequent chapters.

Spectral Measure

Let $[a, b] \subset \mathbf{R}$ be a finite or infinite interval and E be a function on $[a, b]$ with values $E(\lambda)$ for $\lambda \in [a, b]$, which are operators in a Hilbert space $\mathcal{H}$. E is called a *spectral function* if it satisfies the following conditions:
i) $E(\lambda)$ is self-adjoint for each λ,
ii) $E(\lambda)E(\mu) = E(\min(\lambda, \mu))$,
iii) The operator function E is strongly right continuous, i.e. one has

$$\lim_{\epsilon \to 0+} E(\lambda + \epsilon)u = E(\lambda)u \quad \text{for all } u \in \mathcal{H},$$

iv) $E(-\infty) = 0$, $E(\infty) = Id$, i.e. one has

$$\lim_{\lambda \to -\infty} E(\lambda)u = 0, \quad \lim_{\lambda \to \infty} E(\lambda)u = u \quad \text{for all } u \in \mathcal{H}.$$

Conditions i) and ii) mean that $E(\lambda), \lambda \in [a, b]$, are (refining Definition 1.7) orthogonal projection operators. For an interval $\Delta = [\lambda_1, \lambda_2] \subset [a, b]$ one denotes the difference $E(\lambda_2) - E(\lambda_1)$ by $E(\Delta)$.
For $\Delta_1, \Delta_2 \subset [a, b]$ one has

$$E(\Delta_1)E(\Delta_2) = E(\Delta_1 \cap \Delta_2),$$

in particular

$$E(\Delta_1)E(\Delta_2) = 0 \text{ if } \Delta_1 \cap \Delta_2 = \emptyset,$$

i.e. the subspaces $\mathcal{H}_1 = E(\Delta_1)\mathcal{H}$ and $\mathcal{H}_2 = E(\Delta_2)\mathcal{H}$ are orthogonal.

Example 5.4: We take $G = \mathbf{R}, \mathcal{H} = L^2(\mathbf{R})$ and $\pi(a)u(x) := u(x + a)$ for $u \in \mathcal{H}$. Then

$$\mathbf{R} \ni \lambda \longmapsto E(\lambda) := (2\pi)^{-1} \int_{-\infty}^{\lambda} d\lambda' \int_{-\infty}^{\infty} e^{-i\lambda'a} \pi(a) da$$

is a spectral function.

The properties of the spectral function imply that, for any $u \in \mathcal{H}$, the function

$$\sigma_u(\lambda) := \, < E(\lambda)u, u >$$

is a right continuous, non-decreasing function of bounded variation with

$$\sigma_u(-\infty) = 0, \, \sigma_u(\infty) = \| \, u \, \|^2 \, .$$

Moreover it is denumerable additive: For a pairwise disjoint decomposition $\{\Delta_n\}$ with $\Delta = \cup_{n=1}^{\infty} \Delta_n$, one has

$$E(\Delta) = \sum_{n=1}^{\infty} E(\Delta_n) \text{ and } \sigma_u(\Delta) = \sum_{n=1}^{\infty} \sigma_u(\Delta_n).$$

The function σ_u is called the *spectral measure*. We now can state the fundamental theorem of spectral decomposition theory as follows:

Theorem 5.1: Every self-adjoint operator A in $\mathcal{H}$ has the representation

$$A = \int_{-\infty}^{\infty} \lambda dE(\lambda).$$

Continuous Sums of Hilbert Spaces

The operation of direct sum of Hilbert spaces admits a generalization: Let there be given a set X with measure μ and a family $\{\mathcal{H}_x\}_{x \in X}$ of Hilbert spaces. It is natural to try to define a new space

$$\mathcal{H} = \int_X \mathcal{H}_x d\mu(x),$$

the elements of which are to be functions f on X, assuming values in $\mathcal{H}_x$ for all $x \in X$, and the scalar product being defined by the formula

$$< f_1, f_2 > := \int_X < f_1(x), f_2(x) >_{\mathcal{H}_x} d\mu(x).$$

The difficulty consists in the fact that the integrand in this expression may be non-measurable. In the case where all of the spaces $\mathcal{H}_x$ are separable, one can introduce an appropriate definition of measurable vector-functions f, which will guarantee this measurability of the numerical function $x \longmapsto < f_1(x), f_2(x) >$: In the special example that all $\mathcal{H}_x$ coincide with a fixed separable $\mathcal{H}_0$ with basis $\{e_j\}$, the numerical functions $x \longmapsto < f(x), e_j >$ are measurable (for more general cases see [Ki] p.57). Once an appropriate concept of measurable vector-functions has been adopted one defines the *continuous sum* (or *direct integral*) $\mathcal{H} = \int_X \mathcal{H}_x d\mu(x)$ as the set of equivalence classes of measurable vector-functions f with summable square of norms.

An operator A in $\mathcal{H} = \int_X \mathcal{H}_x d\mu(x)$ is called *decomposable* iff there exists a family $\{A(x)\}$ of operators in the spaces $\mathcal{H}_x$ such that

$$(Af)(x) = A(x)f(x) \text{ almost everywhere on } X.$$

By a suitable generalization of the notion of spectral measure to the dual $\hat{G}$ (which for an abelian locally compact group G can be given the structure of a topological space), one has the general statement:

Theorem 5.2: Let π be a unitary continuous representation of an abelian locally compact group G in a Hilbert space $\mathcal{H}$. Then there exists a spectral measure $E(.)$ on the character group $\hat{G}$ such that

$$\pi(g) = \int_{\hat{G}} \chi(g) dE(\chi).$$

This theorem is called SNAG-theorem, as it goes back to Stone, Naimark, Ambrose, and Godement. One can find a proof in [BR] p.160ff. We also recommend the Chapter on locally compact commutative groups in [Ma] p.37ff, which is centered on the notion of *projection valued measure*. This is a function $E \longrightarrow P_E$ associating a projection operator P_E in $\mathcal{H}$ to each Borel set E from a Borel space S with the properties
i) $P_\emptyset = 0$, $P_S = Id$,
ii) $P_{E \cap F} = P_E P_F$ for all E and F,
iii) $P_E = \sum_{j \in \mathbf{N}} P_{E_j}$ for every disjoint decomposition $E = \sqcup_{j \in \mathbf{N}} E_j$.

Chapter 6

The Infinitesimal Method

Up to now, with exception of the Schrödinger representation of the Heisenberg group, all irreducible representations treated here were finite-dimensional. Before we go further into the construction of infinite-dimensional representations, we discuss a method which allows a linearization of our objects and hence simplifies the task to determine the structure of the representations and classify them. As we shall see, this is helpful for compact and noncompact groups as well. The idea is to associate to the linear topological group G a linear object, its *Lie algebra* $\mathfrak{g} = \operatorname{Lie} G$, and study representations $\hat{\pi}$ of these algebras, which are open to purely algebraic and combinatorial studies. One can associate to each representation π of G an *infinitesimal representation* $d\pi$ of $\mathfrak{g}$. Conversely, one can ask which representations $\hat{\pi}$ may be integrated to a unitary representation π of G, i.e., which are of the form $\hat{\pi} = d\pi$, $\pi \in \hat{G}$. As we will see in several examples, this method allows us to classify the (unitary) irreducible representations and, hence, gives a parametrization of $\hat{G}$. It will further prove to be helpful for the construction of explicit models for representations $(\pi, \mathcal{H})$.

6.1 Lie Algebras and their Representations

We introduce some algebraic notions, which will become important in the sequel. The interested reader can learn more about these in any text book on Algebra (e.g. Lang's *Algebra* [La]).

Definition 6.1: Let $\mathbf{K}$ be a field. A $\mathbf{K}$-*algebra* is a $\mathbf{K}$-vector space $\mathcal{A}$ provided with a $\mathbf{K}$-bilinear map (*multiplication*)

$$\mathcal{A} \times \mathcal{A} \longrightarrow \mathcal{A}, \quad (x, y) \longmapsto xy.$$

$\mathcal{A}$ is called

    – *associative* iff one has

$$(xy)z = x(yz) \quad \text{for all } x, y, z \in \mathcal{A},$$

    – *commutative* iff one has

$$xy = yx \quad \text{for all } x, y \in \mathcal{A},$$

– a *Lie algebra* iff the multiplication xy, usually written as $xy =: [x,y]$ (the *Lie bracket*), is anti-symmetric, i.e.

$$[x,y] = -[y,x] \quad \text{for all } x,y \in \mathcal{A},$$

and fulfills the *Jacobi identity*, i.e. one has

$$[x,[y,z]] + [y,[z,x]] + [z,[x,y]] = 0 \quad \text{for all } x,y,z \in \mathcal{A}.$$

– a *Jordan algebra* iff one has

$$xy = yx \text{ and } x^2(xy) = x(x^2y) \quad \text{for all } x,y \in \mathcal{A},$$

– a *Poisson algebra* iff $\mathcal{A}$ is a commutative ring with multiplication

$$\mathcal{A} \times \mathcal{A} \longrightarrow \mathcal{A}, \ (x,y) \longmapsto xy$$

and one has a bilinear map

$$\mathcal{A} \times \mathcal{A} \longrightarrow \mathcal{A}, \ (x,y) \longmapsto \{x,y\}$$

(the *Poisson bracket*) fulfilling the identities

$$
\begin{aligned}
\{x,y\} &= -\{y,x\}, \\
\{x,yz\} &= \{x,y\}z + y\{x,z\}, \\
\{x,\{y,z\}\} &= \{\{x,y\},z\} + \{y,\{x,z\}\} \quad \text{for all } x,y,z \in \mathcal{A}.
\end{aligned}
$$

The dimension of $\mathcal{A}$ as a **K**-vector space is called the *dimension of the algebra*.

A map $\varphi : \mathcal{A} \longrightarrow \mathcal{A}'$ is an *algebra homomorphism* iff it is a **K**-linear map and respects the respective composition, i.e. if one has

$$\varphi(xy) = \varphi(x)\varphi(y) \quad \text{for all } x,y \in \mathcal{A}.$$

Exercise 6.1: Show that $\mathcal{C}^\infty(\mathbf{R}^{2n})$ is a Poisson algebra with

$$\{f,g\} := \sum_{i=1}^{n} \left(\frac{\partial f}{\partial q_i} \frac{\partial g}{\partial p_i} - \frac{\partial f}{\partial p_i} \frac{\partial g}{\partial q_i} \right) \quad \text{for all } f,g \in \mathcal{C}^\infty(\mathbf{R}^{2n}),$$

where the coordinates are denoted by $(q_1, \ldots, q_n, p_1, \ldots, p_n) \in \mathbf{R}^{2n}$.

Exercise 6.2: Verify that each Poisson algebra is also a Lie algebra.

In the following, we are mainly interested in Lie algebras over $\mathbf{K} = \mathbf{R}$ or $\mathbf{K} = \mathbf{C}$, which we usually denote by the letter $\mathfrak{g}$. We recommend as background any text book on this topic, for instance Humphreys' *Introduction to Lie Algebras and Representation Theory* [Hu].

Example 6.1: Every $M_n(\mathbf{K})$ is an associative noncommutative algebra with composition XY of two elements $X,Y \in M_n(\mathbf{K})$ given by matrix multiplication.

Example 6.2: Every associative algebra $\mathcal{A}$ is a Lie algebra with

$$[X, Y] = XY - YX \quad \text{for all } X, Y \in \mathcal{A},$$

the *commutator* of X and Y. In particular, one denotes

$$\mathfrak{gl}(n, \mathbf{K}) := M_n(\mathbf{K}).$$

Example 6.3: If V is any $\mathbf{K}$-vector space, the space $\text{End}\, V$ of linear maps F from V to V is a Lie algebra with

$$[F, G] := FG - GF \quad \text{for all } F, G \in \text{End}\, V.$$

Example 6.4: If $\mathfrak{g}$ is a Lie algebra and $\mathfrak{g}_0$ is a subspace with $[X, Y] \in \mathfrak{g}_0$ for all $X, Y \in \mathfrak{g}_0$, $\mathfrak{g}_0$ is again a Lie algebra. One easily verifies this and that

$$
\begin{aligned}
\mathfrak{sl}(n, \mathbf{K}) \;&:= \{X \in \mathfrak{gl}(n, \mathbf{K}); \text{Tr}\, X = 0\}, \\
\mathfrak{so}(n) \;&:= \{X \in \mathfrak{gl}(n, \mathbf{K}); X = -{}^t X\}, \\
\mathfrak{su}(n) \;&:= \{X \in \mathfrak{gl}(n, \mathbf{K}); X = -{}^t \bar{X}, \text{Tr}\, X = 0\}
\end{aligned}
$$

are examples of Lie algebras.

Example 6.5: If $\mathcal{A}$ is a $\mathbf{K}$-algebra,

$$\text{Der}\mathcal{A} := \{D : \mathcal{A} \longrightarrow \mathcal{A}\ \ \mathbf{K} - \text{linear}; D(ab) = a\, Db + Da\, b \quad \text{for all } a, b \in \mathcal{A}\}$$

is a Lie algebra, the *algebra of derivations*.

Exercise 6.3: Verify that the matrices

$$F := \begin{pmatrix} & 1 \\ & \end{pmatrix}, G := \begin{pmatrix} \\ 1 & \end{pmatrix}, H := \begin{pmatrix} 1 & \\ & -1 \end{pmatrix}$$

are a basis of $\mathfrak{sl}_2(\mathbf{K})$ with the relations

$$[H, F] = 2F, \quad [H, G] = -2G, \quad [F, G] = H.$$

Exercise 6.4: Verify that the matrices

$$X_1 := (1/2)\begin{pmatrix} & -i \\ -i & \end{pmatrix}, X_2 := (1/2)\begin{pmatrix} & -1 \\ 1 & \end{pmatrix}, X_3 := (1/2)\begin{pmatrix} i & \\ & -i \end{pmatrix}$$

are a basis of $\mathfrak{su}(2)$ with the relations

$$[X_i, X_j] = -X_k, (i, j, k) = (1, 2, 3), (2, 3, 1), (3, 1, 2).$$

Exercise 6.5: Verify that the matrices

$$X_1 := \begin{pmatrix} 0 & \\ & -1 \\ & 1 \end{pmatrix}, X_2 := \begin{pmatrix} & & -1 \\ & 0 & \\ 1 & & \end{pmatrix}, X_3 := \begin{pmatrix} & -1 \\ 1 & \\ & & 0 \end{pmatrix}$$

are a basis of $\mathfrak{so}(3)$ with the relations

$$[X_i, X_j] = -X_k, (i, j, k) = (1, 2, 3), (2, 3, 1), (3, 1, 2).$$

Exercise 6.6: Show that $\mathfrak{g} = \mathbf{R}^3$ provided with the usual vector product as composition is a Lie algebra isomorphic to $\mathfrak{so}(3)$.

Exercise 6.7: Show that $\mathfrak{g} = \mathbf{R}^3$ provided with the composition

$$((p, q, r), (p', q', r')) \longmapsto (0, 0, 2(pq' - p'q))$$

is a Lie algebra, the *Heisenberg algebra* $\mathrm{heis}(\mathbf{R})$, with basis

$$P = (1, 0, 0), Q = (0, 1, 0), R = (0, 0, 1)$$

and relations

$$[R, P] = [R, Q] = 0, [P, Q] = 2R.$$

Remark 6.1: It will soon be clear that $\mathfrak{gl}(n, \mathbf{R}), \mathfrak{sl}(n, \mathbf{R}), \mathfrak{so}(n), \mathfrak{su}(n)$, and $\mathrm{heis}(\mathbf{R})$ are associated to the groups $\mathrm{GL}(n, \mathbf{R}), \mathrm{SL}(n, \mathbf{R}), \mathrm{SO}(n), \mathrm{SU}(n)$, resp. $\mathrm{Heis}(\mathbf{R})$.

In analogy with the concept of the linear representation of a group G in a vector space V, one introduces the following notion.

Definition 6.2: A *representation of the Lie algebra* $\mathfrak{g}$ in V is a Lie algebra homomorphism $\hat{\pi} : \mathfrak{g} \longrightarrow \mathrm{End}\, V$, i.e. $\hat{\pi}$ is $\mathbf{K}$-linear and one has

$$\hat{\pi}([X, Y]) = [\hat{\pi}(X), \hat{\pi}(Y)] \quad \text{for all } X, Y \in \mathfrak{g}.$$

Here we have no topology and, hence, no continuity condition.

Each Lie algebra has a *trivial representation* $\hat{\pi} = \hat{\pi}^0$ with $\hat{\pi}^0(X) = 0$ for all $X \in \mathfrak{g}$ and an *adjoint representation* $\hat{\pi} = \mathrm{ad}$ given by

$$\mathrm{ad}\, X(Y) := [X, Y] \quad \text{for all } X, Y \in \mathfrak{g}.$$

Exercise 6.8: Verify this.

In analogy with the respective notions for group representations, we have the following concepts.
− $(\hat{\pi}_0, V_0)$ is a *subrepresentation* of $(\hat{\pi}, V)$ iff $V_0 \subset V$ is a $\hat{\pi}$-invariant subspace, i.e. $\hat{\pi}(X)v_0 \in V_0$ for all $v_0 \in V_0$, and $\hat{\pi}_0 = \hat{\pi}\,|_V$.
− $\hat{\pi}$ is *irreducible* iff $\hat{\pi}$ has no nontrivial subrepresentation.
If $\hat{\pi}_1$ and $\hat{\pi}_2$ are representations of $\mathfrak{g}$ we have
− the *direct sum* $\hat{\pi}_1 \oplus \hat{\pi}_2$ as the representation on $V_1 \oplus V_2$ given by

$$(\hat{\pi}_1 \oplus \hat{\pi}_2)(X)(v_1, v_2) := (\hat{\pi}_1(X)v_1, \hat{\pi}_2(X)v_2) \quad \text{for all } v_1 \in V_1, v_2 \in V_2,$$

and
− the *tensor product* $\hat{\pi}_1 \otimes \hat{\pi}_2$ as the representation on $V_1 \otimes V_2$ given by

$$(\hat{\pi}_1 \otimes \hat{\pi}_2)(X)(v_1 \otimes v_2) := \hat{\pi}_1(X)v_1 \otimes v_2 + v_1 \otimes \hat{\pi}_2(X)v_2 \quad \text{for all } v_1 \in V_1, v_2 \in V_2.$$

Exercise 6.9: Verify that these are in fact representations.

Exercise 6.10: Find the prescription to define a *contragredient representation* $(\hat{\pi}^*, V^*)$ to a given representation $(\hat{\pi}, V)$ of $\mathfrak{g}$.

There is a lot of material about the structure and representations of Lie algebras, much of it going back to Elie and Henri Cartan. As already indicated, a good reference for this is the book by Humphreys [Hu]. We will later need some of this and develop parts of it. Here we only mention that one comes up with facts like this: If $\mathfrak{g}$ is semisimple (i.e. has no nontrivial ideals), then each representation $\hat{\pi}$ of $\mathfrak{g}$ is completely reducible.

Before we discuss more examples, we establish the relation between linear groups and *their* Lie algebras.

6.2 The Lie Algebra of a Linear Group

In Section 3.1 we introduced the concept of a linear group G as a closed subgroup of a matrix group $GL(n, \mathbf{R})$ or $GL(n, \mathbf{C})$ for a suitable $n \in \mathbf{N}$. To such a group we associate a Lie algebra $\mathfrak{g} = \operatorname{Lie} G$ using as essential tool the *exponential function for matrices*. We rely again on some experience from calculus in several complex variables and we define the exponential function exp by

$$\exp X := \sum_{k=0}^{\infty} (1/k!) X^k \quad \text{for all } X \in M_n(\mathbf{C}).$$

Remark 6.2: One has the following facts:

1. $\exp X$ is absolutely convergent for each $X \in M_n(\mathbf{C})$.
2. $\exp X : M_n(\mathbf{C}) \longrightarrow M_n(\mathbf{C})$ is continuously differentiable.
3. There is an open neighbourhood U of $0 \in M_n(\mathbf{R})$, which is diffeomorphic to an open neighbourhood V of $E \in GL(n, \mathbf{R})$.
4. $\log X := \Sigma((-1)^k/k)(X - E)^k$ converges for $\| X - E \| < 1$ and we have

$$\begin{aligned} \exp \log X &= X \text{ for } \| X - E \| < 1, \\ \log \exp X &= X \text{ for } \| \exp X - E \| < 1. \end{aligned}$$

5. One has

i)	$\exp(X + Y)$	$= \exp X \exp Y$	for *commuting* $X, Y \in M_n(\mathbf{C})$,
ii)	$(\exp X)^{-1}$	$= \exp(-X)$	for $X \in M_n(\mathbf{C})$,
iii)	$A \exp X A^{-1}$	$= \exp(AXA^{-1})$	for $X \in M_n(\mathbf{C}), A \in GL(n, \mathbf{C})$,
iv)	$\det(\exp X)$	$= \exp \operatorname{Tr} X$	for $X \in M_n(\mathbf{C})$.

The concientious reader should prove all this as **Exercise 6.11**. She/he should remember that we introduced in (3.1) in Section 3.1

$$< X, Y > = \operatorname{Re} \operatorname{Tr} {}^t \bar{X} Y, \ \| X \| = < X, X >^{1/2} \quad \text{for all } X, Y \in M_n(\mathbf{C}),$$

and that one has the fundamental inequalities

$$\begin{aligned} \| X + Y \| &\leq \| X \| + \| Y \|, \quad \text{the *triangle inequality*,} \\ |< X, Y >| &\leq \| X \| \| Y \|, \quad \text{the *Cauchy Schwarz inequality*.} \end{aligned}$$

Everyone interested in concrete examples could do the following **Exercise 6.12** preparing the ground for the next definition: Verify

$$\exp(tX) = \begin{pmatrix} 1 & t \\ & 1 \end{pmatrix}, \qquad \text{for } X = \begin{pmatrix} & 1 \\ & \end{pmatrix},$$

$$\exp(tX) = \begin{pmatrix} \cos t & \sin t \\ -\sin t & \cos t \end{pmatrix}, \qquad \text{for } X = \begin{pmatrix} & 1 \\ -1 & \end{pmatrix},$$

$$\exp(tX) = \begin{pmatrix} e^t & \\ & e^{-t} \end{pmatrix}, \qquad \text{for } X = \begin{pmatrix} 1 & \\ & -1 \end{pmatrix},$$

$$\exp(tX) = \begin{pmatrix} \cos t & \sin t & \\ -\sin t & \cos t & \\ & & 1 \end{pmatrix}, \qquad \text{for } X = \begin{pmatrix} & 1 & \\ -1 & & \\ & & 0 \end{pmatrix}.$$

Exercise 6.13: Verify that one has the following relations

$$\begin{aligned}
\{\exp(tX), X \in \mathfrak{so}(3); t \in \mathbf{R}\} &= \mathrm{SO}(3), \\
\{\exp(tX), X \in \mathfrak{su}(2); t \in \mathbf{R}\} &= \mathrm{SU}(2), \\
\{\exp(tX), X \in \mathfrak{sl}(2,\mathbf{R}); t \in \mathbf{R}\} &\neq \mathrm{SL}(2,\mathbf{R}).
\end{aligned}$$

This exercise shows that there are groups, which are examples for *groups of exponential type*, i.e. images under the expontial map of *their* Lie algebras. For $G = \mathrm{SL}(2,\mathbf{R})$ one has

$$\{\exp(tX), X \in \mathfrak{sl}(2,\mathbf{R}), t \in \mathbf{R}\} \neq \mathrm{SL}(2,\mathbf{R}),$$

but the matrices $\exp(tX)$, $X \in \mathfrak{sl}(2,\mathbf{R})$, $t \in \mathbf{R}$, generate the identity component of $G = \mathrm{SL}(2,\mathbf{R})$.

We take this as motivation to associate a Lie algebra to each linear group by a kind of inversion of the exponential map:

Definition 6.3: Let G be a linear group contained in $\mathrm{GL}(n,\mathbf{K})$, $\mathbf{K} = \mathbf{R}$ or $\mathbf{K} = \mathbf{C}$. Then we denote

$$\mathfrak{g} = \mathrm{Lie}\, G := \{X \in M_n(\mathbf{K}); \exp(tX) \in G \quad \text{for all } t \in \mathbf{R}\}.$$

Theorem 6.1: $\mathfrak{g}$ is a real Lie algebra.

A hurried or less curious reader may simply believe this and waive the proof. (See for instance [He] p.114: it is, in principle, not too difficult but lengthy. It shows why (in our approach) we had to restrict ourselves to closed matrix groups.) We emphasize that it is essentially the heart of the whole theory that not only a linear group but also every Lie group has an associated Lie algebra. This is based on another important notion, which we now introduce because it gives us a tool to determine explicitly the Lie algebras for those groups we are particularly interested in (even if we we have skipped the proof of the central theorem above).

Definition 6.4: A *one-parameter subgroup* γ of a topological group G is a continuous homomorphism

$$\gamma : \mathbf{R} \longrightarrow G, \quad t \longmapsto \gamma(t).$$

This notation is a bit misleading: to be precise, $G_0 = \gamma(\mathbf{R})$ is indeed a subgroup of G. γ is often abbreviated as a 1-PUG (from the German term "Untergruppe").

Remark 6.3: Each $X \in \mathfrak{g}$ produces a 1-PUG γ, namely the subgroup given by

$$\gamma(t) := \exp(tX).$$

This is obvious since $\exp(tX)$ is continuous (and even differentiable) in t and fulfills the functional equation i) in part 5 of the Remark 6.2 above.

If the 1-PUG $\gamma = \gamma(t)$ is differentiable in t, we assign to it an *infinitesimal generator* $X = X_\gamma$ by

$$X := \frac{d}{dt}\gamma(t)\,|_{t=0}\,.$$

If γ is given as $\gamma(t) = \exp(tX)$, we have X as infinitesimal generator.

This suggests the following procedure to determine the Lie algebra $\mathfrak{g}$ of a given matrix group G (which is also the background of our Exercise 6.13 above): Look for "sufficiently many independent" one-parameter subgroups $\gamma_1, \ldots, \gamma_d$ and determine their generators $X_1, \ldots, X_d$. Then $\mathfrak{g}$ will come out as the **R**-vector space generated by these matrices X_i. Obviously, one needs some explanation:

– Two 1-PUGs γ and γ' are *independent* iff their infinitesimal generators are linearly independent over **R** as matrices in $M_n(\mathbf{C})$.

– The number d of the *sufficiently many* independent (γ_i) is just the *dimension* of our group G as a *topological manifold*: Up to now we could work without this notion, but later on we will even have to consider *differentiable* manifolds. Therefore, let us recall (or introduce) that the notion of a real manifold M of *dimension d* includes the condition that each point $m \in M$ has a neighbourhood which is homeomorphic to an open set in $\mathbf{R}^d$. This amounts to the practical recipe that the dimension of a group G is the number of real parameters needed to describe the elements of G.

For instance for $G = \mathrm{SU}(2)$ and $\mathrm{SO}(3)$ we have as parameters the three Euler angles α, β, γ, which we introduced in 4.3 and 4.2. Hence, here we have $d = 3$, as also in the next example.

Example 6.6: For $G = \mathrm{SL}(2, \mathbf{R})$ we take as a standard parametrization (a special instance of the important *Iwasawa decomposition*)

$$(6.1) \qquad \mathrm{SL}(2, \mathbf{R}) \ni g = \begin{pmatrix} a & b \\ c & d \end{pmatrix} = n(x)t(y)r(\vartheta)$$

with

$$n(x) := \begin{pmatrix} 1 & x \\ & 1 \end{pmatrix},\ t(y) := \begin{pmatrix} y^{1/2} & \\ & y^{-1/2} \end{pmatrix},\ r(\vartheta) := \begin{pmatrix} \cos\vartheta & \sin\vartheta \\ -\sin\vartheta & \cos\vartheta \end{pmatrix},$$

where $x \in \mathbf{R}, y > 0, \vartheta \in [0, 2\pi)$. Therefore, we take

$$\gamma_1(t) := \begin{pmatrix} 1 & t \\ & 1 \end{pmatrix},\ \gamma_2(t) := \begin{pmatrix} e^t & \\ & e^{-t} \end{pmatrix},\ \gamma_3(t) := r(t),$$

and get by differentiating and then putting $t = 0$

$$X_1 = \begin{pmatrix} & 1 \\ & \end{pmatrix},\ X_2 = \begin{pmatrix} 1 & \\ & -1 \end{pmatrix},\ X_3 = \begin{pmatrix} & 1 \\ -1 & \end{pmatrix}.$$

Obviously, we have with the notation introduced in Exercise 6.3 in 6.1

$$\mathrm{Lie}\ \mathrm{SL}(2, \mathbf{R}) = <X_1, X_2, X_3> = <F, G, H> = \mathfrak{sl}_2(\mathbf{R}).$$

Example 6.7: For

$$G = \text{Heis}'(\mathbf{R}) := \left\{ \begin{pmatrix} 1 & x & z \\ & 1 & y \\ & & 1 \end{pmatrix} ; \ x, y, z \in \mathbf{R} \right\}$$

we take

$$\gamma_1(t) := \begin{pmatrix} 1 & t & \\ & 1 & \\ & & 1 \end{pmatrix}, \ \gamma_2(t) := \begin{pmatrix} 1 & & \\ & 1 & t \\ & & 1 \end{pmatrix}, \ \gamma_3(t) := \begin{pmatrix} 1 & & t \\ & 1 & \\ & & 1 \end{pmatrix},$$

and get

$$X_1 = \begin{pmatrix} & 1 & \\ & & \\ & & \end{pmatrix}, \ X_2 = \begin{pmatrix} & & \\ & & 1 \\ & & \end{pmatrix}, \ X_3 = \begin{pmatrix} & & 1 \\ & & \\ & & \end{pmatrix}.$$

By an easy computation of the commutators we get the *Heisenberg commutation relations*

$$[X_1, X_2] = X_3, \ [X_1, X_3] = [X_2, X_3] = 0.$$

Exercise 6.14: Repeat this for $G = \text{Heis}(\mathbf{R})$ as a subgroup of $\text{GL}(4, \mathbf{R})$.

Exercise 6.15: Determine $\text{Lie}\, G$ for $G = \{g = n(x)t(y); \ x \in \mathbf{R}, y > 0\}$.

6.3   Derived Representations

Now that we have associated a Lie algebra $\mathfrak{g} = \text{Lie}\, G$ to a given linear group (using a differentiation procedure), we want to associate an algebra representation $\hat{\pi}$ of $\mathfrak{g}$ to a given representation $(\pi, \mathcal{H})$ of G. Here we will have to use differentiation again. To make this possible, we have to modify the notion of a group representation:

Definition 6.5: Let $(\pi, \mathcal{H})$ be a continuous representation of a linear group G. Then its associated *smooth* representation $(\pi, \mathcal{H}^\infty)$ is the restriction of π to the space $\mathcal{H}^\infty$ of *smooth* vectors, i.e. those $v \in \mathcal{H}$ for which the map

$$G \ni g \longmapsto \pi(g)v \in \mathcal{H}$$

is differentiable.

This definition makes sense if we accept that a linear group is a differentiable manifold and extend the notion of differentiability to vector valued functions. In our examples things are fairly easy as we have functions depending on the parameters of the group where differentiability is not difficult to check.

Definition 6.6: Let $(\pi, \mathcal{H})$ be a continuous representation of a linear group G. Then its associated *derived representation* is the representation $d\pi$ of $\mathfrak{g}$ given on the space $\mathcal{H}^\infty$ of smooth vectors by the prescription

$$d\pi(X)v := \frac{d}{dt} \pi(\exp(tX))v \mid_{t=0} \quad \text{for all } v \in \mathcal{H}^\infty.$$

If it is clear, which representation π is meant, one abbreviates $\hat{X}v := d\pi(X)v$.
It is not difficult to verify that $d\pi(X)$ stays in $\mathcal{H}^\infty$ and rather obvious that $d\pi(X)$ is
linear. But there is some more trouble to prove (see [Kn] p.53, [BR] p.320, or [FH] p.16)
that one has

$$[d\pi(X), d\pi(Y)] = d\pi([X,Y]) \quad \text{for all } X, Y \in \mathfrak{g}.$$

Example 6.8: We take $G = \mathrm{SU}(2)$ and $\pi = \pi_1$ given by

$$\pi(g)f(x,y) := f(\bar{a}x - by, \bar{b}x + ay)$$

for

$$f \in V^{(1)} = <f_{-1}, f_0, f_1>$$

with

$$f_{-1}(x,y) = (1/\sqrt{2})y^2, \ f_0(x,y) = xy, \ f_1(x,y) = (1/\sqrt{2})x^2.$$

As we did in Exercise 6.4 in 6.1, we take as basis of $\mathfrak{su}(2)$

$$X_1 := (1/2)\begin{pmatrix} & -i \\ -i & \end{pmatrix}, \ X_2 := (1/2)\begin{pmatrix} & -1 \\ 1 & \end{pmatrix}, \ X_3 := (1/2)\begin{pmatrix} i & \\ & -i \end{pmatrix}.$$

Then we have the one-parameter subgroups

$$\exp(tX_1) = \begin{pmatrix} \cos(t/2) & -i\sin(t/2) \\ -i\sin(t/2) & \cos(t/2) \end{pmatrix},$$

$$\exp(tX_2) = \begin{pmatrix} \cos(t/2) & -\sin(t/2) \\ \sin(t/2) & \cos(t/2) \end{pmatrix},$$

$$\exp(tX_3) = \begin{pmatrix} e^{it/2} & \\ & e^{-it/2} \end{pmatrix}.$$

It is clear that $V^{(1)}$ consists of smooth vectors and hence we can compute

$$\begin{aligned}
\hat{X}_1 f_{-1} &= \tfrac{d}{dt}\pi(\exp tX_1)f_{-1}\,|_{t=0} \\
&= \tfrac{d}{dt}(1/\sqrt{2})(i\sin(t/2)x + \cos(t/2)y)^2\,|_{t=0} \\
&= (i/\sqrt{2})xy = (i/\sqrt{2})f_0,
\end{aligned}$$

and

$$\begin{aligned}
\hat{X}_1 f_0 &= \tfrac{d}{dt}\pi(\exp tX_1)f_0\,|_{t=0} \\
&= \tfrac{d}{dt}(\cos(t/2)x + i\sin(t/2)y)(i\sin(t/2)x + \cos(t/2)y)\,|_{t=0} \\
&= (i/2)(x^2 + y^2) = (i/\sqrt{2})(f_{-1} + f_1), \\
\hat{X}_1 f_1 &= \tfrac{d}{dt}\pi(\exp tX_1)f_1\,|_{t=0} \\
&= \tfrac{d}{dt}(1/\sqrt{2})(\cos(t/2)x + i\sin(t/2)y)^2\,|_{t=0} \\
&= (i/\sqrt{2})xy = (i/\sqrt{2})f_0.
\end{aligned}$$

Similarly we get

$$\begin{aligned}
\hat{X}_2 f_{-1} &= -(1/\sqrt{2})f_0, \quad \hat{X}_2 f_0 = (1/\sqrt{2})(f_{-1} - f_1), \quad \hat{X}_2 f_1 = (1/\sqrt{2})f_0, \\
\hat{X}_3 f_{-1} &= if_{-1}, \quad\quad\quad\ \ \hat{X}_3 f_0 = 0, \quad\quad\quad\quad\quad\quad\ \ \hat{X}_3 f_1 = -if_{-1}.
\end{aligned}$$

Exercise 6.16: i) Verify this and show that for

$$X_0 := iX_3, \ X_\pm := (1/\sqrt{2})(-iX_1 \mp X_2)$$

one has

$$\hat{X}_0 f_p = p f_p, \quad \hat{X}_\pm f_p = f_{p\pm 1}, \ p = -1, 0, 1 \ (f_2 = f_{-2} = 0).$$

ii) Do the same for $\pi = \pi_j, j > 1$.

Example 6.9: We take the Heisenberg group $G = \text{Heis}(\mathbf{R}) \subset GL(4, \mathbf{R})$ and the Schrödinger representation $\pi = \pi_m, m \in \mathbf{R} \setminus \{0\}$, given by

$$\pi(g)f(x) = e^m(\kappa + (2x + \lambda)\mu)f(x + \lambda)$$

with

$$g = (\lambda, \mu, \kappa) \in G, x \in \mathbf{R}, f \in \mathcal{H} = L^2(\mathbf{R}), e^m(u) := e^{2\pi i m u}.$$

Here $\mathcal{H}^\infty$ is the (dense) subspace of differentiable functions in $L^2(\mathbf{R})$. We realize the Lie algebra $\mathfrak{g} = \text{heis}(\mathbf{R})$ by the space spanned by the four-by-four matrices

$$P = \begin{pmatrix} & & & \\ 1 & & & \\ & & & -1 \\ & & & \end{pmatrix}, \quad Q = \begin{pmatrix} & & 1 & \\ 1 & & & \\ & & & \\ & & & \end{pmatrix}, \quad R = \begin{pmatrix} & & & 1 \\ & & & \\ & & & \\ & & & \end{pmatrix}.$$

Then we have the Heisenberg commutation relations

$$[P, Q] = 2R, \quad [R, P] = [R, Q] = 0$$

and (using again the notation introduced in 0.1)

$$\exp tP = (t, 0, 0), \ \exp tQ = (0, t, 0), \ \exp tR = (0, 0, t).$$

For differentiable f we compute

$$\begin{array}{lclcl}
\hat{P}f &=& \frac{d}{dt}\pi(t, 0, 0)f \mid_{t=0} &=& \frac{d}{dt}f(x + t) \mid_{t=0} = \partial_x f, \\
\hat{Q}f &=& \frac{d}{dt}\pi(0, t, 0)f \mid_{t=0} &=& \frac{d}{dt}e^m(2xt)f(x) \mid_{t=0} = 4\pi i m x f(x), \\
\hat{R}f &=& \frac{d}{dt}\pi(0, 0, t)f \mid_{t=0} &=& \frac{d}{dt}e^m(t)f(x) \mid_{t=0} = 2\pi i m f(x),
\end{array}$$

i.e. we have

$$\hat{P} = \partial_x, \quad \hat{Q} = 4\pi i m x, \quad \hat{R} = 2\pi i m.$$

Exercise 6.17: i) Verify that these operators really fulfill the same commutation relations as P, Q, R.

ii) Take $f_0(x) := e^{-\pi m x^2}$ and

$$\hat{Y}_0 := -i\hat{R}, \quad \hat{Y}_\pm := (1/2)(\hat{P} \pm i\hat{Q})$$

and determine

$$f_p := \hat{Y}_+^p f_0, \ \hat{Y}_- f_p \text{ and } \hat{Y}_0 f_p \text{ for } p = 0, 1, 2, \ldots.$$

These Examples show how to associate a Lie algebra representation to a given group representation. Now the following question arises naturally: Is there always a nontrivial subspace $\mathcal{H}^\infty$ of smooth vectors in a given representation space $\mathcal{H}$? We would even want that it is a dense subspace. The answer to this question is positive. This is easy in our examples but rather delicate in general and one has to rely on work by Garding, Nelson and Harish Chandra. It is not very surprising that equivalent group representations are *infinitesimally equivalent*, i.e. their derived representations are equivalent algebra representations. The converse is not true in general. But as stated in [Kn] p.209, at least for semisimple groups one has that infinitesimally equivalent irreducible representations are unitarily equivalent.

Taking these facts for granted, a discussion of the unitary irreducible representations of a given linear group G can proceed as follows: We try to classify all irreducible representations of the Lie algebra $\mathfrak{g} = \text{Lie}\,G$ of G and determine those, which can be realized by derivation from a unitary representation of the group. In general, this demands for some more information about the structure of the Lie algebra at hand. Before going a bit into this structure theory, we discuss in extenso three fundamental examples whose understanding prepares the ground for the more general cases.

6.4 Unitarily Integrable Representations of $\mathfrak{sl}(2,\mathbf{R})$

A basic reference for this example is the book "SL$(2,\mathbf{R})$" [La1] by Serge Lang but in some way or other our discussion can be found in nearly any text treating the representation theory of semisimple algebras or groups.

As we already know from Exercise 6.3 in 6.1 and Example 6.5 in 6.2, $\mathfrak{g} = \mathfrak{sl}(2,\mathbf{R})$ is the Lie algebra of $G = \text{SL}(2,\mathbf{R})$ and we have

$$\mathfrak{g} = \mathfrak{sl}(2,\mathbf{R}) = \{X \in M_2(\mathbf{R}); \text{Tr}\,X = 0\} = \,<F,G,H>$$

with

$$F = (\begin{smallmatrix} & 1 \\ & \end{smallmatrix}),\, G = (\begin{smallmatrix} \\ 1 & \end{smallmatrix}),\, H = (\begin{smallmatrix} 1 & \\ & -1 \end{smallmatrix})$$

and the relations

$$[H,F] = 2F,\, [H,G] = -2G,\, [F,G] = H.$$

The exercises at the end of the examples in the previous section should give a feeling that it is convenient to complexify the Lie algebra, i.e. here we go over to

$$\mathfrak{g}_c := \mathfrak{g} \otimes_{\mathbf{R}} \mathbf{C} =\, <F,G,H>_{\mathbf{C}} =\, <X_+,X_-,Z>_{\mathbf{C}}$$

where

$$(6.2) \quad X_\pm := (1/2)(H \pm i(F+G)) = (1/2)(\begin{smallmatrix} 1 & \pm i \\ \pm i & -1 \end{smallmatrix}),\, Z := -i(F-G) = (\begin{smallmatrix} & -i \\ i & \end{smallmatrix})$$

with

$$[Z,X_\pm] = \pm 2X_\pm,\, [X_+,X_-] = Z.$$

This normalization will soon prove to be adequate and the operators $X_\pm$ will get a meaning as *ladder operators*: We consider a representation $\hat{\pi}$ of $\mathfrak{g}_c =\, <X_\pm, Z>$ on a $\mathbf{C}$-vector space $V =< v_i >_{i \in I}$ where we abbreviate

$$Xv := \hat{\pi}(X)v \quad \text{for all } v \in V, X \in \mathfrak{g}_c.$$

We are looking for representations $(\hat{\pi},V)$ of $\mathfrak{g}_c$, which may be *integrated*, i.e. which are complexifications of derived representations $d\pi$ for a representation π of SL$(2,\mathbf{R})$.

This has the following consequence: $\pi|_K$ is a representation of $K := SO(2)$. From Example 3.1 in 3.2 we know that all irreducible unitary representations of $SO(2)$ are given by

$$\chi_k(r(\vartheta)) = e^{ik\vartheta}, \quad k \in \mathbf{Z},$$

hence one has the derived representation $d\chi_k$ with

$$d\chi_k(Y) = ik \text{ for } Y = \begin{pmatrix} & 1 \\ -1 & \end{pmatrix} = iZ.$$

This motivates the idea to look for each $k \in \mathbf{Z}$ at the subspace of V consisting of the eigenvectors with eigenvalue k with respect to the action of Z, namely

$$V_k := \{v \in V; \ Zv = kv\}.$$

This space V_k is called a *K-isotypic subspace* of V. Each $v \in V_k$ is said to have *weight* k. Now we can explain the meaning of the term *ladder operator*. The isotypic spaces V_k may be seen as rungs of a ladder numerated by the weight k and the operators $X_\pm$ change the weights in a controlled way:

Remark 6.4: We have

$$X_\pm V_k \subset V_{k\pm 2}.$$

Proof: For $v \in V_k$ we put $v' := X_\pm v$ and check that v' is again a Z-eigenvector of weight $k \pm 2$:

$$Zv' = ZX_\pm v = (X_\pm Z \pm 2X_\pm)v = X_\pm(k \pm 2)v = (k \pm 2)v'.$$

Here we used the Lie algebra relation $[Z, X_\pm] = \pm 2X_\pm$, which for the operators acting on V is transformed into the relation

$$ZX_\pm - X_\pm Z = \pm 2X_\pm.$$

This remark has immediate consequences:

i) If $V = \Sigma V_k$ is the space of an irreducible $\mathfrak{g}_c$-representation, there are only nontrivial V_k where k is throughout odd or even.

ii) If there is a $V_{k_0} = \{0\}$, then all V_k with $k \geq k_0$ or all V_k with $k \leq k_0$ have to be zero. Hence, there can be only four different configurations for our irreducible $\mathfrak{g}_c$-representations. The nontrivial weights k, i.e. those for which $V_k \neq \{0\}$, compose

– case 1: a **finite interval**

$$I_m^n := \{m \leq k \leq n; \ m \equiv k \equiv n \mod 2\}$$

of even or odd integers,

– case 2: a **half line with an upper bound**

$$I^n := \{k \leq n; \ k \equiv n \mod 2\}$$

consisting of even or odd integers,

– case 3: a **half line with lower bound**

$$I_m := \{m \leq k; \ m \equiv k \mod 2\}$$

consisting of even or odd integers,

– case 4: a **full chain of even or odd integers**

$$I_+ := \{k \in 2\mathbf{Z}\} \text{ or } I_- := \{k \in 2\mathbf{Z} + 1\}.$$

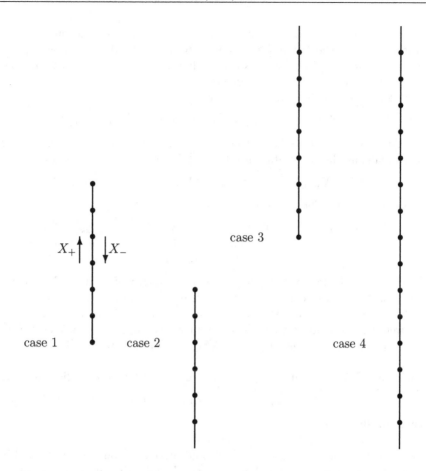

It is quite natural to call an element $0 \neq v \in V_k$ with $k = n$ in configuration 1 or 2 a vector of *highest weight* and similarly for $k = m$ in configuration 1 or 3 a vector of *lowest weight*.

The following statement offers the decisive point for the reasoning that we really get **all** representations by our procedure. It is a special case of a more general theorem by Godement on *spherical* functions. It also is a first example for the famous and important *multiplicity-one* statements.

Lemma: If $\hat{\pi}$ is irreducible, we have $\dim V_k \leq 1$.

For a proof we refer to Lang [La1] p.24 or the article by van Dijk [vD].

We look at the configurations 1 or 2: Let $0 \neq v \in V_n$ be a vector of highest weight, i.e. with

$$Zv_n = nv_n, \ X_+v_n = 0,$$

and put

$$v_{n-2k} := X_-^k v_n.$$

As we want an irreducible representation and we have accepted to take the muliplicity-one statement from the Lemma for granted, we can take this v_{n-2k} as a generator for V_{n-2k} with $v_{n-2k} \neq 0$ for all $k \in \mathbf{N}_0$ in configuration 1 and all k with $n - 2k \geq m$ in configuration 2. We check what X_+ does to v_{n-2k}:

Remark 6.5: For k as fixed above, we have

$$X_+ v_{n-2k} = a_k v_{n-2k+2} \text{ with } a_k := kn - k(k - 1).$$

Proof: Since $X_+ v_n = 0$, we know that $a_0 = 0$. As in the proof of Remark 6.4, for the operators realizing the commutation relation $[X_+, X_-] = Z$ we calculate

$$X_+ v_{n-2} = X_+ X_- v_n = (X_- X_+ + Z)v_n = nv_n.$$

Hence we get $a_1 = n$. Then we verify by induction

$$
\begin{aligned}
X_+ v_{n-2(k+1)} &= X_+ X_- v_{n-2k} = (X_- X_+ + Z)v_{n-2k} \\
&= X_-((kn - k(k - 1))v_{n-2k+2} + (n - 2k)v_{n-2k} \\
&= ((k + 1)n - k(k + 1))v_{n-2k} = a_{k+1} v_{n-2k}.
\end{aligned}
$$

We see that the irreducibility of $\hat{\pi}$ forces a_k to be nonzero for all $k \in \mathbf{N}$ in configuration 2 and for $2k \leq n - m$ in configuration 1. Since $a_k = 0$ iff $n - (k - 1) = 0$, we conclude that n has to be negative in case 2 and $m = -n$ with $n \in \mathbf{N}$ in case 1.

Just as well, we can start to treat the configurations 1 and 3 with a vector $0 \neq v_m \in V_m$ of lowest weight. Here we put $v_{m+2k} := X_+^k v_m$. In parallel to Remark 6.5 we get:

Remark 6.6: For $k \in \mathbf{N}_0$ in configuration 3 and $m + 2k \leq n$ in configuration 1, we have

$$X_- v_{m+2k} = b_k v_{m+2k-2} \text{ with } b_k = -(km + k(k - 1)).$$

Exercise 6.18: Prove this.

As above, we see that we have $m = -n$ with negative n in configuration 1 and that n has to be positive in configuration 3. For instance, the following are possible configurations.

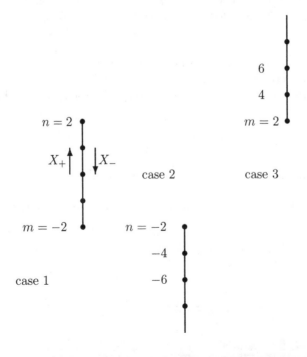

In configuration 4 we have no distinguished vector of highest or lowest weight. In this case we choose for each even (or odd) integer n a nonzero $v_n \in V_n$, i.e. with $Zv_n = nv_n$, such that they are related as follows:

Remark 6.7: For all even (resp. odd) $n \in \mathbf{Z}$ we have

$$X_{\pm}v_n = a_n^{\pm}v_{n\pm2} \text{ with } a_n^{\pm} = (1/2)(s+1\pm n),$$

where $s \in \mathbf{C}$, such that $a_n^{\pm} \neq 0$ for all $n \in 2\mathbf{Z}$ (resp. $2\mathbf{Z}+1$), i.e. $s \notin 2\mathbf{Z}+1$ for $n \in 2\mathbf{Z}$ and $s \notin 2\mathbf{Z}$ for $n \in 2\mathbf{Z}+1$.

Proof: From irreducibility and multiplicity-one we conclude that each $X_{\pm}v_n$ has to be a nonzero multiple of $v_{n\pm2}$. We verify that the choice of $a_n^{\pm}$ is consistent with the commutation relations:

$$
\begin{aligned}
[Z, X_{\pm}]v_n &= (ZX_{\pm} - X_{\pm}Z)v_n = Za_n^{\pm}v_{n\pm2} - a_n^{\pm}nv_{n\pm2} \\
&= \pm2a_n^{\pm}v_{n\pm2} \\
&= \pm2X_{\pm}v_n,
\end{aligned}
$$

and

$$[X_+, X_-]v_n = Zv_n = nv_n.$$

Verify this as **Exercise 6.19.**

The conditions for s are obvious since we have $a_n^{\pm} = 0$ iff $s+1 = \mp n$.

We chose a symmetric procedure for our configuration 4. But as well we could have chosen any nonzero $v_n \in V_n$, and then $v_{n+2j} := X_+^j v_n, v_{n-2j} := X_-^j v_n$ for $j \in \mathbf{N}$ would constitute an equivalent representation.
Exercise 6.20: Check this.

We denote the representations obtained above as follows

- $\hat{\pi}_{s,+}$ or $\hat{\pi}_{s,-}$ for the even resp. odd case in configuration 4 with $s \neq 2\mathbf{Z}+1$ for $\hat{\pi}_{s,+}$ and $s \neq 2\mathbf{Z}$ for $\hat{\pi}_{s,-}, s \in \mathbf{C}$,
- $\hat{\pi}_n^+$ for configuration 3 with lowest weight $n, n \in \mathbf{N}_0$,
- $\hat{\pi}_n^-$ for configuration 2 with highest weight $-n, n \in \mathbf{N}_0$,
- $\hat{\sigma}_n$ for configuration 1 with highest weight n and lowest weight $-n, n \in \mathbf{N}_0$.

Remark 6.8: If in $\hat{\pi}_{s,+}$ and $\hat{\pi}_{s,-}$ we do not exclude the values of s as above, we get all representations as sub - resp. quotient representations of these $\hat{\pi}_{s,\pm}$.

Proposition 6.1: We have the following equivalences

$$\hat{\pi}_{s,+} \sim \hat{\pi}_{-s,+} \text{ and } \hat{\pi}_{s,-} \sim \hat{\pi}_{-s,-}.$$

All other representations listed above are inequivalent.

Proof: i) Let $V = <v_n>$ and $V' = <v_n'>$ be representation spaces for $\hat{\pi}_{s,+}$ resp. $\hat{\pi}_{s',+}$ and $F : V \longrightarrow V'$ an intertwining operator. Because of $Zv_n = nv_n$, we have for each n

$$ZFv_n = FZv_n = nFv_n.$$

Hence one has $Fv_n \in V_n'$, i.e. $Fv_n = b_n v_n'$ with $b_n \in \mathbf{C}$. Moreover, from

$$X_\pm F v_n = F X_\pm v_n$$

we deduce

$$(s' + 1 \pm n) b_n v_{n\pm 2}' = b_{n\pm 2}(s + 1 \pm n) v_{n\pm 2}',$$

i.e.

$$b_n/b_{n-2} = (s + 1 - n)/(s' + 1 - n)$$

and

$$b_{n+2}/b_n = (s' + 1 + n)/(s + 1 + n)$$

resp.

$$b_n/b_{n-2} = (s' + n - 1)/(s + n - 1).$$

Now we see that we have

$$(s + 1 - n)(s + n - 1) = (s' + n - 1)(s' + 1 - n), \text{ i.e. } s' = \pm s.$$

The same conclusions can be made for $\hat{\pi}_{s,-}$.

ii) The inequivalence of the $\hat{\sigma}_n$'s among themselves and with respect to all the other representations is clear because the dimensions are different. The other inequivalences have to be checked individually. As an example let $V = <v_n>$ and $V' = <v_n'>$ be spaces of representations $\hat{\pi}_k^+$ and $\hat{\pi}_{s,+}$ and $F : V \longrightarrow V'$ an intertwining operator. As above, for $k \geq n$ we have the relation $Fv_n = b_n v_n'$ with $b_n \in \mathbf{C}$. Hence F is not surjective and both representations are not equivalent. The same reasoning goes through in all the other cases.

Exercise 6.21: Realize this.

Classification of the Unitarily Integrable Representations of $\mathfrak{sl}(2,\mathbf{R})$

At first, we derive a necessary condition for a general representation $\hat{\pi}$ of a Lie algebra $\mathfrak{g} = \mathrm{Lie}\,G$ to be integrable to a unitary representation π of G.

Proposition 6.2: If $(\pi, \mathcal{H})$ is a unitary representation of G and $X \in \mathfrak{g} = \mathrm{Lie}\,G$, the operator $\hat{X} := d\pi(X)$ on $\mathcal{H}^\infty$ is skew-Hermitian, i.e. one has $\hat{X} = -\hat{X}^*$.

Proof: For $v, w \in \mathcal{H}^\infty$, unitarity of π implies

$$< v, \pi(g)w > = < \pi(g^{-1})v, w > \quad \text{for all } g \in G$$

and hence, for $t \in \mathbf{R}, t \neq 0$,

$$< v, i\frac{\pi(\exp(tX))w - w}{t} > = < i\frac{\pi(\exp(-tX))v - v}{-t}, w > .$$

Recalling the definition

$$d\pi(X)v := \frac{d}{dt}\pi(\exp(tX))v \mid_{t=0} = \lim_{t \to 0} \frac{\pi(\exp tX)v - v}{t},$$

we get from the last equation in the limit $t \longrightarrow 0$

$$< v, i\,d\pi(X)w > = - < i\,d\pi(X)v, w > .$$

Therefore we have

$$< v, i\, d\pi(X)w > = - < v, i\, d\pi(X)^* w >,$$

i.e. $\hat{X} = d\pi(X)$ is skew-Hermitian.

Now, we apply this to $\mathfrak{g} = \mathfrak{sl}(2, \mathbf{R})$. The Proposition leads to the condition:

Remark 6.9: We have $d\pi(X_\pm)^* = -d\pi(X_\mp)$.

Proof: Using the notation from the beginning of this section we have

$$X_\pm = (1/2)(H \pm i(F + G)) = (1/2)\begin{pmatrix} 1 & \pm i \\ \pm i & -1 \end{pmatrix} =: (X_1 \pm iX_2)$$

with $X_1 = H, X_2 = F + G \in \mathfrak{sl}(2, \mathbf{R})$. Hence, the skew-Hermiticity of $\hat{X}_i$ leads to

$$(\hat{X}_1 \pm i\hat{X}_2)^* = -(\hat{X}_1 \mp i\hat{X}_2).$$

Now, if we ask for unitarity, we have to check whether we find a scalar product $<,>$ defined on the representation space V such that we get

$$< \hat{X}_\pm v, w > = - < v, \hat{X}_\mp w > \quad \text{for all } v, w \in V.$$

For $V = < v_j >_{j \in J}$ this condition implies

(6.3) $\qquad < \hat{X}_+ v_j, v_{j+2} > = - < v_j, \hat{X}_- v_{j+2} > \quad \text{for all } j \in J.$

i) In configuration 1 and 2 we have

$$\hat{X}_- v_{j+2} = v_j$$

and from Remark 6.5 for $j = n - 2k$

$$\hat{X}_+ v_j = a_k v_{j+2} \text{ with } a_k = k(n - k + 1),$$

i.e.

$$a_k \parallel v_{j+2} \parallel^2 = - \parallel v_j \parallel^2 .$$

An equation like this is only possible for $a_k < 0$, which is true for the a_k in configuration 2 but not for configuration 1, where we have seen that all a_k are positive.

Similarly Remark 6.6 shows that in configuration 3 the condition (6.3) is no obstacle.

Exercise 6.22: Verify this.

ii) Configuration 4 is more delicate: From Remark 6.7 we have

$$\hat{X}_\pm v_j = a_j^\pm v_{j\pm2} \text{ with } a_j^\pm = (1/2)(s + 1 \pm j),$$

i.e. the condition (6.3) has the form

(6.4) $\qquad (\bar{s} + 1 + j) \parallel v_{j+2} \parallel^2 = -(s - 1 - j) \parallel v_j \parallel^2 .$

At first we treat the even case $V = < v_j >_{j \in 2\mathbf{Z}}$. Condition (6.4) requires for $j = 0$

$$(\bar{s} + 1) \parallel v_2 \parallel^2 = -(s - 1) \parallel v_0 \parallel^2 .$$

As the norm has to be positive, a relation like this is only possible if one has

$$(1 - s)/(\bar{s} + 1) > 0, \text{ i. e. } (1 - s)(1 + s) > 0.$$

We put $s = \sigma + i\tau, \sigma, \tau \in \mathbf{R}$, and get the condition

$$s^2 = \sigma^2 - \tau^2 + 2\sigma\tau i < 1,$$

i.e. unitarity demands for

$$\sigma = 0 \quad \text{and} \quad \tau \in \mathbf{R} \text{ arbitrary}$$

or

$$\tau = 0 \quad \text{and} \quad \sigma^2 < 1.$$

In the odd case $V = < v_j >_{j \in 2\mathbf{Z}+1}$ we get for $j = -1$ from (j')

$$\bar{s} \parallel v_1 \parallel^2 = -s \parallel v_{-1} \parallel^2 .$$

This is possible only for

$$-\bar{s}/s > 0, \text{ i. e. } \bar{s}^2 = \sigma^2 - \tau^2 - 2\sigma\tau i < 0.$$

Therefore, here we have only one possibility, namely

$$\sigma = 0 \quad \text{and} \quad \tau \in \mathbf{R}.$$

Putting all this together, we have proved the following statement.

Theorem 6.2: Unitary irreducible representations π of $G = \mathrm{SL}(2, \mathbf{R})$ can exist only if their derived representation $\hat{\pi} = d\pi$ is among the following list:

– $\hat{\pi}$ the trivial representation,
– $\hat{\pi} = \hat{\pi}_k^{\pm}, k \in \mathbf{N}$,
– $\hat{\pi} = \hat{\pi}_{is,\pm}, s \in \mathbf{R}$ ($s \neq 0$ for $\hat{\pi}_{is,-}$)
– $\hat{\pi} = \hat{\pi}_s$, $s \in \mathbf{R}, 0 < s^2 < 1$.

In the next chapter we will prove that, in fact, all these representations $\hat{\pi}$ can be integrated to unitary representations π of G. They get the following names
– $\pi = \pi_k^{\mp}$ *discrete series representation of highest (lowest) weight $\mp k$,*
– $\pi = \pi_{is,\pm}$ *even (odd) principal series representation,*
– $\pi = \pi_s$ *complementary (or supplementary) series representation.*

Because of the equivalences stated in Proposition 6.1, we have for the unitary dual of $G = \mathrm{SL}(2, \mathbf{R})$

$$\hat{G} = \{\pi_0; \pi_k^{\pm}, k \in \mathbf{N}; \pi_{is,\pm}, s \in \mathbf{R}_{>0}; \pi_s, 0 < s < 1\}.$$

We will study this more closely in the next chapter but already indicate a kind of visualization.

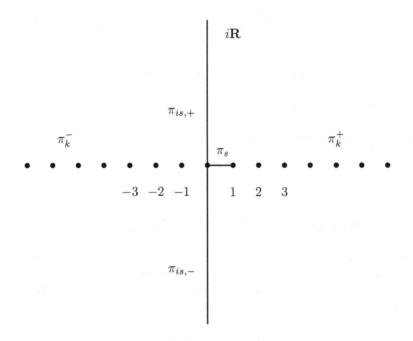

We can observe a very nice fact, which will become important later and is the starting point for a general procedure in the construction of representations:

Exercise 6.23: Show that the operator

$$\Omega := X_+X_- + X_-X_+ + (1/2)Z^2$$

acts as a scalar for the three types of representations in Theorem 6.2 and determine these scalars.

Our infinitesimal considerations showed that there is no nontrivial finite-dimensional unitary representation of $\mathrm{SL}(2,\mathbf{R})$. There are other proofs for this fact. Because it is so elegant and instructive, we repeat the one from [KT] p.16.

Theorem 6.3: Every finite-dimensional representation π of $G = \mathrm{SL}(2,\mathbf{R})$ is trivial.

Proof: If G has an n-dimensional representation space, we have a homomorphism

$$\pi : \mathrm{SL}(2,\mathbf{R}) \longrightarrow \mathrm{U}(n).$$

Since

$$\begin{pmatrix} a & \\ & a^{-1} \end{pmatrix}\begin{pmatrix} 1 & b \\ & 1 \end{pmatrix}\begin{pmatrix} a & \\ & a^{-1} \end{pmatrix}^{-1} = \begin{pmatrix} 1 & a^2b \\ & 1 \end{pmatrix}$$

all $n(b) = \begin{pmatrix} 1 & b \\ & 1 \end{pmatrix}$ are conjugate. So all $\pi(n(b))$ with $b > 0$ are conjugate. It is a not too difficult topological statement that in a compact group all conjugacy classes are closed. As $\mathrm{U}(n)$ is compact, we can deduce

$$\lim_{b \to 0^+} \pi(n(b)) = \pi(E_2) = E_n,$$

and therefore $\pi(n(b)) = E_n$ for all $b > 0$. But we can repeat this conclusion for $b < 0$
and as well for $\bar{n}(b) = (\begin{smallmatrix} 1 & \\ b & 1 \end{smallmatrix})$ and get

$$\pi(n(b)) = \pi(\bar{n}(b)) = E_n \quad \text{for all } b \in \mathbf{R}.$$

Now the proof is finished if we recall the following

Lemma: $\mathrm{SL}(2, \mathbf{R})$ is generated by the elements $n(b)$ and $\bar{n}(b), b \in \mathbf{R}$.

Exercise 6.24: Prove this.

6.5 The Examples $\mathfrak{su}(2)$ and $\mathrm{heis}(\mathbf{R})$

In section 6.3 we computed derived representations for $G = \mathrm{SU}(2)$ and $\mathrm{Heis}(\mathbf{R})$. We look
at this again on the ground of what we did in the last section.
We refer to Exercise 6.4 in 6.1 and Example 6.8 in 6.3 and (having in mind later applications) change the basis (X_j) of $\mathfrak{su}(2)$ we used there to

(6.5) $H_1 = (\begin{smallmatrix} & -i \\ i & \end{smallmatrix})$, $H_2 = (\begin{smallmatrix} & -i \\ -i & \end{smallmatrix})$, $H_3 = (\begin{smallmatrix} & -1 \\ 1 & \end{smallmatrix})$,

with
$$[H_j, H_k] = 2H_l, \quad (j, k, l) = (1, 2, 3), (2, 3, 1), (3, 1, 2).$$

$\mathfrak{su}(2)$ and $\mathfrak{sl}(2, \mathbf{R})$ have the same complexifications. Hence we can express the matrices
$Z, X_\pm$ from (6.2) in the previous section as

$$Z = iH_3, \ X_\pm = (1/2)(iH_1 \mp H_2)$$

and repeat the discussion of the representation space $V = <v_j>_{j \in J}$ to get the configurations 1 to 4. Let us look at configuration 1: We take a vector $v_n \in V$ of highest weight
$n \in \mathbf{N}$, put

$$v_{n-2k} = \hat{X}_-^k v_n, \ \hat{X}_+ v_{n-2k} =: a_k v_{n-2k+2},$$

get $a_k = kn - k(k - 1)$ and $m = -n$ for the lowest weight. But in this case the skew-
hermiticity of the H_j changes things and Remark 6.9 in 6.4 is to be replaced by:

Remark 6.10: For a unitary representation π of $\mathrm{SU}(2)$ we must have

$$d\pi(X_\pm)^* = d\pi(X_\mp).$$

Hence, the discussion in 6.4 shows that for unitarity in this case the a_k have to be
positive as, in fact, they are for $n \geq k \geq 0$. We see that our discussion rediscovers the
representation space $V = V^{(j)}$ from section 4.2 with $n = 2j$. By the way, we get here for
free another proof that the representation π_j from 4.2 is irreducible.
Configurations 2 to 4 are not unitarizable.

For $G = \mathrm{Heis}(\mathbf{R})$, as in Example 6.8 in 6.3, we have

$$\mathfrak{g} = \mathrm{heis}(\mathbf{R}) = <P, Q, R>$$

with the Heisenberg commutation relations

$$[P, Q] = 2R, \ [R, P] = [R, Q] = 0.$$

We complexify $\mathfrak{g}$ and take as a $\mathbf{C}$-basis

$$Y_0 := iR, \ Y_\pm := (1/2)(P \pm iQ)$$

to get the relations $[Y_+, Y_-] = Y_0$, $[Y_0, Y_\pm] = 0$.

We want to construct a representation $\hat{\pi}$ of $\mathfrak{g}$ on a space $V = <v_j>_{j \in J}$, (which we continue linearly to a representation of $\mathfrak{g}_c$,) such that $\hat{\pi}$ can be unitarily integrated, i.e. we can find a unitary representation π of Heis($\mathbf{R}$) with $d\pi = \hat{\pi}$: In every group one has as a distinguished subgroup, the *center* $C(G)$, defined by

$$C(G) := \{g \in G; gg_0 = g_0 g \ \text{ for all } \ g_0 \in G\}.$$

Obviously for $G = \text{Heis}(\mathbf{R})$ we have

$$C := C(\text{Heis}(\mathbf{R})) = \{(0, 0, \kappa); \ \kappa \in \mathbf{R}\} \simeq \mathbf{R}.$$

Hence, a representation π of our G restricted to the center C is built up (see Chapter 5) from the characters of $\mathbf{R}$. We write here

$$\psi(\kappa) := \psi_m(\kappa) := \exp(2\pi i m\kappa) = e^m(\kappa), \ m \in \mathbf{R} \setminus \{0\}$$

with derived representation $d\psi(\kappa) = 2\pi i m$. Having this in mind (and the discussion of Example 6.9 in 6.3), we propose the following construction of a representation $(\hat{\pi}, V)$, where we abbreviate again $\hat{\pi}(Y)v =: Yv$.

We suppose that we have a *vacuum vector*, i.e. an element $v_0 \in V$ with

$$Y_0 v_0 = \mu v_0, \ Y_- v_0 = 0, \ \mu = 2\pi m.$$

The first relation is inspired by differentiation of the prescription of the Schrödinger representation

$$\pi((0, 0, \kappa))f = \psi(\kappa)f$$

where we have $(0, 0, \kappa) = \exp \kappa R$ and $Y_0 = iR$.

We put $v_j := Y_+^j v_0$. As Y_0 commutes with $Y_\pm$, we have $Y_0 v_j = \mu v_j$. Then we try to realize a relation

$$Y_- v_j = a_j v_{j-1} :$$

From the commutation relation $[Y_+, Y_-] = Y_0$ we deduce

$$\begin{aligned} Y_- v_1 &= Y_- Y_+ v_0 = (Y_+ Y_- - Y_0)v_0 = -\mu, \\ Y_- v_j &= Y_- Y_+ v_{j-1} = (Y_+ Y_- - Y_0)v_{j-1} = (a_{j-1} - \mu)v_{j-1} = a_j v_{j-1}, \end{aligned}$$

and, hence, by induction

$$a_j = -j\mu.$$

Thus we have V spanned by $v_0, v_1, v_2, \ldots$ with

$$Y_0 v_j = \mu v_j, \ Y_+ v_j = v_{j+1}, \ Y_- v_j = -j\mu v_{j-1}.$$

We check whether this representation can be unitarized: We look at the necessary condition in Proposition 6.2 in 6.4. The skew-Hermiticity of P, Q, R leads to

$$\hat{Y}_\pm^* = d\pi(Y_\pm)^* = (1/2)(P^* \mp iQ^*) = -(1/2)(P \mp iQ) = -\hat{Y}_\mp.$$

Hence, as in the $\mathfrak{sl}$-discussion, we get the necessary condition for a scalar product $<,>$ on V

$$< \hat{Y}_\pm v, w > = - < v, \hat{Y}_\pm w >,$$

in particular, for $v = v_{j-1}, w = v_j$, we have

$$\| v_j \|^2 = j\mu \| v_{j-1} \|^2 .$$

We see that unitarity is possible iff $\mu = 2\pi m > 0$. A *model*, i.e. a realization of this representation $\hat{\pi}$ by integration, is given by the Schrödinger representation π_m, which we discussed already in several occasions (see Example 6.9 and Exercise 6.17 in 6.3).

Remark 6.11: Our infinitesimal considerations show that π_m is irreducible.

Remark 6.12: This discussion above is already a good part of the way to a proof of the famous *Stone-von Neumann Theorem* stating that up to equivalence there is no other irreducible unitary representation π of Heis($\mathbf{R}$) with $\pi |_C = \psi_m$ for $m \neq 0$ (for a complete proof see e.g. [LV] p.19ff).

Remark 6.13: What we did here is essentially the same as that appears in the physics literature under the heading of *harmonic oscillator*. Our $v_0 = f_0$ with $f_0(x) = e^{-2\pi m x^2}$ describes the *vacuum*, and Y_+ and Y_- are *creation* resp. *annihilation operators* producing higher resp. lower excitations.

Exercise 6.25: Use Exercise 6.17 in 6.3 to verify the equation

$$(Y_+Y_- + Y_-Y_+)f = (1/2)(f'' - (4\pi m x)^2 f) = -2\pi m f$$

and show that f_0 is a solution. Recover the Hermite polynomials.

6.6   Roots, Weights and some Structure Theory

In the examples discussed above in 6.4 and 6.5 we realized the representations of the complexification $\mathfrak{g}_c$ of the given Lie algebra $\mathfrak{g} = \text{Lie } G$ by *ladder operators* $X_\pm$ resp. $Y_\pm$ changing generators v_j of our representation space V by going to $v_{j\pm2}$ resp. $v_{j\pm1}$ in a controlled way. For $G = \text{SL}(2, \mathbf{R})$ and SU(2) the indices of the generators were related to the *weights*, i.e. the eigenvalues of the operator Z. For $G = \text{Heis}(\mathbf{R})$ we observe a different behaviour: We have an operator Y_0 producing the same eigenvalue when applied to all generators of our representation space. This is expression of the fact that $\mathfrak{sl}(2, \mathbf{R})$ and $\mathfrak{su}(2)$ resp. the groups SL(2, $\mathbf{R}$) and SU(2) on the one hand and heis($\mathbf{R}$) resp. Heis($\mathbf{R}$) on the other are examples of two different types for which we have to expect different ways of generalizations. Let us already guess that we expect to generalize the machinery of the weights in the first case (the *semisimple* one) by assuming to have a greater number of commuting operators whose eigenvalues constitute the weights and to have two different kinds of operators raising or lowering the weights (ordered, say, lexicographically).

We have to introduce some more notions to get the tools for an appropriate structure theory. The reader interested mainly in special examples and eager to see the representation theory of the other groups mentioned in the introduction may want to skip this section. Otherwise she or he is strongly recommended to a parallel study of a more detailed source like [Hu], [HN] or [Ja]. Here we follow [Ki] p.92f, [KT], and [Kn1], which don't give proofs neither (for these see [Kn] p.113ff, and/or [Kn2] Ch.I - VI).

6.6.1 Specifications of Groups and Lie Algebras

In this section we mainly treat Lie algebras. But we start with some definitions for groups, which often correspond to related definitions for algebras.

The reports [Kn1] and [KT] take as their central objects real or complex linear connected *reductive* groups. The definition of a reductive group is not the same with all authors but the following definition ([Kn] p.3, [KT] p.25) is very comprehensive and practical for our purposes.

Definition 6.7: A linear connected *reductive* group is a closed connected group of real or complex matrices that is stable under conjugate transpose, i.e. under the *Cartan involution*

$$\Theta : M_n(\mathbf{C}) \longrightarrow M_n(\mathbf{C}), \quad X \longmapsto ({}^t\bar{X})^{-1}.$$

A linear connected *semisimple* group is a linear connected reductive group with finite center.

More standard is the following definition: A group is called *simple* iff it is non-trivial, and has no normal subgroups other than $\{e\}$ and G itself.

Example 6.10: The following groups are reductive

a) $G = \mathrm{GL}(n, \mathbf{C})$,

b) $G = \mathrm{SL}(n, \mathbf{C})$,

c) $G = \mathrm{SO}(n, \mathbf{C})$,

d) $G = \mathrm{Sp}(n, \mathbf{C}) := \{g \in \mathrm{SL}(2n, \mathbf{C}); \ {}^t g J g = J := (\begin{smallmatrix} & E_n \\ -E_n & \end{smallmatrix})\}$.

The center of $\mathrm{GL}(n, \mathbf{C})$ is isomorphic to $\mathbf{C}^*$, so $\mathrm{GL}(n, \mathbf{C})$ is not semisimple. The other groups are semisimple (with exception $n = 2$ for c)). More examples come up by the groups of real matrices in the above complex groups. $\mathrm{GL}(n, \mathbf{R})$ is disconnected and therefore not reductive in the sense of the above definition, however its identity component is. We will come back to a classification scheme behind these examples in the discussion of the corresponding notions for Lie algebras.

For the moment we follow another approach (as in [Ki] p.17). Non-commutative groups can be classified according to the degree of their non-commutativity: Given two subsets A and B of the group G, let $[A, B]$ denote the set of all elements of the form $aba^{-1}b^{-1}$ as a runs through A and b runs through B. We define two sequences of subgroups:

Let $G_0 := G$ and for $n \in \mathbf{N}$ let G_n be the subgroup generated by the set $[G_{n-1}, G_{n-1}]$. G_n is called the n-th *derived group* of G.

Let $G^0 := G$ and for $n \in \mathbf{N}$ assume G^n to be the subgroup generated by the set $[G, G^{n-1}]$. We obviously obtain the following inclusions

$$G = G_0 \supset G_1 \supset \cdots \supset G_n \supset \cdots,$$

$$G = G^0 \supset G^1 \supset \cdots \supset G^n \supset \cdots.$$

For a commutative group these sequences are trivial, we have $G_n = G^n = \{e\}$ for all $n \in \mathbf{N}$.

Definition 6.8: G is called *solvable of class k* iff $G_n = \{e\}$ beginning with $n = k$. G is called *nilpotent of class k* iff $G^n = \{e\}$ beginning with $n = k$.

Exercise 6.26: Show that $\mathrm{Heis}(\mathbf{R})$ is nilpotent and solvable of class 1.

Now we pass to the corresponding concepts for Lie algebras (following [Ki] p.89).

Definition 6.9: A linear subspace $\mathfrak{g}_0$ in a Lie algebra is called a *subalgebra* iff

$$[X, Y] \subset \mathfrak{g}_0 \quad \text{for all } X, Y \in \mathfrak{g}_0.$$

A linear subspace $\mathfrak{a}$ in a Lie algebra is called an *ideal* iff

$$[X, Y] \subset \mathfrak{a} \quad \text{for all } X \in \mathfrak{g} \text{ and } Y \in \mathfrak{a}.$$

If $\mathfrak{a}$ is an ideal in $\mathfrak{g}$, then the factor space $\mathfrak{g}/\mathfrak{a}$ is provided in a natural way with the structure of a Lie algebra, which is called the *factor algebra* of $\mathfrak{g}$ by $\mathfrak{a}$.

In every Lie algebra $\mathfrak{g}$ we can define two sequences of subspaces, where here the expression $[\mathfrak{a}, \mathfrak{b}]$ denotes the linear hull of all $[X, Y]$, $X \in \mathfrak{a}, Y \in \mathfrak{b}$:

$$\mathfrak{g}_1 := [\mathfrak{g}, \mathfrak{g}], \quad \mathfrak{g}_2 := [\mathfrak{g}, \mathfrak{g}_1], \quad \ldots, \quad \mathfrak{g}_{n+1} := [\mathfrak{g}, \mathfrak{g}_n],$$
$$\mathfrak{g}^1 := [\mathfrak{g}, \mathfrak{g}], \quad \mathfrak{g}^2 := [\mathfrak{g}^1, \mathfrak{g}^1], \quad \ldots, \quad \mathfrak{g}^{n+1} := [\mathfrak{g}^n, \mathfrak{g}^n].$$

It is clear that one has the following inclusions

$$\mathfrak{g}_n \supset \mathfrak{g}_{n+1}, \quad \mathfrak{g}^n \supset \mathfrak{g}^{n+1}, \quad \mathfrak{g}_n \supset \mathfrak{g}^n, \quad n \in \mathbf{N}.$$

Exercise 6.27: Prove that all $\mathfrak{g}^n$ and $\mathfrak{g}_n$ are ideals in $\mathfrak{g}$ and that $\mathfrak{g}_n/\mathfrak{g}_{n+1}$ and $\mathfrak{g}^n/\mathfrak{g}^{n+1}$ are commutative.

For $\dim \mathfrak{g} < \infty$ the sequences $(\mathfrak{g}_n)$ and $(\mathfrak{g}^n)$ must stabilize, beginning with a certain n we have $\mathfrak{g}_n = \mathfrak{g}_{n+1} = \cdots = \mathfrak{g}_\infty$ and $\mathfrak{g}^n = \mathfrak{g}^{n+1} = \cdots = \mathfrak{g}^\infty$.

Definition 6.10: $\mathfrak{g}$ is called *solvable* resp. *nilpotent* iff $\mathfrak{g}^\infty = \{0\}$ resp. $\mathfrak{g}_\infty = \{0\}$. Obviously every nilpotent algebra is solvable.

Example 6.11: $\mathrm{heis}(\mathbf{R})$ is nilpotent, and hence solvable as we have more generally

$$\begin{aligned}
\mathfrak{b}_n(\mathbf{K}) &:= \{X = (x_{ij}) \in M_n(\mathbf{K}); \, x_{ij} = 0 \text{ for } i > j\} \text{ is solvable,} \\
\mathfrak{n}_n(\mathbf{K}) &:= \{X = (x_{ij}) \in M_n(\mathbf{K}); , \, x_{ij} = 0 \text{ for } i \geq j\} \text{ is nilpotent.}
\end{aligned}$$

It is clear that every $X \in \mathfrak{n}_n$ is *nilpotent*, i.e. has the property $X^n = 0$. This is generalized by the following statement justifying the name nilpotent algebra:

Theorem 6.4 (Engel): $\mathfrak{g}$ is nilpotent iff the operator $\operatorname{ad} X$ is nilpotent for all $X \in \mathfrak{g}$, i.e. for every X there is an $n \in \mathbf{N}$ with $(\operatorname{ad} X)^n = 0$.
We recall that $\operatorname{ad} X$ is defined by $\operatorname{ad} X(Y) := [X, Y]$ for all $Y \in \mathfrak{g}$.

Now we pass to the other side of Lie algebra structure theory. As in [Kn1] p.1 ff we restrict ourselves to the treatment of finite dimensional real or complex algebras $\mathfrak{g}$.

Definition 6.11: $\mathfrak{g}$ is said to be *simple* iff $\mathfrak{g}$ is non-abelian and $\mathfrak{g}$ has no proper non-zero ideals.

In this case one has $[\mathfrak{g}, \mathfrak{g}] = \mathfrak{g}$, which shows that we are as far from solvability as possible.

Definition 6.12: $\mathfrak{g}$ is said to be *semisimple* iff $\mathfrak{g}$ has no non-zero abelian ideal.

There are other (equivalent) definitions, for instance $\mathfrak{g}$ is said to be *semisimple* iff one has $\operatorname{rad} \mathfrak{g} = 0$. In this definition the radical $\operatorname{rad} \mathfrak{g}$ is the sum of all solvable ideals of $\mathfrak{g}$.

Semisimple and simple algebras are related a follows.

Theorem 6.5: $\mathfrak{g}$ is semisimple iff $\mathfrak{g}$ is the sum of simple ideals. In this case there are no other simple ideals, the direct sum decomposition is unique up to order of summands and every ideal is the sum of the simple ideals. Also in this case $[\mathfrak{g}, \mathfrak{g}] = \mathfrak{g}$.

Finally, from [KT] p.29 we take over

Definition 6.13: A *reductive* Lie algebra is a Lie algebra that is the direct sum of two ideals, one equal to a semisimple algebra and the other to an abelian Lie algebra.

We have a practical criterium:

Theorem 6.6: If $\mathfrak{g}$ is a real Lie algebra of real or complex (even quaternion) matrices closed under conjugate transposition, then $\mathfrak{g}$ is reductive. If moreover the center of $\mathfrak{g}$ is trivial, i.e. $Z_\mathfrak{g} := \{X \in \mathfrak{g}; [X, Y] = 0 \ \text{ for all } \ Y \in \mathfrak{g}\} = 0$, then $\mathfrak{g}$ is semisimple.

Reductive Lie algebras have a very convenient property:

Proposition 6.3: A Lie algebra $\mathfrak{g}$ is reductive iff each ideal $\mathfrak{a}$ in $\mathfrak{g}$ has a *complementary* ideal, i.e. an ideal $\mathfrak{b}$ with $\mathfrak{g} = \mathfrak{a} \oplus \mathfrak{b}$.

As to be expected, to some extent the notions just defined for groups and algebras fit together and one has statements like the following:

Proposition 6.4 ([KT] p.29): If G is a linear connected semisimple group, then $\mathfrak{g} = \operatorname{Lie} G$ is semisimple. More generally, if G is linear connected reductive, then $\mathfrak{g}$ is reductive with $\mathfrak{g} = Z_\mathfrak{g} \oplus [\mathfrak{g}, \mathfrak{g}]$ as a direct sum of ideals. Here $Z_\mathfrak{g}$ denotes the center of $\mathfrak{g}$, and the commutator ideal is semisimple.

Example 6.12: $\mathfrak{gl}(n, \mathbf{R}) = \{\text{scalars}\} \oplus \mathfrak{sl}(n, \mathbf{R})$.

The main tool for the structure theory of semisimple Lie algebras is the following bilinear form that was first defined by Killing and then extensively used by E. Cartan.

Definition 6.14: The *Killing form* B is the symmetric bilinear form on $\mathfrak{g}$ defined by

$$B(X,Y) := \mathrm{Tr}\,(\mathrm{ad}X\,\mathrm{ad}Y) \quad \text{for all } X,Y \in \mathfrak{g}.$$

Remark 6.14: B is invariant in the sense that

$$B([X,Y],Z]) = B(X,[Y,Z]) \quad \text{for all } X,Y,Z \in \mathfrak{g}.$$

Exercise 6.28: Determine the matrix of B with respect to the basis
a) F, G, H (from Exercise 6.3 in 6.1) for $\mathfrak{g} = \mathfrak{sl}(2, \mathbf{R})$,
b) P, Q, R (from Exercise 6.7 in 6.1) for $\mathfrak{g} = \mathrm{heis}(\mathbf{R})$.

The starting point for structure theory of semisimple Lie algebras is *Cartan's Criterium for Semisimplicity.*

Theorem 6.7: $\mathfrak{g}$ is semisimple if and only if the Killing form is *nondegenerate*, that is $B(X,Y) = 0$ for all $Y \in \mathfrak{g}$ implies $X = 0$.

The proof uses the remarkable fact that $\ker B$ is an ideal in $\mathfrak{g}$.
There is another important and perhaps a bit more accessible bilinear form on $\mathfrak{g}$: The *trace form* B_0 is given by

$$B_0(X,Y) := \mathrm{Tr}\,(XY) \quad \text{for all } X,Y \in \mathfrak{g}.$$

The trace form is invariant in the same sense as the Killing form. Both forms are related (see e.g. [Fog] IV.4):

Remark 6.15: We have

$$
\begin{aligned}
B(X,Y) &= 2n\,\mathrm{Tr}\,(XY) & \text{for } \mathfrak{g} &= \mathfrak{sl}(n, \mathbf{K}) \text{ and } n \geq 2, \\
&= (n-2)\mathrm{Tr}\,(XY) & \text{for } \mathfrak{g} &= \mathfrak{so}(n, \mathbf{K}) \text{ and } n \geq 3, \\
&= 2(n+1)\mathrm{Tr}\,(XY) & \text{for } \mathfrak{g} &= \mathfrak{sp}(2n, \mathbf{K}) \text{ and } n \geq 1.
\end{aligned}
$$

A variant of the trace form already appeared in our definition of topology in matrix spaces. We introduced for $X, Y \in M_n(\mathbf{C})$

$$< X,Y > := \mathrm{Re}\,\mathrm{Tr}\,{}^t\bar{X}Y, \text{ and } \| X \|^2 := \mathrm{Re}\,\mathrm{Tr}\,{}^t\bar{X}X.$$

As infinitesimal object corresponding to the Cartan involution for our matrix groups G

$$\Theta : \mathrm{GL}(n, \mathbf{C}) \longrightarrow \mathrm{GL}(n, \mathbf{C}), \quad g \longmapsto ({}^t\bar{g})^{-1},$$

we have the map for the Lie algebra $\mathfrak{g} = Lie\,G$

$$\theta : M_n(\mathbf{C}) \longrightarrow M_n(\mathbf{C}), \quad X \longmapsto -{}^t\bar{X}.$$

Hence we have

$$< X,Y > := -\mathrm{Re}\,B_0(X, \theta Y)$$

as a scalar product on $\mathfrak{g}$ as a real vector space. This will become important later on.

6.6.2 Structure Theory for Complex Semisimple Lie Algebras

A complete classification exists for semisimple Lie algebras, as one knows their building blocks, the simple algebras: Over $\mathbf{C}$ there exist four infinite series of *classical* simple Lie algebras and five *exceptional* simple Lie algebras. Over $\mathbf{R}$ we have 12 infinite series and 23 exceptional simple algebras. From all these, here we reproduce only the following standard list of classical complex simple algebras

1) $A_n := \{X \in M_{n+1}(\mathbf{C}); \operatorname{Tr} X = 0\}$, $n = 1, 2, 3, \ldots$

2) $B_n := \{X \in M_{2n+1}(\mathbf{C}); X = -{}^t X\}$, $n = 2, 3, \ldots$

3) $C_n := \{X \in M_{2n}(\mathbf{C}); X J_{2n} + J_{2n} X = 0\}$, $n = 3, 4, \ldots, J_{2n} := \begin{pmatrix} & E_n \\ -E_n & \end{pmatrix}$,

4) $D_n := \{X \in M_{2n}(\mathbf{C}); X = -{}^t X\}$, $n = 4, 5, \ldots$.

We have the isomorphisms

$$B_1 \simeq A_1 \simeq C_1, C_2 \simeq B_2, D_2 \simeq A_1 \oplus A_1, D_3 \simeq A_3$$

and D_1 is commutative. In the sequel we shall in most cases return to our previous notation and write

$$A_n = \mathfrak{sl}(n+1, \mathbf{C}), B_n = \mathfrak{so}(2n+1, \mathbf{C}), C_n = \mathfrak{sp}(n, \mathbf{C}), D_n = \mathfrak{so}(2n, \mathbf{C}).$$

All these algebras are also simple algebras over $\mathbf{R}$. The remaining real classical algebras remain simple upon complexification. For a complete list see for instance [Ki] p.92/3. By the way, we introduce here the following also otherwise important concept:

Definition 6.15: Given a complex Lie algebra $\mathfrak{g}$, a real algebra $\mathfrak{g}_0$ is called a *real form* of $\mathfrak{g}$ iff $(\mathfrak{g}_0)_c := \mathfrak{g}_0 \otimes \mathbf{C} = \mathfrak{g}$.

Example 6.13: $\mathfrak{g} = \mathfrak{sl}(2, \mathbf{C})$ has just two real forms $\mathfrak{g}_0 = \mathfrak{sl}(2, \mathbf{R})$ and $\mathfrak{su}(2)$. More about this is to be found in [Kn1] p.15.

The main ingredients for a classification scheme are the *root* and *Dynkin diagrams*. We give a sketch starting by the case of complex semisimple algebras where the theory is most accessible, and then go to compact real and finally briefly to noncompact real algebras. The central tool to define roots and their diagrams is a certain distinguished abelian subalgebra.

Definition 6.16: Let $\mathfrak{g}$ be a complex semisimple Lie algebra. A *Cartan subalgebra* $\mathfrak{h}$ is a maximal abelian subspace of $\mathfrak{g}$ in which every $\operatorname{ad} H, H \in \mathfrak{h}$, is diagonizable.

There are other equivalent ways to characterize a Cartan subalgebra $\mathfrak{h}$, for instance (see [Kn1] p.2): $\mathfrak{h}$ is a nilpotent subalgebra whose *normalizer* $N_{\mathfrak{g}}(\mathfrak{h})$ satisfies

$$N_{\mathfrak{g}}(\mathfrak{h}) := \{X \in \mathfrak{g}, [X, H] \in \mathfrak{h} \quad \text{for all } H \in \mathfrak{h}\} = \mathfrak{h}.$$

Each semisimple complex algebra has a Cartan subalgebra. Any two are conjugate via $\operatorname{Int}\mathfrak{g}$ ([Kn1] p.24), where $\operatorname{Int}\mathfrak{g}$ is a certain *analytic subgroup* of $\operatorname{GL}(\mathfrak{g})$ with Lie algebra $\operatorname{ad}\mathfrak{g}$. We refer to [Kn2] p.69/70 for the notion of an analytic subgroup, which comes up quite naturally when one analyzes the relation between Lie algebras and Lie groups.

The theory of Cartan subalgebras for the complex semisimple case extends to a complex reductive Lie algebra $\mathfrak{g}$ by just saying that the center of $\mathfrak{g}$ is to be adjoined to a Cartan subalgebra of the semisimple part of $\mathfrak{g}$.

Example 6.14: For $\mathfrak{g} = \mathfrak{sl}(n, \mathbf{C})$ we have as Cartan subalgebra

$$\mathfrak{h} := \{\text{diagonal matrices in } \mathfrak{g}\}.$$

Using our Cartan subalgebra $\mathfrak{h}$, we generalize the decomposition we introduced at the beginning of Section 6.4 for $\mathfrak{g} = \mathfrak{sl}(2, \mathbf{C})$

$$\mathfrak{g} = <Z_0> + <X_+> + <X_->,$$

$$[Z_0, X_\pm] = \pm 2 X_\pm, \quad [X_+, X_-] = Z_0,$$

to the *root space decomposition*

$$\mathfrak{g} = \mathfrak{h} \oplus \bigoplus_{\alpha \in \Delta} \mathfrak{g}_\alpha$$

as follows: For all $H \in \mathfrak{h}$ the maps $\operatorname{ad} H$ are diagonizable, and, as the elements of $\mathfrak{h}$ all commute, they are simultaneously diagonizable (by the finite-dimensional spectral theorem from Linear Algebra). Let $V_1, \ldots, V_\ell$ be the eigenspaces in $\mathfrak{g}$ for the different systems of eigentupels. If $\mathfrak{h} := <H_1, \ldots, H_r>$ and $\operatorname{ad} H_i$ acts as $\lambda_{ij} \cdot id$ on V_j, define a linear functional λ_j on $\mathfrak{h}$ by $\lambda_j(H_i) = \lambda_{ij}$. If $H := \Sigma c_i H_i$, then $\operatorname{ad} H$ acts on V_j by multiplication with

$$\sum_i c_i \lambda_{ij} = \sum_i c_i \lambda_j(H_i) =: \lambda_j(H).$$

In other words, $\operatorname{ad} \mathfrak{h}$ acts in simultaneously diagonal fashion on $\mathfrak{g}$ and the simultaneous eigenvalues are members of the dual vector space $\mathfrak{h}^* := \operatorname{Hom}_{\mathbf{C}}(\mathfrak{h}, \mathbf{C})$. There are finitely many such simultaneous eigenvalues and we write

$$\mathfrak{g}_\lambda := \{X \in \mathfrak{g}; \ [H, X] = \lambda(H)X \quad \text{for all } H \in \mathfrak{h}\}$$

for the eigenspace corresponding to $\lambda \in \mathfrak{h}^*$. The nonzero such λ are called *roots* and the corresponding $\mathfrak{g}_\lambda$ a *root space*. The (finite) set of all roots is denoted by Δ.

The following are some elementary properties of root space decompositions (for proofs see for instance [Kn2] II.4):

Proposition 6.5: We have

a) $[\mathfrak{g}_\alpha, \mathfrak{g}_\beta] \subset \mathfrak{g}_{\alpha+\beta}$.

b) If $\alpha, \beta \in \Delta \setminus \{0\}$ and $\alpha + \beta \neq 0$, then $B(\mathfrak{g}_\alpha, \mathfrak{g}_\beta) = 0$, i.e. root spaces are orthogonal with respect to the Killing form.

c) B is nonsingular on $\mathfrak{g}_\alpha \times \mathfrak{g}_{-\alpha}$ if $\alpha \in \Delta$.

d) $-\alpha \in \Delta$ if $\alpha \in \Delta$.

e) $B|_{\mathfrak{h} \times \mathfrak{h}}$ is nondegenerate. We define H_α to be the element of $\mathfrak{h}$ paired with α.

f) Δ spans $\mathfrak{h}^*$.

Some deeper properties of root space decompositions are assembled in the following statement (see again [Kn2] II.4).

Theorem 6.8: Root space decompositions have the following properties:

a) dim $\mathfrak{g}_\alpha = 1$ for $\alpha \in \Delta$.
b) $n\alpha \notin \Delta$ for $\alpha \in \Delta$ and any integer $n \geq 2$.
c) $[\mathfrak{g}_\alpha, \mathfrak{g}_\beta] = \mathfrak{g}_{\alpha+\beta}$ for $\alpha + \beta \neq 0$.
d) The real subspace $\mathfrak{h}_0$ of $\mathfrak{h}$ on which all roots are real is a real form of $\mathfrak{h}$ and $B\big|_{\mathfrak{h}_0 \times \mathfrak{h}_0}$ is an inner product.

We transfer $B\big|_{\mathfrak{h}_0 \times \mathfrak{h}_0}$ to the real span $\mathfrak{h}_0^*$ of the roots obtaining a scalar product $<,>$ and a norm $\| \cdot \|$. It is not too difficult to see that for $\alpha \in \Delta$ there is an orthogonal transformation s_α given on $\mathfrak{h}_0^*$ by

$$s_\alpha(\varphi) := \varphi - \frac{2 <\varphi, \alpha>}{\| \alpha \|^2} \alpha \quad \text{for } \varphi \in \mathfrak{h}_0^*.$$

s_α is called a *root reflection* in α and the hyperplane $\alpha^\perp$ a *mirror*.

The analysis of the root space shows that it has very nice geometrical and combinatorical properties. The abstraction of these is the following:

Definition 6.17: An *abstract root system* is a finite set Δ of nonzero elements in a real inner product space V such that

a) Δ spans V,
b) all reflections s_α for $\alpha \in \Delta$ carry Δ to itself,
c) $2 <\beta, \alpha> / \| \alpha \|^2 \in \mathbf{Z}$ for all $\alpha, \beta \in \Delta$.

The abstract root system is called *reduced* iff $\alpha \in \Delta$ implies $2\alpha \notin \Delta$.
And it is called *reducible* iff $\Delta = \Delta' \cup \Delta''$ with $\Delta' \perp \Delta''$, otherwise it is *irreducible*.

The root system of a complex semisimple Lie algebra $\mathfrak{g}$ with respect to a Cartan subalgebra $\mathfrak{h}$ forms a reduced abstract root system in $\mathfrak{h}_0^*$. And a semisimple Lie algebra $\mathfrak{g}$ is simple if the corresponding root system is irreducible.

Definition 6.18: The dimension of the underlying space V of an abstract root system Δ is called its *rank*, and if Δ is the root system of a semisimple Lie algebra $\mathfrak{g}$, we also refer to $r = \dim \mathfrak{h}$ as the *rank* of $\mathfrak{g}$.

To give an illustration, we sketch the reduced root systems of rank 2.

      case 1   case 2   case 3   case 4

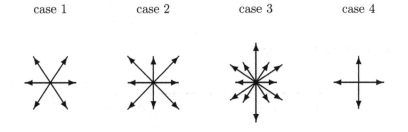

Ordering of the Roots, Cartan Matrices, and Dynkin Diagrams

We fix an ordered basis $\alpha_1, \ldots, \alpha_r$ of $\mathfrak{h}_0^*$ and define $\lambda = \Sigma c_i \alpha_i$ to be *positive* if the first nonzero c_i is positive. The ordering comes from saying $\lambda > \mu$ if $\lambda - \mu$ is positive. Let Δ^+ be the set of positive members in Δ.

A root α is called *simple* if $\alpha > 0$ and α does not decompose as $\alpha = \beta_1 + \beta_2$ with β_1 and β_2 positive roots. And a root α is called *reduced* if $(1/2)\alpha$ is not a root.
Relative to a given simple system $\alpha_1, \ldots, \alpha_r$, the *Cartan matrix* C is the $r \times r$-matrix with entries

$$c_{ij} = 2 < \alpha_i, \alpha_j > / \parallel \alpha_i \parallel^2 .$$

It has the following properties
a) $c_{ij} \in \mathbf{Z}$ for all i, j,
b) $c_{ii} = 2$ for all i,
c) $c_{ij} \leq 0$ for all $i \neq j$,
d) $c_{ij} = 0$ iff $c_{ji} = 0$,
e) there exists a diagonal matrix D with positive diagonal entries such that DCD^{-1} is symmetric positive definite.

An *abstract Cartan matrix* is a square matrix satisfying properties a) through e) as above. To such a matrix we associate a diagram usually called a *Dynkin diagram* (historically more correct is the term *CDW-diagram*, C indicating Coxeter and W pointing to Witt): To the elements $\alpha_1, \ldots, \alpha_r$ of a basis of a root system correspond bijectively r points of a plane, which are also called $\alpha_1, \ldots, \alpha_r$. For $i \neq j$ one joins the point α_i to α_j by $c_{ij}c_{ji}$ lines, which do not touch any $\alpha_k, k \neq i, j$. For $c_{ij} \neq c_{ji}$ the lines are drawn as arrows pointing in the direction to α_j if $c_{ij} < c_{ji}$.

The main facts are that a Cartan matrix and equivalently the CDW-diagram determine the Lie algebra uniquely (up to isomorphism). In particular this correspondence does not depend on the choice of a basis of the root system.

Example 6.15: For $\mathfrak{g} = \mathfrak{sl}(r + 1, \mathbf{C}), r \geq 1$ and its standard Cartan subalgebra

$$\mathfrak{h} := \{D(d_1, \ldots, d_r); \ d_1, \ldots, d_r \in \mathbf{C}, \Sigma d_i = 0\}$$

we have the Cartan matrix

$$C := \begin{pmatrix} 2 & -1 & & & & \\ -1 & 2 & & & & \\ & & \bullet & & & \\ & & & \bullet & & \\ & & & & 2 & -1 \\ & & & & -1 & 2 \end{pmatrix}$$

and the corresponding CDW-diagram

The diagrams for the other classical algebras are the following

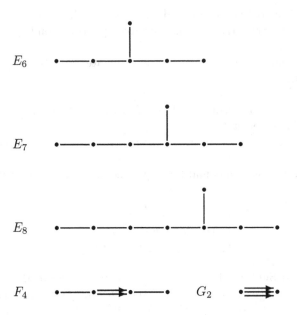

The remaining *exceptional* graphs of indecomposable reduced root systems are the following

6.6.3   Structure Theory for Compact Real Lie Algebras

Up to now, we treated complex Lie algebras. We go over to a real Lie algebra $\mathfrak{g}_0$ and (following [Kn1] p.16) we call a subalgebra $\mathfrak{h}_0$ of $\mathfrak{g}_0$ a *Cartan subalgebra* if its complexification is a Cartan subalgebra of the complex algebra $\mathfrak{g} = \mathfrak{g}_0 \otimes \mathbf{C}$. At first we look at the compact case: If $\mathfrak{g}_0$ is the Lie algebra of a compact Lie group G and if $\mathfrak{t}_0$ is a maximal abelian subspace of $\mathfrak{g}_0$, then $\mathfrak{t}_0$ is a Cartan subalgebra. We have already discussed the example $\mathfrak{g}_0 = \mathfrak{su}(2)$ in 4.2 and we will discuss $\mathfrak{g}_0 = \mathfrak{su}(3)$ in the next section and see that we have $\dim \mathfrak{t}_0 = 2$ in this case. To illustrate the general setting, we collect some more remarks about the background notions (from [Kn1] p.15):

Theorem 6.9: If $\mathfrak{g}_0$ is semisimple, then the following conditions are equivalent:
a) $\mathfrak{g}_0$ is the Lie algebra of some compact Lie group.
b) Int $\mathfrak{g}_0$ is compact.
c) The Killing form of $\mathfrak{g}_0$ is negative definite.

If $\mathfrak{g}$ is semisimple complex, a real form $\mathfrak{g}_0$ of $\mathfrak{g}$ is said to be *compact* if the equivalent conditions of the theorem hold. Our main example $\mathfrak{su}(n)$ is a compact form of $\mathfrak{sl}(n, \mathbf{C})$. The fundamental result is here:

Theorem 6.10: Each complex semisimple Lie algebra has a compact real form.

Another important topic is maximal tori.

Definition 6.19: Let G be a compact connected linear group. A *maximal torus* in G is a subgroup T maximal with respect to the property of being compact connected abelian.

From [Kn2] Proposition 4.30, Theorem 4.34 and 4.36 we take over:

Theorem 6.11: If G is a compact connected linear group, then
a) the maximal tori in G are exactly the analytic subgoups corresponding to the maximal abelian subalgebras of $\mathfrak{g}_0 = \operatorname{Lie} G$;
b) any two maximal abelian subalgebras of $\mathfrak{g}_0$ are conjugate via $\operatorname{Ad} G$ and hence any two maximal tori are conjugate via G.

Theorem 6.12: If G is compact connected and T a maximal torus, then each element of G is conjugate to a member of T.

Example 6.16:
1. For $G = \mathrm{SU}(n)$ one has as a maximal torus, its Lie algebra and its complexified Lie algebra

$$
\begin{aligned}
T &= \{D(e^{i\vartheta_1}, \ldots, e^{i\vartheta_n}); \Sigma \vartheta_j = 0\}, \\
\mathfrak{t}_0 &= \{D(i\vartheta_1, \ldots, i\vartheta_n); \Sigma \vartheta_j = 0\}, \\
\mathfrak{t} &= \{D(z_1, \ldots, z_n); \Sigma z_j = 0\}.
\end{aligned}
$$

2. For $G = \mathrm{SO}(2n)$ and $\mathrm{SO}(2n + 1)$ one has as a maximal torus T the diagonal block matrices with 2×2-blocks $r(\vartheta_j), j = 1, \ldots, n$, in the diagonal, resp. additionally 1 in the second case.

We go back to the general case and use the notation we have introduced. In this setting, we can form a root-space decomposition

$$
\mathfrak{g} = \mathfrak{t} \oplus \bigoplus_{\alpha \in \Delta} \mathfrak{g}_\alpha.
$$

Each root is the complexified differential of a multiplicative character χ_α of the maximal torus T that corresponds to $\mathfrak{t}_0$, with

$$
\operatorname{Ad}(t)X = \chi_\alpha(t)X \quad \text{for all } X \in \mathfrak{g}_\alpha.
$$

Another central concept is the one of the Weyl group. We keep the notation used above: G is compact connected, $\mathfrak{g}_0 = \operatorname{Lie} G$, T a maximal torus, $\mathfrak{t}_0 = \operatorname{Lie} T$, $\mathfrak{t} = \mathfrak{t}_0 \otimes \mathbf{C}$, $\Delta(\mathfrak{g}, \mathfrak{t})$ is the set of roots, and B is the negative of a G invariant scalar product on $\mathfrak{g}_0$. We define $\mathfrak{t}_\mathbf{R} = i\mathfrak{t}_0$. As roots are real on $\mathfrak{t}_\mathbf{R}$, they are in $\mathfrak{t}_\mathbf{R}^*$. The form B, when extended to be complex bilinear, is positive definite on $\mathfrak{t}_\mathbf{R}$, yielding a scalar product $<,>$ on $\mathfrak{t}_\mathbf{R}^*$. Now, the *Weyl group* $W = W(\Delta(\mathfrak{g}, \mathfrak{t}))$ is in this context defined as the group generated by the root reflections $s_\alpha, \alpha \in \Delta(\mathfrak{g}, \mathfrak{t})$, given (as already fixed above) on $\mathfrak{t}_\mathbf{R}^*$ by $s_\alpha(\lambda) = \lambda - \frac{2<\lambda,\alpha>}{|\alpha|^2}\alpha$. This is a finite group, which also can be characterized like this:

One defines $W(G, T)$ as the quotient of the normalizer by the centralizer

$$W(G, T) := N_G(T)/Z_G(T).$$

By Corollary 4.52 in [Kn2] p.260 one has $Z_G(T) = T$ and hence also the formula $W(G, T) = N_G(T)/T$.

Theorem 6.13: The group $W(G, T)$, when considered as acting on $\mathfrak{t}_\mathbf{R}^*$, coincides with $W(\Delta(\mathfrak{g}, \mathfrak{t}))$.

6.6.4   Structure Theory for Noncompact Real Lie Algebras

Now, in the final step, we briefly treat the general case of a real Lie algebra $\mathfrak{g}_0$ of a noncompact semisimple group G. We already introduced the Killing form B, the Cartan involution θ with $\theta X = -^t\bar{X}$ for a matrix X, and the fact that if $\mathfrak{g}_0$ consists of real, complex or quaternion matrices and is closed under conjugate transpose, then it is reductive. More generally we call here a *Cartan involution* θ any involution of $\mathfrak{g}_0$ such that the symmetric bilinear form

$$B_\theta(X, Y) := -B(X, \theta Y)$$

is positive definite. Then we have the *Cartan decomposition*

$$\mathfrak{g}_0 = \mathfrak{k}_0 \oplus \mathfrak{p}_0, \quad \mathfrak{k}_0 := \{X \in \mathfrak{g}_0;\, \theta X = X\}, \quad \mathfrak{p}_0 := \{X \in \mathfrak{g}_0;\, \theta X = -X\}$$

with the bracket relations

$$[\mathfrak{k}_0, \mathfrak{k}_0] \subseteq \mathfrak{k}_0, \quad [\mathfrak{k}_0, \mathfrak{p}_0] \subseteq \mathfrak{p}_0, \quad [\mathfrak{p}_0, \mathfrak{p}_0] \subseteq \mathfrak{k}_0.$$

One has the following useful facts ([Kn 2] VI.2):
a) $\mathfrak{g}_0$ has a Cartan involution.
b) Any two Cartan involutions of $\mathfrak{g}_0$ are conjugate via $\operatorname{Int} \mathfrak{g}_0$.
c) If $\mathfrak{g}$ is a complex semisimple Lie algebra, then any two compact real forms of $\mathfrak{g}$ are conjugate via $\operatorname{Int} \mathfrak{g}$.
d) If $\mathfrak{g}$ is a complex semisimple Lie algebra and is considered as a real Lie algebra, then the only Cartan involutions of $\mathfrak{g}$ are the conjugations with respect to the compact real forms of $\mathfrak{g}$.

The Cartan decomposition of Lie algebras has a global counterpart (see for instance [Kn1] Theorem 4.3). Its most rudimentary form is the following:

Proposition 6.6: For every $A \in \mathrm{GL}(n, \mathbf{C})$ resp. $\mathrm{GL}(n, \mathbf{R})$, there are $S_1, S_2 \in \mathrm{O}(n)$ resp. $\mathrm{U}(n)$ and a diagonal matrix D with real positive elements in the diagonal such that $A = S_1 D S_2$. This decomposition is not unique.

Proof as **Exercise 6.29** (use the fact that to each positive definite matrix B there is a unique positive definite matrix P with $B = P^2$ and/or see [He] p.73).

Restricted Roots

We want to find a way to describe Cartan subalgebras in our general situation: Let G be semisimple linear group with Lie $G = \mathfrak{g}_0$, $\mathfrak{g} = \mathfrak{g}_0 \otimes \mathbf{C}$, θ a Cartan involution of $\mathfrak{g}_0$, and $\mathfrak{g}_0 = \mathfrak{k}_0 \oplus \mathfrak{p}_0$ the corresponding Cartan decomposition. Let B be the Killing form (or more generally any nondegenerate symmetric invariant bilinear form on $\mathfrak{g}_0$ with $B(\theta X, \theta Y) = B(X, Y)$ such that $B_\theta(X, Y) = -B(X, \theta Y)$ is positive definite).

Definition 6.20: Let $\mathfrak{a}_0$ be a maximal abelian subspace of $\mathfrak{p}_0$. *Restricted roots* are the nonzero $\lambda \in \mathfrak{a}_0^*$ such that

$$(\mathfrak{g}_0)_\lambda := \{X \in \mathfrak{g}_0;\ (\mathrm{ad}\,H)X = \lambda(H)X \quad \text{for all } H \in \mathfrak{a}_0\} \neq \{0\}.$$

Let Σ be the set of restricted roots and $\mathfrak{m}_0 := Z_{\mathfrak{k}_0}$. We fix a basis of $\mathfrak{a}_0$ and an associated lexicographic ordering in $\mathfrak{a}_0^*$ and define Σ^+ as the set of *positive resticted* roots. Then

$$\mathfrak{n}_0 = \bigoplus_{\lambda \in \Sigma^+} (\mathfrak{g}_0)_\lambda$$

is a nilpotent Lie subalgebra, and we have the following *Iwasawa decomposition*:

Theorem 6.14: The semisimple Lie algebra $\mathfrak{g}_0$ is a vector space direct sum

$$\mathfrak{g}_0 = \mathfrak{k}_0 \oplus \mathfrak{a}_0 \oplus \mathfrak{n}_0.$$

Here $\mathfrak{a}_0$ is abelian, $\mathfrak{n}_0$ is nilpotent, $\mathfrak{a}_0 \oplus \mathfrak{n}_0$ is a solvable subalgebra of $\mathfrak{g}_0$, and $\mathfrak{a}_0 \oplus \mathfrak{n}_0$ has $[\mathfrak{a}_0 \oplus \mathfrak{n}_0, \mathfrak{a}_0 \oplus \mathfrak{n}_0] = \mathfrak{n}_0$.

For a proof see for instance [Kn2] Proposition 6.43. There is also a global version ([Kn2] Theorem 6.46) stating (roughly) that one has a diffeomorphism of the corresponding groups $K \times A \times N \longrightarrow G$ given by $(k, a, n) \longmapsto kan$. We already know all this in our standard example $G = \mathrm{SL}(2, \mathbf{R})$: If we take $\theta X = -{}^t X$, we have

$$\mathfrak{g}_0 = \mathfrak{sl}(2, \mathbf{R}) = \mathfrak{k}_0 \oplus \mathfrak{a}_0 \oplus \mathfrak{n}_0,$$

with

$$\mathfrak{k}_0 = < \begin{pmatrix} & 1 \\ -1 & \end{pmatrix} >, \ \mathfrak{a}_0 = < \begin{pmatrix} 1 & \\ & -1 \end{pmatrix} >, \ \mathfrak{n}_0 = < \begin{pmatrix} & 1 \\ & \end{pmatrix} >$$

and as in Example 6.6 in 6.2

$$G = KAN \text{ with } K = \mathrm{SO}(2), \ A = \{t(y);\, y > 0\}, \ N = \{n(x);\, x \in \mathbf{R}\}.$$

As to be expected, roots and restricted roots are related to each other. If $\mathfrak{t}_0$ is a maximal abelian subspace of $\mathfrak{m}_0 := Z_{\mathfrak{t}_0}(\mathfrak{a}_0)$, then $\mathfrak{h}_0 := \mathfrak{a}_0 \oplus \mathfrak{t}_0$ is a Cartan subalgebra of $\mathfrak{g}_0$ in the sense we defined at the beginning ([Kn2] Proposition 6.47). Roots are real valued on $\mathfrak{a}_0$ and imaginary valued on $\mathfrak{t}_0$. The nonzero restrictions to $\mathfrak{a}_0$ of the roots turn out to be restricted roots. Roots and restricted roots can be ordered compatibly by taking $\mathfrak{a}_0$ before $i\mathfrak{t}_0$. Cartan subalgebras in this setting are not always unique up to conjugacy.

Exercise 6.30: Show that $\mathfrak{h}_0 :=< \left(\begin{smallmatrix} & 1 \\ 1 & \end{smallmatrix} \right) >$ and $\mathfrak{h}_0' :=< \left(\begin{smallmatrix} & 1 \\ -1 & \end{smallmatrix} \right) >$ are Cartan subagebras of $\mathfrak{g}_0 = \mathfrak{sl}(2, \mathbf{R})$ and determine the corresponding Iwasawa decomposition.

Every Cartan subalgebra of $\mathfrak{g}_0$ is conjugate (via $\mathrm{Int}\mathfrak{g}_0$) to this $\mathfrak{h}_0'$ or the $\mathfrak{h}_0$ above.

6.6.5   Representations of Highest Weight

After this excursion into structure theory of complex and real Lie algebras, we finally come back to show a bit more of the general representation theory, which is behind the examples discussed in our sections 6.4 and 6.5 and which we shall apply in our next section to the example $\mathfrak{g}_0 = \mathfrak{su}(3)$. We start by following the presentation in [Kn1] p.8. Let at first $\mathfrak{g}$ be a complex Lie algebra, $\mathfrak{h}$ a Cartan subalgebra, $\Delta = \Delta(\mathfrak{g}, \mathfrak{h})$ the set of roots, $\mathfrak{h}_0$ the real form of $\mathfrak{h}$ where roots are real valued, B the Killing form (or a more general form as explained above), and $H_\lambda \in \mathfrak{h}_0$ corresponding to $\lambda \in \mathfrak{h}_0^*$. Let $\hat{\pi} : \mathfrak{g} \longrightarrow \mathrm{End}\, V$ be a representation. For $\lambda \in \mathfrak{h}^*$ we put

$$V_\lambda := \{v \in V; (\hat{\pi}(H) - \lambda(H)1)^n v = 0 \quad \text{for all } H \in \mathfrak{h} \text{ and some } n = n(H, V) \in \mathbf{N}\}.$$

If $V_\lambda \neq \{0\}$, V_λ is called a *generalized weight space* and λ a *weight*. If V is finite dimensional, V is the direct sum of its generalized weight spaces. This is a generalization of the fact from linear algebra about eigenspace decompositions of a linear transformation on a finite-dimensional vector space. If λ is a weight, then the subspace

$$V_\lambda^0 := \{v \in V; \hat{\pi}(H)v = \lambda(H)v \quad \text{for all } H \in \mathfrak{h}\}$$

is nonzero and called the *weight space* corresponding to λ. One introduces a lexicographic ordering among the weights and hence has the notion of *highest* and *lowest weights*. The set of weights belonging to a representation $\hat{\pi}$ is denoted by $\Gamma(\hat{\pi})$. For dim $\mathfrak{h} = 1$ we have $\mathfrak{h} \simeq \mathfrak{h}^* \simeq \mathbf{C}$ and so the weights are simply the complex numbers we met in our examples in 6.4 and 6.5, for instance for $\mathfrak{g} = \mathfrak{sl}(2, \mathbf{C})$ we found representations $\hat{\sigma}_N$ with dim $V_N = N + 1$ and weights $-N, -N + 2, \ldots, N - 2, N$. In continuation of this, we treat a more general example.

Example 6.17: Let $G = \mathrm{SU}(n)$ and, hence, $\mathfrak{g} = \mathfrak{su}(n)_c = \mathfrak{sl}(n, \mathbf{C})$ and the Cartan subalgebra the diagonal subalgebra $\mathfrak{h} :=< H; \mathrm{Tr}\, H = 0 >$. We choose as generators of $\mathfrak{h}$ the matrices $H_j := E_{jj} - E_{j+1,j+1}$ and (slightly misusing) we write also $H_j = e_j - e_{j+1}$ where e_j denote the canonical basis vectors of $\mathbf{C}^{n-1} \simeq \mathfrak{h}$. We have three natural types of representations :
At first, let V be the space of homogeneous polynomials P of degree N in $z_1, \ldots, z_n$ and their conjugates and take the action of $g \in \mathrm{SL}(n, \mathbf{C})$ given by

$$(\pi(g)P)(z, \bar{z}) := P(g^{-1}z, \bar{g}^{-1}\bar{z}), \quad z = {}^t(z_1, \ldots, z_n).$$

Hence for $H = D(i\vartheta_1, \ldots . i\vartheta_n)$ with $\vartheta_j \in \mathbf{R}, \Sigma \vartheta_j = 0$, we come up with

$$(d\pi(H)P)(z, \bar{z}) = \sum_{j=1}^{n}(-i\vartheta_j z_j)\partial_{z_j}P(z, \bar{z}) + \sum_{j=1}^{n}(-i\vartheta_j \bar{z}_j)\partial_{\bar{z}_j}P(z, \bar{z}).$$

If P is a monomial

$$P(z, \bar{z}) = z_1^{k_1} \cdot \ldots \cdot z_n^{k_n} \cdot \bar{z}_1^{l_1} \cdot \ldots \cdot \bar{z}_n^{l_n} \text{ with } \sum_{j=1}^{n}(k_j + l_j) = N,$$

then we get

$$d\pi(H)P = \sum_{j=1}^{n}(l_j - k_j)(i\vartheta_j)P.$$

Exercise 6.31: Describe the weights for this representation with respect to the lexico-graphic ordering for the basis elements of $\mathfrak{h}_0 = <iH_1, \ldots, iH_{n-1}>_{\mathbf{R}}$. Do this also for the two other "natural" representations, namely on the subspaces V_1 of holomorphic and V_2 of antiholomorphic polynomials in V (i.e. polyomials only in $z_1, \ldots, z_n$ resp. $\bar{z}_1, \ldots, \bar{z}_n$). We will come back to this for the case $n = 3$ in the next section.

In our examples $\mathfrak{g}_0 = \mathfrak{sl}(2, \mathbf{R})$ and $\mathfrak{g}_0 = \mathfrak{su}(2)$ in 6.4 resp. 6.5 we saw that the finite-dimensionality of the representation space V was equivalent to an integrality condition for the weight. Now we want to look at the general case where the weights are l-tupels if one fixes a basis of a Cartan subalgebra $\mathfrak{h}$ resp. its dual $\mathfrak{h}^*$ and the real form $\mathfrak{h}_0^*$.

Definition 6.21: $\lambda \in \mathfrak{h}^*$ is said to be *algebraically integral* if

$$2 < \lambda, \alpha > /|\alpha|^2 \in \mathbf{Z} \quad \text{for all } \alpha \in \Delta.$$

Then as a generalization of what we saw in 6.4 and 6.5, we have the elementary properties of the weights for a finite-dimensional representation $\hat{\pi}$ on a vector space V:

Proposition 6.7: a) $\hat{\pi}(\mathfrak{h})$ acts diagonally on V, so that every generalized weight vector is a weight vector and V is the direct sum of all weight spaces.
b) Every weight is real valued on $\mathfrak{h}_0$ and algebraically integral.
c) Roots and weights are related by $\hat{\pi}(\mathfrak{g}_\alpha)V_\lambda \subset V_{\lambda+\alpha}$.

We fix a lexicographical ordering and take Δ^+ to be the set of positive roots with $\Pi = \{\alpha_1, \ldots, \alpha_l\}$ as the corresponding simple system of roots. We say that $\lambda \in V$ is *dominant* if $< \lambda, \alpha_j > \geq 0$ for all $\alpha_j \in \Pi$. The central fact is here the beautiful *Theorem of the Highest Weight* ([Kn2] Th. 5.5):

Theorem 6.15: The irreducible finite-dimensional representations $\hat{\pi}$ of $\mathfrak{g}$ stand (up to equivalence) in one-one correspondence with the algebraically integral dominant linear functionals λ on $\mathfrak{h}$, the correspondence being that λ is the highest weight of $\hat{\pi}_\lambda$.
The highest weight λ of $\hat{\pi}_\lambda$ has these additional properties:
a) λ depends only on the simple system Π and not on the ordering used to define Π.
b) The weight space V_λ for λ is one-dimensional.
c) Each root vector E_α for arbitrary $\alpha \in \Delta^+$ annihilates the members of V_λ, and the members of V_λ are the only vectors with this property.

d) Every weight of $\hat{\pi}_\lambda$ is of the form $\lambda - \sum_{i=1}^{l} n_i \alpha_i$ with $n_i \in \mathbf{N}_0$ and $\alpha_i \in \Pi$.

e) Each weight space V_μ for $\hat{\pi}_\lambda$ has $\dim V_{w\mu} = \dim V_\mu$ for all w in the Weyl group $W(\Delta)$, and each weight μ has $|\lambda| \geq |\mu|$ with equality only if μ is in the orbit $W(\Delta)\lambda$.

We already introduced in 3.2 the concept of complete reducibility. Here we can state the following fact ([Kn2] Th. 5.29).

Theorem 6.16: Let $\hat{\pi}$ be a complex linear representation of $\mathfrak{g}$ on a finite-dimensional complex vector space V. Then V is completely reducible in the sense that there exist invariant subspaces $U_1, \ldots, U_r$ of V such that $V = U_1 \oplus \cdots \oplus U_r$ and such that the restriction of the representation to each U_i is irreducible.

The proofs of these two theorems use three tools, which are useful also in other contexts:
- Universal enveloping algebras,
- Casimir elements,
- Verma modules.

Again following [Kn1] p.10f, we briefly present these as they will help us to a better understanding in our later chapters.

The Universal Enveloping Algebra

This is a general and far reaching concept applicable for any complex Lie algebra $\mathfrak{g}$:

We take the *tensor algebra*

$$T(\mathfrak{g}) := \mathbf{C} \oplus \mathfrak{g} \oplus (\mathfrak{g} \otimes \mathfrak{g}) \oplus \ldots.$$

and the two-sided ideal $\mathfrak{a}$ in $T(\mathfrak{g})$ generated by all

$$X \otimes Y - Y \otimes X - [X, Y], \ X, Y \in T^1(\mathfrak{g}).$$

Here $T^1(\mathfrak{g})$ denotes the space of first order tensors.

Then the *universal enveloping algebra* is the associative algebra with identity given by

$$U(\mathfrak{g}) := T(\mathfrak{g})/\mathfrak{a}.$$

This formal definition has the practical consequence that $U(\mathfrak{g})$ consists of sums of monomials usually written (slightly misusing) as $a_{(j)} X_1^{j_1} \ldots X_n^{j_n}$, $a_{(j)} \in \mathbf{C}$, $X_1, \ldots, X_n \in \mathfrak{g}$ with the usual addition and a multiplication coming up from the Lie algebra relations by $X_i X_j = X_j X_i + [X_i, X_j]$. More carefully and formally correct, one has the following: Let $\iota : \mathfrak{g} \longrightarrow U(\mathfrak{g})$ be the composition of natural maps

$$\iota : \mathfrak{g} \simeq T^1(\mathfrak{g}) \hookrightarrow T(\mathfrak{g}) \longrightarrow U(\mathfrak{g}),$$

so that

$$\iota([X, Y]) = \iota(X)\iota(Y) - \iota(Y)\iota(X).$$

ι is in fact injective as a consequence of the fundamental *Poincaré-Birkhoff-Witt Theorem*:

Theorem 6.17: Let $\{X_i\}_{i \in I}$ be a basis of $\mathfrak{g}$, and suppose that a simple ordering has been imposed on the index set I. Then the set of all monomials

$$(\iota X_{i_1})^{j_1} \cdot \ldots \cdot (\iota X_{i_n})^{j_n}$$

with $i_1 < \cdots < i_n$ and with all $j_k \geq 0$, is a basis of $U(\mathfrak{g})$. In particular the canonical map $\iota : \mathfrak{g} \longrightarrow U(\mathfrak{g})$ is one-to-one.

The name universal enveloping algebra stems from the following *universal mapping property*.

Theorem 6.18: Whenever $\mathcal{A}$ is a complex associative algebra with identity and we have a linear mapping $\varphi : \mathfrak{g} \longrightarrow \mathcal{A}$ such that

$$\varphi(X)\varphi(Y) - \varphi(Y)\varphi(X) = \varphi([X,Y]) \quad \text{for all } X, Y \in \mathfrak{g},$$

then there exists a unique algebra homomorphism $\tilde{\varphi} : U(\mathfrak{g}) \longrightarrow \mathcal{A}$ such that $\tilde{\varphi}(1) = 1$ and $\varphi = \tilde{\varphi} \circ \iota$.

The map $\tilde{\varphi}$ from the theorem may be thought of as an extension of φ from $\mathfrak{g}$ to all of $U(\mathfrak{g})$. This leads to the following useful statement.

Theorem 6.19: Representations of $\mathfrak{g}$ on complex vector spaces stand in one-one correspondence with left $U(\mathfrak{g})$ modules in which 1 acts as the identity.
This fact is essential for the construction of representations and implicit in our examples in 6.4 and 6.5.

We now come back to the case that $\mathfrak{g}$ is semisimple and to the notation introduced above. We enumerate the positive roots as $\alpha_1, \ldots, \alpha_m$ and we let $H_1, \ldots, H_l$ be a basis of $\mathfrak{h}$. Then for the construction of highest weight representations it is appropriate to use the ordered basis

$$E_{-\alpha_1}, \ldots, E_{-\alpha_m}, H_1, \ldots, H_l, E_{\alpha_1}, \ldots, E_{\alpha_m}$$

in the Poincaré-Birkhoff-Witt Theorem. The theorem says that

$$E_{-\alpha_1}^{p_1} \ldots E_{-\alpha_m}^{p_m} H_1^{k_1} \ldots H_l^{k_l} E_{\alpha_1}^{q_1} \ldots E_{\alpha_m}^{q_m}$$

is a basis of $U(\mathfrak{g})$. If one applies members of this basis to a nonzero highest weight vector v_0 of V, one gets control of a general member of $U(\mathfrak{g})v_0$: $E_{\alpha_1}^{q_1} \ldots E_{\alpha_m}^{q_m}$ will act as 0 if $q_1 + \cdots + q_m > 0$, and $H_1^{k_1} \ldots H_l^{k_l}$ will act as a scalar. Thus one has only to sort out the effect of $E_{-\alpha_1}^{p_1} \ldots E_{-\alpha_m}^{p_m}$ and most of the conclusions of the Theorem of the Highest Weight follow readily. If one looks at our examples in 6.4 and 6.5, one can get an idea how this works even in the case of representations, which are not finite-dimensional, and how the integrality of the weight leads to finite-dimensionality.

The Casimir Element

For a complex semisimple Lie algebra $\mathfrak{g}$ with Killing form B, the *Casimir element* Ω is the element

$$\Omega_0 := \sum_{i,j} B(X_i, X_j) \tilde{X}_i \tilde{X}_j$$

of $U(\mathfrak{g})$, where (X_i) is a basis of $\mathfrak{g}$ and $(\tilde{X}_i)$ is the dual basis relative to B. One can show that Ω_0 is defined independently of the basis (X_i) and is an element of the center $Z(\mathfrak{g})$ of $U(\mathfrak{g})$.

Exercise 6.32: Check this for the case $\mathfrak{g} = \mathfrak{sl}(2, \mathbf{C})$ and determine the Casimir element.

In the general case one has the following statement.

Theorem 6.20: Let Ω_0 be the Casimir element, $(H_i)_{i=1,..,l}$ an orthogonal basis of $\mathfrak{h}_0$ relative to B, and choose root vectors E_α so that $B(E_\alpha, E_{-\alpha}) = 1$ for all roots α. Then
a) $\Omega_0 = \sum_{i=1}^{l} H_i^2 + \sum_{\alpha \in \Delta} E_\alpha E_{-\alpha}$.
b) Ω_0 operates by the scalar $|\lambda|^2 + 2 < \lambda, \delta > = |\lambda + \delta|^2 - |\delta|^2$ in an irreducible finite-dimensional representation of $\mathfrak{g}$ of highest weight λ, where δ is half the sum of the positive roots.
c) The scalar by which Ω_0 operates in an irreducible finite-dimensional representation of $\mathfrak{g}$ is nonzero if the representation is not trivial.

The main point is that $\ker \Omega_0$ is an invariant subspace of V if V is not irreducible.

Remark 6.16: The center of $U(\mathfrak{g})$ is important also in the context of the determination of infinite-dimensional representations. We observed in Exercise 6.23 in 6.4 that Ω, a multiple of the Casimir element Ω_0, acts as a scalar for the unitary irreducible representations of $\mathfrak{sl}(2, \mathbf{R})$. For more general information we recommend [Kn] p.214 where as a special case of Corollary 8.14 one finds the statement: If π is unitary, then each member of the center $Z(\mathfrak{g}_c)$ of $U(\mathfrak{g}_c)$ acts as a scalar operator on the $K-$finite vectors of π. We will come back to this in 7.2 while explicitly constructing representations of $SL(2, \mathbf{R})$.

The Verma Module

We fix a lexicographic ordering and introduce

$$\mathfrak{b} := \mathfrak{h} \oplus \bigoplus_{\alpha > 0} \mathfrak{g}_\alpha.$$

For $\nu \in \mathfrak{h}^*$, make $\mathbf{C}$ into a one-dimensional $U(\mathfrak{b})$ module $\mathbf{C}_\nu$ by defining the action of $H \in \mathfrak{h}$ by $Hz = \nu(H)z$ for $z \in \mathbf{C}$ and the action of $\bigoplus_{\alpha>0} \mathfrak{g}_\alpha$ by zero. For $\mu \in \mathfrak{h}^*$, we define the *Verma module* $V(\mu)$ by

$$V(\mu) := U(\mathfrak{g}) \otimes_{U(\mathfrak{h})} \mathbf{C}_{\mu-\delta}.$$

where δ is again half the sum of the positive roots and this term is introduced to simplify calculations with the Weyl group.

Verma modules are essential for the construction of representations. They have the following elementary properties:

a) $V(\mu) \neq 0$.

b) $V(\mu)$ is a universal highest weight module for highest weight modules of $U(\mathfrak{g})$ with highest weight $\mu - \delta$.

c) Each weight space of $V(\mu)$ is finite-dimensional.

d) $V(\mu)$ has a unique irreducible quotient $L(\mu)$.

If λ is dominant and algebraically integral, then $L(\lambda + \delta)$ is the irreducible representation of highest weight λ looked for in the theorem of the Highest Weight (Theorem 6.15).

In our treatment of finite and compact groups we already saw the effectiveness of the theory of characters of representations. As for the moment we look at finite-dimensional representations we can use characters here too. To allow for more generalization, we treat them for now as formal exponential sums (again following [Kn1] p.12/3):

Let again $\mathfrak{g}$ be a semisimple Lie algebra, $\mathfrak{h}$ a Cartan subalgebra, Δ a set of roots provided with a lexicographic ordering, $\alpha_1, \ldots, \alpha_l$ the simple roots, and $W(\Delta)$ the Weyl group. We regard the set $\mathbf{Z}^{\mathfrak{h}^*}$ of functions f from $\mathfrak{h}^*$ to $\mathbf{Z}$ as an abelian group under pointwise addition. We write

$$f = \sum_{\lambda \in \mathfrak{h}^*} f(\lambda) e^{\lambda}.$$

The *support* of f is defined to be the set of $\lambda \in \mathfrak{h}^*$ for which $f(\lambda) \neq 0$. Within $\mathbf{Z}^{\mathfrak{h}^*}$, let $\mathbf{Z}[\mathfrak{h}^*]$ be the subgroup of all f of finite support. The subgroup $\mathbf{Z}[\mathfrak{h}^*]$ has a natural commutative ring structure, which is determined by $e^{\lambda} e^{\mu} = e^{\lambda + \mu}$. Moreover, we introduce a larger ring $\mathbf{Z} < \mathfrak{h}^* >$ with

$$\mathbf{Z}[\mathfrak{h}^*] \subseteq \mathbf{Z} < \mathfrak{h}^* > \subseteq \mathbf{Z}^{\mathfrak{h}^*}$$

consisting of all $f \in \mathbf{Z}^{\mathfrak{h}^*}$ whose support is contained in the union of finitely many sets $\nu_i - Q^+$, $\nu_i \in \mathfrak{h}^*$ and

$$Q^+ := \{\sum_{i=1}^{l} n_i \alpha_i;\ n_i \in \mathbf{N}_0\}.$$

Multiplication in $\mathbf{Z} < \mathfrak{h}^* >$ is given by

$$\left(\sum_{\lambda \in \mathfrak{h}^*} c_\lambda e^\lambda\right)\left(\sum_{\mu \in \mathfrak{h}^*} c_\mu e^\mu\right) := \sum_{\lambda \in \mathfrak{h}^*} \left(\sum_{\lambda + \mu = \nu} c_\lambda c_\mu\right) e^\nu.$$

If V is a representation of $\mathfrak{g}$ (not necessarily finite-dimensional), one says that V *has a character* if V is the direct sum of its weight spaces under $\mathfrak{h}$, i.e., $V = \oplus_{\mu \in \mathfrak{h}^*} V_\mu$, and if $\dim V_\mu < \infty$ for $\mu \in \mathfrak{h}^*$. In this case the *character* is

$$\operatorname{char}(V) := \sum_{\mu \in \mathfrak{h}^*} (\dim V_\mu) e^\mu$$

as an element of $\mathbf{Z}^{\mathfrak{h}^*}$. This definition is meaningful if V is finite-dimensional or if V is a Verma module.

We have two more important notions:
The *Weyl denominator* is the element $d \in \mathbf{Z}[\mathfrak{h}^*]$ given by

$$d := e^\delta \Pi_{\alpha \in \Delta^+}(1 - e^{-\alpha}).$$

Here δ is again half the sum of the positive roots.
The *Kostant partition function* $\mathcal{P}$ is the function from Q^+ to $\mathbf{N}$ that tells the number of ways, apart from order, that a member of Q^+ can be written as the sum of positve roots. We put $\mathcal{P}(0) = 1$ and define $K := \Sigma_{\gamma \in Q^+}\mathcal{P}(\gamma)e^{-\gamma} \in \mathbf{Z} < \mathfrak{h}^* >$. Then one can prove that one has $Ke^{-\delta}d = 1$ in the ring $\mathbf{Z} < \mathfrak{h}^* >$, hence $d^{-1} \in \mathbf{Z} < \mathfrak{h}^* >$. Then we have as the last main theorem in this context the famous *Weyl Character Formula*:

Theorem 6.21: Let (π, V) be an irreducible finite-dimensional representation of the complex semisimple Lie algebra $\mathfrak{g}$ with highest weight λ. Then

$$\operatorname{char}(V) = d^{-1} \sum_{w \in W(\Delta)} (\det w)e^{w(\lambda+\delta)}.$$

Now we leave the treatment of the complex semisimple case and go over to the following situation: G is compact connected, $\mathfrak{g}_0 := \operatorname{Lie} G$, $\mathfrak{g}$ the complexification of $\mathfrak{g}_0$, T a maximal torus, $\mathfrak{t}_0 := \operatorname{Lie} T$, $\Delta(\mathfrak{g}, \mathfrak{t})$ the set of roots, and B the negative of a G invariant inner product on $\mathfrak{g}_0$, and $\mathfrak{t}_{\mathbf{R}} := i\mathfrak{t}_0$. As we know, roots are real on $\mathfrak{t}_{\mathbf{R}}$, hence are in $\mathfrak{t}_{\mathbf{R}}^*$. The form B, when extended to be complex bilinear, is positive definite on $\mathfrak{t}_{\mathbf{R}}$, yielding an inner product $<,>$ on $\mathfrak{t}_{\mathbf{R}}^*$. $W(\Delta(\mathfrak{g}, \mathfrak{t}))$ is the Weyl group generated by the root reflections s_α for $\alpha \in \Delta(\mathfrak{g}, \mathfrak{t})$. Besides the notion of algebraic integrality already exploited in the complex semisimple case above, we have here still another notion of integrality:
We say that $\lambda \in \mathfrak{t}^*$ is *analytically integral* if the following equivalent conditions hold:
1) Whenever $H \in \mathfrak{t}_0$ satisfies $\exp H = 1$, then $\lambda(H)$ is in $2\pi i\mathbf{Z}$.
2) There is a multiplicative character ψ_λ of T with $\psi_\lambda(\exp H) = e^{\lambda(H)}$ for all $H \in \mathfrak{t}_0$.
In [Kn1] p.18 one finds a list of properties of these notions. We cite part of it:
a) Weights of finite-dimensional representations of G are analytically integral. In particular every root is analytically integral.
b) Analytically integral implies algebraically integral.
c) If G is simply connected and semisimple, then algebraically integral implies analytically integral.

For instance, the half sum δ of positve roots is algebraically integral but not analytically if $G = \operatorname{SO}(3)$.

In our situation the Theorem 6.15 (Theorem of the Highest Weight) comes in the following form:

Theorem 6.22: Let G be a compact connected Lie group with complexified Lie algebra $\mathfrak{g}$, let T be a maximal torus with complexified Lie algebra $\mathfrak{t}$, and let $\Delta^+(\mathfrak{g}, \mathfrak{t})$ be a positive system for the roots. Apart from equivalence the irreducible finite-dimensional representations π of G stand in one-one correspondence with the dominant analytically integral linear functionals λ on $\mathfrak{t}$, the correspondence being that λ is the highest weight of π.

And we restate Theorem 6.21 (Weyl's Character Formula) in the form:

Theorem 6.23: The character χ_λ of the irreducible finite-dimensional representation of G with highest weight λ is given by

$$\chi_\lambda = \frac{\sum_{w \in W} (\det w) \psi_{w(\lambda+\delta)-\delta}(t)}{\Pi_{\alpha \in \Delta^+}(1 - \psi_{-\alpha}(t))}$$

at every $t \in T$ where no ψ_α takes the value 1 on t.

In the next section we shall illustrate this by treating an example, which was a milestone in elementary particle physics.

6.7   The Example $\mathfrak{su}(3)$

In 1962 Gell-Mann proposed in a seminal paper [Gel] a *symmetry scheme* for the description of *hadrons*, i.e. certain elementary particles (defined by interacting by *strong interaction*), which had a great influence in elementary physics and beyond, as it led to the notion of *quarks* and *the eightfold way*. We can not go too much into the physical content but discuss the mathematical background to give another example of the general theory. Respecting the historical context, we adopt the notation introduced by Gell-Mann and used in Cornwell's presentation in [Co] vol II, p.502ff: As $\mathfrak{su}(3)$ consists of traceless skew hermitian three-by-three matrices, one can use as a basis

$$A_1 := \begin{pmatrix} & i & \\ i & & \\ & & \end{pmatrix}, \quad A_2 := \begin{pmatrix} & 1 & \\ -1 & & \\ & & \end{pmatrix}, \quad A_3 := \begin{pmatrix} i & & \\ & -i & \\ & & \end{pmatrix},$$

$$A_4 := \begin{pmatrix} & & i \\ & & \\ i & & \end{pmatrix}, \quad A_5 := \begin{pmatrix} & & 1 \\ & & \\ -1 & & \end{pmatrix}, \quad A_6 := \begin{pmatrix} & & \\ & & i \\ & i & \end{pmatrix},$$

$$A_7 := \begin{pmatrix} & & \\ & & 1 \\ & -1 & \end{pmatrix}, \quad A_8 := \frac{1}{\sqrt{3}} \begin{pmatrix} i & & \\ & i & \\ & & -2i \end{pmatrix}.$$

This basis is also a basis for the complexification $\mathfrak{su}(3)_c = \mathfrak{sl}(3, \mathbf{C})$. One may be tempted to use the elementary matrices E_{ij} with entries $(E_{ij})_{kl} = \delta_{ik}\delta_{jl}$ and take as a basis

$$E_{ij}, i \neq j, (i,j = 1,2,3) \; H_1 := E_{11} - E_{22}, H_2 := E_{22} - E_{33}.$$

One has the commutation relations $[E_{ij}, E_{kl}] = \delta_{jk}E_{il} - \delta_{il}E_{kj}$. For a diagonal matrix $H := D(h_1, h_2, h_3)$ we get

$$[H, E_{ij}] = (e_i(H) - e_j(H))E_{ij}, \; e_i(H) = h_i,$$

and hence

$$\begin{array}{llll} [H_1, E_{12}] & = & 2E_{12}, & [H_2, E_{12}] & = & -E_{12}, \\ [H_1, E_{13}] & = & E_{13}, & [H_2, E_{13}] & = & E_{13}, \\ [H_1, E_{23}] & = & -E_{23}, & [H_2, E_{23}] & = & 2E_{23}, \end{array}$$

and

$$[E_{12}, E_{21}] = H_1, [E_{23}, E_{32}] = H_2, [E_{13}, E_{31}] = E_{11} - E_{33} =: H_3.$$

We see that $\mathfrak{h} :=< H_1, H_2 >$ is a Cartan subalgbra,

$$\mathfrak{g}_{\alpha_1} :=< E_{12} >, \quad \mathfrak{g}_{\alpha_2} :=< E_{13} >, \quad \mathfrak{g}_{\alpha_3} :=< E_{23} >$$

are *root spaces*, i.e. nontrivial eigenspaces for the roots $\alpha_1, \alpha_2, \alpha_3 \in \mathfrak{h}^*$ given by

$$\begin{aligned}
\tilde{\alpha}_1 &:= (\alpha_1(H_1), \alpha_1(H_2)) = (2, -1), \\
\tilde{\alpha}_2 &:= (\alpha_2(H_1), \alpha_2(H_2)) = (1, 1), \\
\tilde{\alpha}_3 &:= (\alpha_3(H_1), \alpha_3(H_2)) = (-1, 2).
\end{aligned}$$

(In the physics literature these tuples $\tilde{\alpha}_j$ themselves sometimes are called roots.)
We choose these roots α_j as *positive roots* and then get in this coordinization a slightly unsymmetric picture:

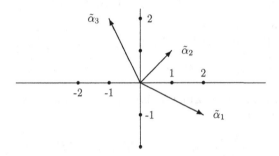

Hence we better use Gell-Mann's matrices and change to $X_j := -iA_j$. Then we get $H_1 = X_3$, $H_2' := X_8$. Moreover, for $X_\pm := (1/2)(X_1 \pm iX_2)$, $Y_\pm := (1/2)(X_6 \pm iX_7)$, and $Z_\pm := (1/2)(X_4 \pm iX_5)$, we have $X_+ = E_{12}, X_- = E_{21}$ etc. and the commutation relations

$$\begin{aligned}
[H_1, X_\pm] &= 2X_\pm, & [H_2', X_\pm] &= 0, \\
[H_1, Y_\pm] &= \mp Y_\pm, & [H_2', Y_\pm] &= \pm\sqrt{3}Y_\pm, \\
[H_1, Z_\pm] &= \pm Z_\pm, & [H_2', Z_\pm] &= \pm\sqrt{3}Z_\pm.
\end{aligned}$$

Now the Cartan subalgebra is $\mathfrak{h} =< H_1, H_2' >$ and the root spaces are

$$\mathfrak{g}_{\pm\alpha_1} =< X_\pm, >, \quad \mathfrak{g}_{\pm\alpha_2} =< Y_\pm >, \quad \mathfrak{g}_{\pm\alpha_3} =< Z_\pm >$$

with positive roots given by

$$\begin{aligned}
\bar{\alpha}_1 &:= (\alpha_1(H_1), \alpha_1(H_2')) = (2, 0), \\
\bar{\alpha}_2 &:= (\alpha_2(H_1), \alpha_2(H_2')) = (-1, \sqrt{3}), \\
\bar{\alpha}_3 &:= (\alpha_3(H_1), \alpha_3(H_2')) = (1, \sqrt{3}).
\end{aligned}$$

α_1 and α_2 are simple roots, we have $\alpha_1 + \alpha_2 = \alpha_3$ and we get a picture with hexagonal symmetry:

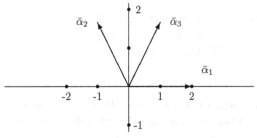

In the literature (see for instance [Co] p.518) we find still another normalization, which is motivated like this: The Killing form B for $\mathfrak{su}(3)$ is given by diagonal matrices

$$B(A_p, A_q) = -12\delta_{pq}, \text{ resp. } B(X_p, X_q) = 12\delta_{pq}.$$

Exercise 6.33: Verify this.

One uses B to introduce a nondegenerate symmetric bilinear form $<,>$ on $\mathfrak{h}^*$ as follows: We define a map

$$\mathfrak{h}^* \supset \Delta \ni \alpha \longmapsto H_\alpha \in \mathfrak{h} \text{ by } B(H_\alpha, H) = \alpha(H) \quad \text{for all } H \in \mathfrak{h},$$

and then

$$< \alpha, \beta >:= B(H_\alpha, H_\beta) \quad \text{for all } \alpha, \beta \in \Delta.$$

This leads to inner products on $\mathfrak{h}_\mathbf{R} = < H_{\alpha_1}, \ldots, H_{\alpha_l} >_\mathbf{R}$ ($\alpha_1, \ldots, \alpha_l$ simple positive roots) and its dual space. For $\mathfrak{su}(3)$ we have $l = 2$ and with α_1, α_2 from above

$$< \alpha_1, \alpha_1 >=< \alpha_2, \alpha_2 >= 1/3, \; < \alpha_1, \alpha_2 >=< \alpha_2, \alpha_1 >= -1/6.$$

It is convenient to introduce an orthonormal basis $\check{H}_1, \ldots, \check{H}_l$ of $\mathfrak{h}$, i.e. with $B(\check{H}_p, \check{H}_q) = \delta_{pq}$. Then we have the simple rule

$$< \alpha, \beta >= \sum_j \check{\alpha}_j \check{\beta}_j, \; \check{\alpha}_j = \alpha(\check{H}_j), \check{\beta}_j = \beta(\check{H}_j).$$

In our case this leads to

$$\check{H}_1 = (1/(2\sqrt{3}))H_1, \; \check{H}_2 = (1/2)H_2' = (1/(2\sqrt{3}))X_8$$

and

$$\check{\alpha}_1 = (\sqrt{3}/6)\bar{\alpha}_1 = (1/\sqrt{3}, 0), \; \check{\alpha}_2 = (\sqrt{3}/6)\bar{\alpha}_2 = (-1/(2\sqrt{3}), 1/2).$$

We see that, for the Cartan matrix $C = (C_{pq}) = (2 < \alpha_p, \alpha_q > / < \alpha_p, \alpha_p >)$ from 6.6.2, we have in this case

$$C = \begin{pmatrix} 2 & -1 \\ -1 & 2 \end{pmatrix}.$$

The Weyl group W (see the end of 6.6.3) is generated by the reflections

$$s_{\alpha_p}(\check{\alpha}_q) = \check{\alpha}_q - C_{pq}\check{\alpha}_p$$

at the planes in $\mathfrak{h}_\mathbf{R}^* \simeq \mathbf{R}^l$ orthogonal to the $\check{\alpha}_p$ with $l = 2$ and $p, q = 1, 2$ in our case. For $\mathfrak{g}_0 = \mathfrak{su}(3)$ we have 6 elements in W.

Exercise 6.34: Determine these explicitely.

Now we proceed to the discussion of the representations $(\hat{\pi}, V)$ of $\mathfrak{su}(3)$. The general theory tells us that we can assume that $\hat{\pi}(H)$ is a diagonal matrix for each $H \in \mathfrak{h}$. The diagonal elements $\hat{\pi}(H)_{jj} =: \lambda_j(H)$ fix the weights λ_j, i.e. linear functionals on $\mathfrak{h}$. We write

$$\vec{\lambda} = (\lambda(\check{H}_1), \ldots, \lambda(\check{H}_l))$$

for an ON-basis of $\mathfrak{h}$. By the Theorem of the Highest Weight (Theorem 6.15 and Theorem 6.22), an irreducible representation is determined by its highest weight Λ and this highest weight is simple, i.e. has multiplicity one. Moreover the general theory says ([Co] p.568) that these highest weights are linear combinations with non negative integer coefficients of *fundamental weights* $\Lambda_1, \ldots, \Lambda_l$ defined by

$$\Lambda_j(H) = \sum_{k=1}^{l}(C^{-1})_{kj}\alpha_k(H)$$

where $\alpha_1, \ldots, \alpha_l$ are the positive simple roots, fixed at the beginning.
And with $\delta = (1/2)\Sigma\alpha_j$ the half sum of these positive simple roots *Weyl's dimensionality formula* says ([Co] p.570) that the irreducible finite-dimensional representation $(\hat{\pi}, V)$ with highest weight Λ has dimension

$$d = \Pi_{j=1}^{l}\frac{<\Lambda + \delta, \alpha_j>}{<\delta, \alpha_j>}.$$

In our case we have

$$C^{-1} = \begin{pmatrix} 2/3 & 1/3 \\ 1/3 & 2/3 \end{pmatrix}.$$

and the fundamental weights

$$\vec{\Lambda}_1 = (2/3)\check{\alpha}_1 + (1/3)\check{\alpha}_2 = (1/6)(\sqrt{3}, 1), \quad \vec{\Lambda}_2 = (1/3)\check{\alpha}_1 + (2/3)\check{\alpha}_2 = (1/6)(0, 2).$$

We write $\hat{\pi}(n_1, n_2)$ for the representation with highest weight $\Lambda = n_1\Lambda_1 + n_2\Lambda_2$ and get by Weyl's formula for its dimension

(6.6) $$d = (n_1 + 1)(n_2 + 1)((1/2)(n_1 + n_2) + 1).$$

In the physics literature one often denotes the representation by its dimension. Then one has

a) $\hat{\pi}(0,0) = \{1\}$, the trivial representation, which has only one weight, namely the highest weight $\Lambda = 0$,
b) $\hat{\pi}(1,0) = \{3\}$ with highest weight $\Lambda = \Lambda_1$ and dimension 3,
c) $\hat{\pi}(0,1) = \{3^*\}$ with highest weight $\Lambda = \Lambda_2$ and dimension 3,
d) $\hat{\pi}(2,0) = \{6\}$ with highest weight $\Lambda = 2\Lambda_1$ and dimension 6,
e) $\hat{\pi}(0,2) = \{6^*\}$ with highest weight $\Lambda = 2\Lambda_2$ and dimension 6,
f) $\hat{\pi}(1,1) = \{8\}$ with highest weight $\Lambda = \Lambda_1 + \Lambda_2$ and dimension 8,
g) $\hat{\pi}(3,0) = \{10\}$ with highest weight $\Lambda = 3\Lambda_1$ and dimension 10,

and so on. The weights appearing in the representations can be determined by the fact that they are of the form

$$\lambda = \Lambda - m_1\alpha_1 - m_2\alpha_2, \ m_1, m_2 \in \mathbf{N}_0,$$

and that the Weyl group transforms weights into weights (preserving the multiplicity). We sketch some of the weight diagrams and later discuss an elementary method to get these diagrams, which generalizes our discussion in 6.4 and 6.5.

$\check{\pi}(1,0) = \{3\}$

$\lambda(\check{H}_2)$

$\Lambda = (1/6)(\sqrt{3},1) = (2/3)\alpha_1 + (1/3)\alpha_2$

$s_{\alpha_1}\Lambda = (1/6)(-\sqrt{3},1) = -(1/3\alpha_1 + (1/3)\alpha_2$

$\lambda(\check{H}_1)$

$s_{\alpha_2}s_{\alpha_1}\Lambda = (1/6)(0,2) = -(1/3)\alpha_1 - (2/3)\alpha_2$

$\check{\pi}(0,1) = \{3^*\}$

$\lambda(\check{H}_2)$

$\Lambda = (1/6)(0,2) = (1/3)\alpha_1 + (2/3)\alpha_2$

$\lambda(\check{H}_1)$

$s_{\alpha_2}\Lambda = (1/6)(\sqrt{3},-1) = -(1/3\alpha_1 - (1/3)\alpha_2$

$s_{\alpha_1}s_{\alpha_2}\Lambda = (1/6)(-\sqrt{3},-1) = -(2/3)\alpha_1 - (1/3)\alpha_2$

$\check{\pi}(2,0) = \{6\}$

$\lambda(\check{H}_2)$

$\Lambda - \alpha_1 = (1/6)(0,2) = (1/3)\alpha_1 + (2/3)\alpha_2$

$s_{\alpha_1}\Lambda = -(2/3)\alpha_1 + (2/3)\alpha_2$

$\Lambda = (1/6)(2\sqrt{3},2) = (4/3)\alpha_1 + (2/3)\alpha_2$

$s_{\alpha_1}s_{\alpha_2}(\Lambda - \alpha_1) = -(2/3)\alpha_1 - (1/3)\alpha_2$

$\lambda(\check{H}_1)$

$s_{\alpha_2}(\Lambda - \alpha_1) = (1/6)(\sqrt{3},-1) = (1/3)\alpha_1 - (1/3)\alpha_2$

$s_{\alpha_2}s_{\alpha_1}\Lambda = (1/6)(0,-4) = -(2/3)\alpha_1 - (4/3)\alpha_2$

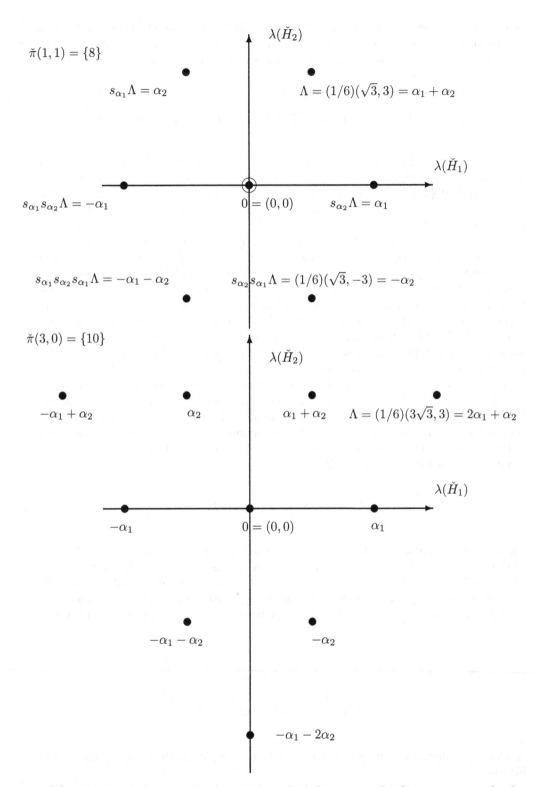

$\check{\pi}(1,1) = \{8\}$

$\lambda(\check{H}_2)$

$s_{\alpha_1}\Lambda = \alpha_2$

$\Lambda = (1/6)(\sqrt{3},3) = \alpha_1 + \alpha_2$

$\lambda(\check{H}_1)$

$s_{\alpha_1}s_{\alpha_2}\Lambda = -\alpha_1$

$0 = (0,0)$ $s_{\alpha_2}\Lambda = \alpha_1$

$s_{\alpha_1}s_{\alpha_2}s_{\alpha_1}\Lambda = -\alpha_1 - \alpha_2$ $s_{\alpha_2}s_{\alpha_1}\Lambda = (1/6)(\sqrt{3},-3) = -\alpha_2$

$\check{\pi}(3,0) = \{10\}$

$\lambda(\check{H}_2)$

$-\alpha_1 + \alpha_2$ α_2 $\alpha_1 + \alpha_2$ $\Lambda = (1/6)(3\sqrt{3},3) = 2\alpha_1 + \alpha_2$

$\lambda(\check{H}_1)$

$-\alpha_1$ $0 = (0,0)$ α_1

$-\alpha_1 - \alpha_2$ $-\alpha_2$

$-\alpha_1 - 2\alpha_2$

In $\{8\}$ we find a first example of a weight, which has not multiplicity one, namely the weight $(0,0)$.

Similar to our discussion for $G = SU(2)$ and $\mathfrak{g} = \mathfrak{su}(2)$ in 4.2, one has here the problem of the explicit decomposition of representations into irreducible components. This goes under the heading of *Clebsch Gordon series* and can be found for instance in [Co] p.611ff. As examples we cite

$$\{3\} \otimes \{3^*\} \simeq \{8\} \oplus \{1\} \text{ and } \{3\} \otimes \{3\} \otimes \{3\} \simeq \{10\} \oplus 2\{8\} \oplus \{1\}.$$

The proofs of these formulae need some skill or at least patience. But it is very tempting to imagine the bigger weight diagrams in our examples as composition of the triangles $\{3\}$ and $\{3^*\}$. This leads to an important interpretation in the theory of elementary particles: We already remarked that the discussion of the irreducible representations of $SU(2)$ resp. $SO(3)$ is useful in the description of particles subjected to $SO(3)$-symmetry (in particular an electron in an hydrogen atom), namely a representation $\pi_j, j = 0, 1/2, 1, 3/2, \ldots$ describes a multiplet of states of the particle with angular momentum resp. spin $2j + 1$ and the members of the multiplet are distinguished by a magnetic quantum number $m \in \mathbf{Z}$ with $m = -(2j + 1), -(2j - 1), \ldots, (2j + 1)$. As one tried to describe and classify the *heavier* elementary particles, one observed patterns, which could be related to the representation theory of certain compact groups, in particular $G = SU(3)$: In Gell-Mann's *eightfold way* we take $SU(3)$ as an *internal symmetry group* to get the ingredients of an atom, the proton p and the neutron n as members of a multiplet of eight states, which correspond to our representation $\{8\}$. To be little bit more precise, one looks at a set of *hadrons*, i.e. particles characterized by an experimentally fixed behaviour (having strong interaction), and describes these by certain *quantum numbers*, namely *electric charge Q, hypercharge Y, baryon number B, strangeness S, isospin I with third component* I_3. The baryon number is $B = 1$ for *baryons* like proton or the neutron, -1 for antibaryons, and zero for other particles like the π-mesons. In this context, one has the following relations

$$Y = B + S, \ Q = I_3 + (1/2)Y$$

and one relates particle multipets to representations of $\mathfrak{su}(3)$ by assigning an I_3 - axis to the axis measuring the weight $\lambda(H_1)$ and an Y-axis to $\lambda(H_2)$. This way, the lowest dimensional nontrivial (three-dimensional) representations correspond to two triples (u, d, s) and $(\bar{u}, \bar{d}, \bar{s})$, named *quarks* resp. *antiquarks* (with u for *up*, d for *down*, and s for *nonzero strangeness*). We give a list of the quantum numbers characterizing the quarks

quark	B	I	I_3	Y	S	Q
u	$1/3$	$1/2$	$-1/2$	$1/3$	0	$2/3$
d	$1/3$	$1/2$	$1/2$	$1/3$	0	$-1/3$
s	$1/3$	0	0	$-2/3$	-1	$-1/3$

and for more information refer to p. 50 f of [FS]. In this picture the weight diagrams look like this:

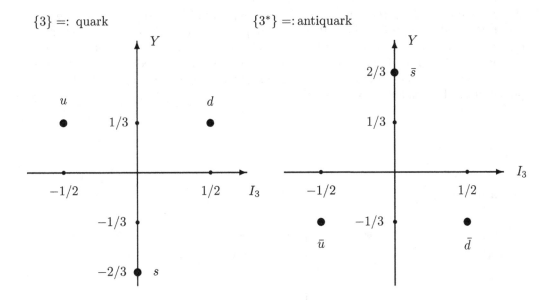

{3} =: quark   {3*} =: antiquark

Up to now the experimental evidence for the existence of the quarks (as far as we know) is still indirect: they are thought of as building blocks of the more observable particles like the proton and the neutron. The following diagram shows the quantum numbers of the baryon octet {8}. By a simple addition process (comparing the (I_3, Y)-coordinates) one can guess the *quark content* of the neutron as $n = (udd)$ and of the proton as $p = (uud)$. One can do as well for the other particles showing up in the diagram.

$\check{\pi}(1,1) = \{8\}$

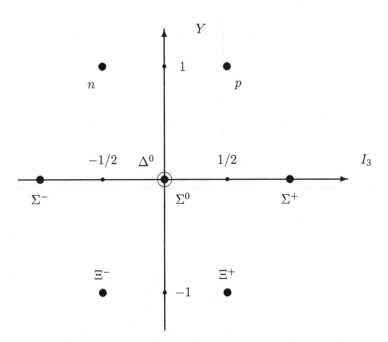

We shall come back to these interpretations of representations later on when we discuss the Euclidean groups and the Poincaré group. A broader discussion of the physical content of these schemes is beyond the scope of this book. But we still reproduce the baryon decuplet $\{10\}$. It became famous in history because it predicted the existence of an Ω^--particle (the bottom of the diagram) with fixed properties, which after this prediction really was found experimentally.

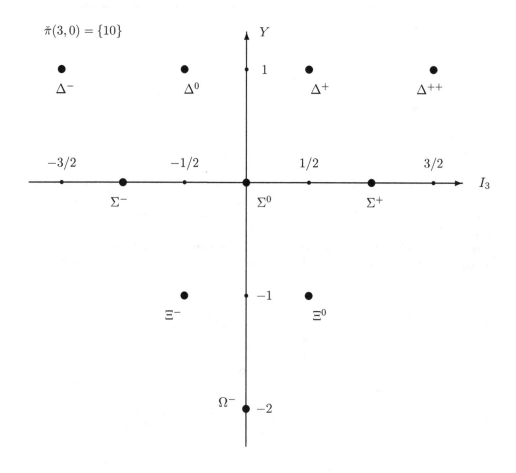

To finish this section, let us look what happens when we apply to our example the procedure, which was successful in our constructions of the representations of $\mathfrak{su}(2)$ and $\mathfrak{sl}(2,\mathbf{R})$ in 6.4 and 6.5. We stay in the unnormalized picture with

$$\mathfrak{g} =< H_1, H_2', X_\pm, Y_\pm, Z_\pm >$$

with the commutation relations

$$[X_+, X_-] = 0, \ [Y_+, Y_-] = (1/2)(\sqrt{3}H_2' - H_1), \ [Z_+, Z_-] = (1/2)(\sqrt{3}H_2' + H_1),$$

$$
\begin{array}{llllll}
[X_+, Y_+] & = & Z_+, & [X_+, Z_-] & = & -Y_-, \\
[X_-, Z_+] & = & Y_+, & [X_-, Y_-] & = & -Z_-, \\
[Y_+, Z_+] & = & X_-, & [Y_-, Z_+] & = & -X_+, \\
[H_1, X_\pm] & = & 2X_\pm, & [H_2', X_\pm] & = & 0, \\
[H_1, Y_\pm] & = & \mp Y_\pm, & [H_2', Y_\pm] & = & \pm\sqrt{3}Y_\pm, \\
[H_1, Z_\pm] & = & \pm Z_\pm, & [H_2', Z_\pm] & = & \pm\sqrt{3}Z_\pm,
\end{array}
$$

(the others are zero) and the roots with coordinates

$$\bar{\alpha}_1 = (2,0), \ \bar{\alpha}_2 = (-1,\sqrt{3}), \ \bar{\alpha}_2 = (1,\sqrt{3}).$$

(We then get the weights from our diagrams above by multiplication with $(\sqrt{3}/6)$.) And as we prefer to think positively, we start our construction of a representation space V by taking a vector $v \in V, v \neq 0$, of **lowest** weight $\Lambda = (k,l)$, i.e. with

$$H_1 v = kv, \ H_2' = lv, \ X_- v = Y_- v = Z_- v = 0.$$

We try to construct V by applying the *plus* or *creation operators* X_+, Y_+, Z_+ to v, i.e. we try

$$V = \sum v_{rst}\mathbf{C}, \ v_{rst} := X_+^r Y_+^s Z_+^t v.$$

From the PBW-Theorem we recall that we have to be careful with the order in which we apply the creation operators. For instance, one has

$$v_{110}' := Y_+ X_+ v = (X_+ Y_+ + [Y_+, X_+])v = v_{110} - v_{001}$$

since $[X_+, Y_+] = Z_+$, but $Z_+ X_+ v = X_+ Z_+ v = v_{101}$ as X_+ and Z_+ commute.
From the commutation relations it is clear that v_{rst} has weight $\Lambda + r\bar{\alpha}_1 + s\bar{\alpha}_2 + t\bar{\alpha}_3$, so, in particular, the elements $v_{100}, v_{010}, v_{001}$ of the *first shell* have the respective weights coordinatized by $(k+2, l), (k-1, l+\sqrt{3}), (k+1, l+\sqrt{3})$. If we apply the *minus* or *annihilation operators* X_-, Y_-, Z_-, using the commutation relations, we get

$$X_- v_{100} = X_- X_+ v = ([X_-, X_+] + X_+ X_-)v = -H_1 v + 0 = -kv,$$

and the same way

$$
\begin{array}{lll}
Y_- v_{100} & = & 0, \\
Z_- v_{100} & = & 0, \\
X_- v_{010} & = & 0, \\
Y_- v_{010} & = & -(1/2)(l\sqrt{3} - k)v, \\
Z_- v_{010} & = & 0, \\
X_- v_{001} & = & v_{010}, \\
Y_- v_{001} & = & -v_{100}, \\
Z_- v_{001} & = & -(1/2)(l\sqrt{3} + k)v,
\end{array}
$$

Before we interpret this and get into danger to lose track in the next shells, let us see how the natural representation of $\mathfrak{su}(3)$ looks in this picture:
Here we have

$$V = \sum_{j=1}^{3} e_j \mathbf{C}, \ e_1 = {}^{t}(1,0,0), e_2 = {}^{t}(0,1,0), e_3 = {}^{t}(0,0,1),$$

and hence

$$H_1 e_1 \ = \ \begin{pmatrix} 1 & & \\ & -1 & \\ & & 0 \end{pmatrix} \begin{pmatrix} 1 \\ 0 \\ 0 \end{pmatrix} = e_1,$$

$$H_2' e_1 \ = \ (1/\sqrt{3}) \begin{pmatrix} 1 & & \\ & 1 & \\ & & -2 \end{pmatrix} \begin{pmatrix} 1 \\ 0 \\ 0 \end{pmatrix} = (1/\sqrt{3}) e_1.$$

In the same way we get

$$\begin{array}{rclcrcl}
H_1 e_2 & = & -e_2, & H_2' e_2 & = & (1/\sqrt{3})\, e_2 \\
H_1 e_3 & = & 0, & H_2' e_3 & = & -(2/\sqrt{3})\, e_3.
\end{array}$$

Recalling $X_+ = E_{12}, X_- = E_{21}$ and the analogous relations for the other matrices, one has

$$\begin{array}{rclcrclcrcl}
X_+ e_1 & = & 0, & Y_+ e_1 & = & 0, & Z_+ e_1 & = & 0, \\
X_- e_1 & = & e_2, & Y_- e_1 & = & 0, & Z_- e_1 & = & e_3, \\
X_+ e_2 & = & e_1, & Y_+ e_2 & = & 0, & Z_+ e_2 & = & 0, \\
X_- e_2 & = & 0, & Y_- e_2 & = & e_3, & Z_- e_2 & = & 0, \\
X_+ e_3 & = & 0, & Y_+ e_3 & = & e_2, & Z_+ e_3 & = & e_1, \\
X_- e_3 & = & 0, & Y_- e_3 & = & 0, & Z_- e_3 & = & 0.
\end{array}$$

These relations show that, in the coordinates fixed above, e_1 is a highest weight vector of weight $(1, (1/\sqrt{3}))$, e_2 has weight $(-1, \sqrt{3})$ and e_3 is a lowest weight vector with weight $(0, -(2/\sqrt{3}))$. Thus, as to be expected, we find the weight diagram of $\{3\}$. This fits into the construction we started above as follows: we put $v_1 := v = e_3$ and get in the first shell translating the equations just obtained

$$X_+ v = v_{100} = 0, Y_+ v = v_{010} = e_2 := v_2, Z_+ v = v_{001} = e_1 := v_3.$$

Moreover, the equations translate to the fact that the creation operators applied to the highest weight vector v_3 and to v_2 give zero with the exception $X_+ v_2 = v_3$. The only non-trivial actions of the annihilation operators are $Y_- v_2 = v_1$, $X_- v_3 = v_2$ and $Z_- v_3 = v_1$. We compare this to the general form for the action obtained at the beginning (now with $v_{100} = 0, v_{010} = v_2, v_{001} = v_3$) and find as neccessary condition for a three-dimensional representation in our general scheme the condition $k = 0, l = -2/\sqrt{3}$, just as it should be.

To show how to come to higher dimensional representations, one has to go to higher shells:

We have $v_{200} = X_+^2 v, v_{110} = X_+ Y_+ v, v_{101} = X_+ Z_+ v$ and so on. Here we remember to pay attention to the order if the creation operators do not commute: As already remarked, we have $v'_{110} := Y_+ X_+ v = v_{110} - v_{001}$ since one has $[X_+, Y_+] = Z_+$. Though it is a bit lengthy we give the list of the action of the annihilation operators (to be obtained parallel to the case $X_- v_{100}$ above and using the actions of the minus operators on the first shell)

$$
\begin{aligned}
X_- v_{110} &= -(k-1)v_{010}, \\
Y_- v_{110} &= -(1/2)(l\sqrt{3} - k)v_{100}, \\
Z_- v_{110} &= -(1/2)(l\sqrt{3} - k)v, \\
X_- v_{101} &= -(k+1)v_{001} + v_{110}, \\
Y_- v_{101} &= -v_{200}, \\
Z_- v_{101} &= -(1/2)(l\sqrt{3} + k + 2)v_{100}, \\
X_- v_{011} &= v_{200}, \\
Y_- v_{011} &= -(1/2)(l\sqrt{3} - k)v_{001} - v_{110}, \\
Z_- v_{011} &= -(1/2)(l\sqrt{3} + k + 2)v_{010}, \\
X_- v_{200} &= -2(k+1)v_{100}, \\
Y_- v_{200} &= 0, \\
Z_- v_{200} &= 0, \\
X_- v_{020} &= 0, \\
Y_- v_{020} &= -(l\sqrt{3} - k + 2)v_{010}, \\
Z_- v_{020} &= 0, \\
X_- v_{002} &= 2v_{001}, \\
Y_- v_{002} &= -2v_{101}, \\
Z_- v_{002} &= -(l\sqrt{3} - k + 2)v_{001}.
\end{aligned}
$$

Just as well one can determine the action of the minus operators in the next shell and so on. As an example, we show how the eight-dimensional representation $\{8\}$ (with highest weight $\Lambda = \alpha_1 + \alpha_2$) comes up in this picture: As above we start with a lowest weight vector $v_1 := v$, which has in our coordinates the weight $\Lambda_0 = (k, l)$. In the first shell we have

$$v_2 := X_+ v = v_{100}, v_3 := Y_+ v = v_{010}, v_4 = Z_+ v = v_{001}$$

of respective weights $\Lambda_0 + \alpha_1, \Lambda_0 + \alpha_2, \Lambda_0 + \alpha_1 + \alpha_2$. For the second shell we put (secretly looking at the weight diagram we already constructed using the results from the general theory, otherwise the reasoning would take a bit more time)

$$X_+^2 v = v_{200} = 0, \ Y_+ X_+ v = v'_{110} =: v_5, \ Z_+ X_+ v =: v_6,$$

and using the relation $v'_{110} = v_{110} - v_{001}$ of weight $\Lambda_0 + \alpha_1 + \alpha_2$ obtained at the beginning

$$X_+ Y_+ v = v_{110} = v_5 + v_4, \ Y_+^2 v = 0, \ Z_+ Y_+ v = v_{011},$$

and, since X_+ and Z_+ commute

$$X_+ Z_+ v = v_{101} = v_6, Y_+ Z_+ v =: v_7, Z_+^2 v = v_{002} =: v_8.$$

If we look at the relations coming from the application of the minus operators to the elements of the second shell listed above, we find in particular the equations

$$X_- v_{200} = -2(k+1)v_{100}, \ Y_- v_{020} = (l\sqrt{3} - k + 2)v_{010}.$$

Since we have fixed $v_{200} = v_{020} = 0$, these equations lead to $k = -1$ and $l = 2/\sqrt{3}$. One verifies easily that all the other relations are fulfilled and the weight diagram for the eightfold way from above in this context leads to the following picture where only part of the relations between the operators and generators are made explicit:

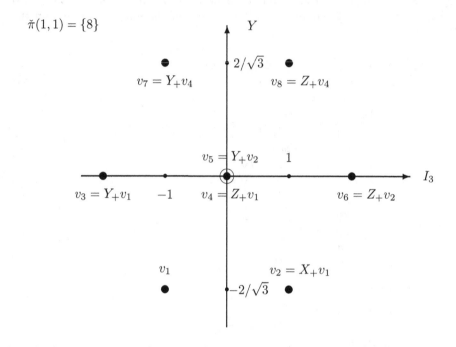

$\check{\pi}(1,1) = \{8\}$

Exercise 6.35: Do the same to construct the representations $\{6\}$ and $\{10\}$.

We hope that these examples help to get a feeling how the discreteness of the weights is related to the finite-dimensionality of the representations. And, as this procedure gets more and more complicated, one gets perhaps some motivation to look into the proofs of the general theorems in the previous section about the integrality of the weights, the existence of fundamental weights, the action of the Weyl group etc.

Chapter 7

Induced Representations

Induction is in this context a method to construct representations of a group starting by a representation of a subgroup. As we shall see, in its ultimate specialization one rediscovers Example 1.5 in our section 1.3, where we constructed representations in function spaces on G-homogeneous spaces. Already in 1898, Frobenius worked with this method for finite groups. Later on, Wigner, Bargman, and Gelfand-Neumark used it to construct representations of special groups, in particular the Poincaré group. But it was Mackey, who, from 1950 on, developed a systematic treatment using essentially elements of functional analysis. Because it is beyond the scope of this book, we do not describe these here as carefully as we should. In most places we simply give recipes so that we can construct the representations for the groups we are striving for. We refer to Mackey [Ma], Kirillov [Ki] p.157ff, Barut-Raczka [BR] p.473ff, or Warner [Wa] p.365ff for the necessary background.

7.1 The Principle of Induction

Let G be a group, H a subgroup, and π_0 a representation of H in a space $\mathcal{H}_0$, all this with certain properties to be specified soon. From these data one can construct representations of G in several ways. The guiding principle is the following. We look at a space of functions $\mathcal{H}$ consisting of functions $\phi : G \longrightarrow \mathcal{H}_0$, which allow for some additional conditions also to be specified below and for the *fundamental functional equation*

(7.1) $$\phi(hg) = \delta(h)^{1/2}\pi_0(h)\phi(g), \quad \text{for all } h \in H, g \in G.$$

Here $\delta : H \longrightarrow \mathbf{R}_{>0}$ is a *normalizing function*, which has several different descriptions. It is identically one in many important cases, so we do not bother with it for the moment (if δ is left out, one has *unnormalized induction*, which later on will appear in connection with an interpretation via line bundles). If $\mathcal{H}$ is closed under right translation, this space is the representation space for the representation

$$\pi = ind_H^G \pi_0$$

given by right translation $\pi(g_0)\phi(g) = \phi(gg_0)$. Before we make all this more precise, in the hope to add some motivation, we treat as an example a representation, which we already treated as one of our first examples in 1.3.

Example 7.1: Let G be the Heisenberg group $\mathrm{Heis}(\mathbf{R}) = \{g := (\lambda, \mu, \kappa);\ \lambda, \mu, \kappa \in \mathbf{R}\}$, H the subgroup consisting of the elements $h := (0, \mu, \kappa)$ and π_0 the one-dimensional representation of H, which for a fixed $m \in \mathbf{R}^*$ is given by

$$\pi_0(h) := e^m(\kappa) = e^{2\pi i m \kappa}.$$

As we know from 3.3 that the Heisenberg group and the subgroup H are unimodular, one has $\delta \equiv 1$ (we shall see this below in 7.1.2) and we have to deal with the functional equation (7.1)

$$\phi(hg) = \pi_0(h)\phi(g) = e^m(\kappa)\phi(g), \quad \text{for all } h \in H, g \in G.$$

By this functional equation the complex function $\phi(g) = \phi(\lambda, \mu, \kappa)$ is forced to have the form

(7.2) $$\phi(\lambda, \mu, \kappa) = e^m(\kappa + \lambda\mu)f(\lambda)$$

where f is a one-variable function since every $g \in \mathrm{Heis}(\mathbf{R})$ has a decomposition

$$g = (\lambda, \mu, \kappa) = (0, \mu, \kappa + \lambda\mu)(\lambda, 0, 0) =: h_0 s(\lambda)$$

with $h_0 \in H$ and $s(\lambda) = (\lambda, 0, 0) \in G$. Application of the right translation with $g_0 \in G$ (via the muliplication law in the Heisenberg group) leads to

$$
\begin{aligned}
(\pi(g_0)\phi)(g) = \phi(gg_0) &= \phi(\lambda + \lambda_0, \mu + \mu_0, \kappa + \kappa_0 + \lambda\mu_0 - \lambda_0\mu) \\
&= e^m(\kappa + \kappa_0 + \lambda\mu + \lambda_0\mu_0 + 2\lambda\mu_0)f(\lambda + \lambda_0) \\
&= e^m(\kappa + \lambda\mu)e^m(\kappa_0 + (\lambda_0 + 2\lambda)\mu_0)f(\lambda + \lambda_0).
\end{aligned}
$$

Comparing with (7.2) above, we rediscover the Schrödinger representation from Section 1.3: If we choose $\mathcal{H}$ such that in the decomposition of the elements ϕ above we have $f \in L^2(\mathbf{R})$, then

$$f \longrightarrow \pi(g_0)f \text{ with } \pi(g_0)f(x) = e^m(\kappa_0 + (\lambda_0 + 2x)\mu_0)f(x + \lambda_0)$$

is the prescription for the unitary representation from Example 1.6 in 1.3. The representation given by right translation on the space of functions ϕ on G of the form (7.2) is called *Heisenberg representation*.

7.1.1 Preliminary Approach

Now we want to see what is behind this example and try a more systematic treatment. At first we do not consider the most general case but (following [La] p.43f) one sufficient for several applications. Let
- G be an unimodular connected linear group,
- H be a closed subgroup,
- K be an unimodular closed subgroup such that the map

$$H \times K \longrightarrow G, (h, k) \longmapsto hk$$

is a topological isomorphism,
- $\delta = \Delta_H$ be the modular function of H, and
- π_0 a representation of H in $\mathcal{H}_0$.

Then the homogeneous space of right H-cosets $\mathcal{X} = H\backslash G$ can be identified with K (provided with an action of K from the right). Let $\mathcal{H}^\pi$ be the space of functions

$$\phi : G \longrightarrow \mathcal{H}_0$$

satisfying the functional condition (7.1)

$$\phi(hg) = \delta(h)^{1/2}\pi_0(h)\phi(g) \quad \text{for all } h \in H, g \in G$$

and the *finiteness condition* that the restriction $\phi|_K =: f$ is in $L^2(K)$. We define a norm for $\mathcal{H}^\pi$ by

$$(7.3) \qquad \| \phi \|^2_{\mathcal{H}^\pi} := \| f \|^2_{L^2(K)} = \int_K |f(k)|^2_{\mathcal{H}_0} dk,$$

and with $\tilde{\phi}\,|_K =: \tilde{f}$ the scalar product by

$$< \phi, \tilde{\phi} > := < f, \tilde{f} > .$$

We denote $\pi = ind^G_H \pi_0$ for π given by right translation $(\pi(g_0)\phi)(g) = \phi(gg_0)$ and call this the *representation of G induced by π_0*. We have to show that this really makes sense and shall see at the same time why the strange normalizing function δ has to be introduced.

Theorem 7.1: If π_0 is bounded, π defines a bounded representation of G on $\mathcal{H}^\pi$. π is unitary if π_0 is.

Proof: We have to show that ϕ^{g_0} given by $\phi^{g_0}(g) := \phi(gg_0)$ is in $\mathcal{H}^\pi$. One has

$$\phi^{g_0}(hg) = \phi(hgg_0) = \delta(h)^{1/2}\pi_0(h)\phi(gg_0) = \delta(h)^{1/2}\pi_0(h)\phi^{g_0}(g),$$

i.e. ϕ^{g_0} satifies the functional equation. Now to the finiteness condition: We write $g = hk$ and $kg_0 =: h'_k k'$ with $h, h'_k \in H, k, k' \in K$. Then we have (with $\delta = \Delta_H$)

$$\phi^{g_0}(k) = \phi(kg_0) = \phi(h'_k k') = \delta(h'_k)^{1/2}\pi_0(h'_k)\phi(k'),$$

and, since π_0 is bounded, $\| \phi^{g_0} \|^2_{\mathcal{H}^\pi}$ is bounded by a constant times

$$\int_K \Delta_H(h'_k) \mid \phi(k') \mid^2 dk,$$

and equality holds if π_0 is unitary. The proof is done if we can show that for $f \in C_c(K)$ one has

$$\int_K f(k)dk = \int_K \Delta_H(h'_k)f(k')dk.$$

We take $\phi \in C_c(G)$ and use Fubini's Theorem. Since G is unimodular, we get

$$
\begin{aligned}
\int_H \int_K \phi(hk)dkdh &= \int_G \phi(g)dg = \int_G \phi(gg_0)dg \\
&= \int_H \int_K \phi(hkg_0)dkdh \\
&= \int_K \int_H \phi(hh'_k k')dhdk \\
&= \int_K \int_H \phi(hk')\Delta_H(h'_k)dhdk \\
&= \int_H \int_K \phi(hk')\Delta_H(h'_k)dkdh.
\end{aligned}
$$

Here we take ϕ such that

$$\phi(hk) = \varphi(h)f(k) \text{ with } f \in C(K), \varphi \in C_c(H), \int_H \varphi(h)dh = 1,$$

and get the desired result.

Exercise 7.1: Verify that the case of the Heisenberg group and its Schrödinger representation is covered by this theorem.

Remark 7.1: Using the notions from our section 6.6 where we introduced a bit more structure theory, one can also state a useful variant of Theorem 7.1 (as in Flicker [Fl] p.39):
If G has an Iwasawa decomposition $G = NAK, K$ a maximal compact subgroup, A the maximal torus in $H = NA$, N the unipotent radical, and χ a character of H, the representation space consists of smooth functions $\phi : G \longrightarrow \mathbf{C}$ with

$$\phi(nak) = (\delta^{1/2}\chi)(a)\phi(k), \quad \text{for all } a \in A, n \in N, k \in K,$$

where the normalizing function can be described by

$$\delta(a) = |\det(\operatorname{Ad} a |_{\operatorname{Lie} N})|.$$

The representation is given again by right translation.

7.1.2   Mackey's Approach

We shall afterwards come back to more easily accessible special cases but now go (following Mackey) to a more general situation (see [Ki] p.187ff, [BR] p.473) where we have to intensify the measure theoretic material from section 3.3. Let
- G be a connected linear group,
- H be a closed subgroup,
- π_0 a unitary representation of H in $\mathcal{H}_0$ (later on, in some cases, we will also treat non-unitary representations),
- $\delta := \Delta_H/\Delta_G$ where Δ_H and Δ_G are the modular functions of H resp. G introduced in 3.3,
- $\mathcal{X}$ the space of right H-cosets $\mathcal{X} = H\backslash G$,
- $p : G \longrightarrow \mathcal{X}$ the natural projection $g \longmapsto Hg =: x \in \mathcal{X}$.

Theorem 7.2: The *normalized induced representation*

$$\pi := \operatorname{ind}_H^G \pi_0$$

is given by right translation on the space $\mathcal{H}^\pi$, which is defined as the completion of its dense subspace spanned by the continuous functions $\phi : G \longrightarrow \mathcal{H}_0$ satisfying the *functional equation* (7.1)

$$\phi(hg) = (\delta(h))^{1/2}\pi_0(h)\phi(g) \quad \text{for all } h \in H, g \in G$$

and the *finite norm condition*

(7.4) $$\| \phi \|_{\mathcal{H}^\pi}^2 := \int_{\mathcal{X}} \| \phi(s(x)) \|_{\mathcal{H}_0}^2 \, d\mu_s(x) < \infty.$$

Before we elaborate on the fact that we really get a unitary representation, this condition introducing the norm $\| \cdot \|_{\mathcal{H}^\pi}$ needs some explanation:

Quasi-invariant Measures, Master Equation and the Mackey Decomposition

As above, we denote by $\mathcal{X}$ the space of left H-cosets, $\mathcal{X} = H \backslash G$, and $p : G \longrightarrow \mathcal{X}$ the natural projection given by $g \longmapsto Hg =: x \in \mathcal{X}$. Then it is a general fact that the homogeneous space $\mathcal{X}$ admits *quasi-invariant* measures μ. These are defined by the condition that, for every $g \in G$, μ and μ_g (with $\mu_g(B) := \mu(Bg)$ for all Borel sets B) have the same null sets, i.e. $\mu(B) = 0$ is equivalent to $\mu_g(B) = 0$. Again we do not prove this general statement but later on give enough examples to illuminate the situation (we hope). For $f \in \mathcal{C}_c(\mathcal{X})$ we have

$$\mu_g(f) = \int_{\mathcal{X}} f(xg)d\mu(x) = \int_{\mathcal{X}} f(x)d\mu(xg^{-1}) \text{ with } d\mu(xg^{-1}) =: d\mu_g(x)$$

and for $g \in G$ we have a *density function* ρ_g, which may be understood as the quotient of $d\mu_g$ and $d\mu$ (in the general theory this goes under the name of a *Radon-Nikodym derivative*). The construction of quasi-invariant measures can be done as follows. We take over (from [BR] p.70:) [Ma1] part I, Lemma 1.1:

Theorem 7.3 (*Mackey's Decomposition Theorem*): Let G be a separable locally compact group and let H be a closed subgroup of G. Then there exists a Borel set B in G such that every element $g \in G$ can be uniquely represented in the form

$$g = hs, \ h \in H, \ s \in B.$$

In our situation where we have

$$p : G \longrightarrow \mathcal{X}, \ g \longmapsto Hg =: x,$$

this leads to the existence of a *Borel section* $s : \mathcal{X} \longrightarrow G$, i.e. a map preserving Borel sets with $p \circ s = id_{\mathcal{X}}$. Hence for $g \in G$, we have a unique Mackey decomposition

(7.5) $$G \ni g = h(x)s(x), \ h(x) \in H, x \in \mathcal{X}.$$

Then the invariant measures $d_r h$ and $d_r g$ on H resp. G are related to a quasi-invariant measure $d\mu_s$ associated to the section φ by

$$d_r g =: \frac{\Delta_G(h)}{\Delta_H(h)} d_r h d\mu_s(x)$$

(by a reasoning, which is a refinement of the last part of the proof of Theorem 7.1, or by solving the problems in [Ki] p.132) and we have the *quasi-measure relation*

(7.6) $$\frac{d\mu_s(xg)}{d\mu_s(x)} = \rho_g(x) = \frac{\Delta_H(h(g,x))}{\Delta_G(h(g,x))} = \delta(h(g,x)).$$

To someone eventually thinking this is complicated, we recommend to look at the case of the Heisenberg group we treated above: We have the section $s : \mathcal{X} \longrightarrow G$ given by $\mathcal{X} \ni x \longmapsto s(x) = (x,0,0)$ and the Mackey decomposition

$$\text{Heis}(\mathbf{R}) \ni g = (\lambda, \mu, \kappa) = h(x)s(x)$$

with

$$h(x) = (0, \mu, \kappa + \lambda\mu), \ s(x) = (\lambda, 0, 0), \ x = \lambda.$$

And, since the groups G and H are both unimodular, we simply have

$$d_r g = d\lambda d\mu d\kappa, \ d_r h = d\mu d\kappa, \ d\mu_s(x) = dx.$$

In this special case, one easily verifies directly that dx is not only quasi-invariant but invariant. And the general finite norm condition above reduces to the condition that f is an L^2-function on $\mathcal{X} \simeq \mathbf{R}$, i.e. the condition we found above while rediscovering the Schrödinger representation.

The First Realization

In the general case we will need the explicit form of the action of G on $\mathcal{X}$ in the framework of the Mackey decomposition: We apply the Mackey decomposition (7.5) to $s(x)g_0$ and get a consistent picture by chosing xg_0 such that, for $h^* := h(xg_0) \in H$ (which, for reasons to be seen later, we also write as $h(g_0, x)$), we have the *Master Equation* (the name is from [Ki1] p.372)

$$(7.7) \qquad\qquad s(x)g_0 = h^* s(xg_0).$$

Now, let us verify that the space $\mathcal{H}^\pi$, as defined above, is in fact the space for our induced representation with $\pi(g_0)\phi = \phi^{g_0}$:
If ϕ satisfies the functional equation (7.1), by the same reasoning as in the proof of Theorem 7.1, so does ϕ^{g_0} for every $g_0 \in G$.
The main point is the norm condition. The more experienced reader will see that, in principle, we do the same as in the proof of Theorem 7.1: Using (7.7) above, the functional equation (7.1) and the unitarity of π_0, one has from (7.4)

$$
\begin{aligned}
\| \phi^{g_0} \|^2_{\mathcal{H}^\pi} &= \int_\mathcal{X} \| \phi^{g_0}(s(x)) \|^2_{\mathcal{H}_0} d\mu_s(x) \\
&= \int_\mathcal{X} \| \phi(s(x)g_0) \|^2_{\mathcal{H}_0} d\mu_s(x) \\
&= \int_\mathcal{X} \| \phi(h^* s(xg_0)) \|^2_{\mathcal{H}_0} d\mu_s(x) \\
&= \int_\mathcal{X} \| \phi(s(xg_0)) \|^2_{\mathcal{H}_0} \delta(h(g_0, x)) d\mu_s(x).
\end{aligned}
$$

Replacing $x \longmapsto xg_0^{-1}$, we see that translation does not change the norm

$$
\begin{aligned}
\| \phi^{g_0} \|^2_{\mathcal{H}^\pi} &= \int_\mathcal{X} \| \phi(s(x)) \|^2_{\mathcal{H}_0} \delta(h(g_0, xg_0^{-1})) d\mu_s(xg_0^{-1}) \\
&= \int_\mathcal{X} \| \phi(s(x)) \|^2_{\mathcal{H}_0} d\mu_s(x) = \| \phi \|^2_{\mathcal{H}^\pi},
\end{aligned}
$$

since from the Master Equation (7.7) (as expression of the associativity of the group operation) one deduces the *cocycle condition*

$$(7.8) \qquad\qquad h(g_1, x)h(g_2, xg_1) = h(g_1 g_2, x).$$

Via the chain rule the quasi-measure relation (7.6) $d\mu_s(xg_0)/d\mu_s(x) = \rho_{g_0}(x) = \delta(h(x, g_0))$ provides for

$$\delta(xg_0^{-1}, g_0) d\mu_s(xg_0^{-1}) = d\mu_s(x).$$

It is now evident that π is unitary if we put

$$< \phi, \psi > := \int_\mathcal{X} < \phi(s(x)), \psi(s(x)) >_{\mathcal{H}_0} d\mu_s(x).$$

Thus we have the essential tools to extend the validity of Theorem 7.1 to the more general hypothesis stated above (we leave out to show that changing the section φ leads to equivalent representations). This realization of the induced representation by functions living on the group is usually called *the first realization* or *induced picture*.

Before we go over to another realization, let us point out that the representation by right translation has an equivalent version where G acts by left translation, i.e. we have $\pi'(g_0)\phi(g) = \phi(g_0^{-1}g)$ where here we have the space $\mathcal{H}'^{\pi}$ consisting of functions ϕ fulfilling the fundamental functional equation

$$(7.9) \qquad \phi(gh) = \delta(h^{-1})^{1/2}\pi_0(h^{-1})\phi(g) \quad \text{for all } h \in H, \ g \in G,$$

and the same finiteness condition.

Exercise 7.2: Fill in missing details.

The Second Realization

We stay with the same hypotheses as above and turn to functions living on the homogeneous space $\mathcal{X} = H\backslash G$. We denote by

$$\mathcal{H}_\pi := L^2(\mathcal{X}, d\mu_s, \mathcal{H}_0)$$

the space of $\mathcal{H}_0$-valued functions f on $\mathcal{X}$, which are quadratic integrable with respect to the quasi-invariant measure $d\mu_s$. In this context we have as fundamental fact the following

Lemma: The map

$$\psi : \mathcal{H}^\pi \longrightarrow \mathcal{H}_\pi, \ \phi \longmapsto f$$

given by $f(x) := \phi(s(x))$ for $x \in \mathcal{X}$ is bijective and an isometry.

Proof: ψ is linear and has as its inverse the map

$$\varphi : \mathcal{H}_\pi \longrightarrow \mathcal{H}^\pi, f \longmapsto \phi_f$$

for $g = h(x)s(x), x \in \mathcal{X}$ given by $\phi_f(g) := \delta(h(x))^{1/2}\pi_0(h(x))f(x)$. One has

$$\| \phi_f \|^2_{\mathcal{H}^\pi} = \int_{\mathcal{X}} \| \phi_f(s(x)) \|^2_{\mathcal{H}_0} d\mu_s(x)$$
$$= \int_{\mathcal{X}} \| f(x) \|^2_{\mathcal{H}_0} d\mu_s(x) = \| f \|^2_{\mathcal{H}_\pi}.$$

Hence the maps preserve the norm.
Both spaces have compatible Hilbert space structure since in a complex inner product space V for $u_1, u_2 \in V$ one has the relation

$$4 < u_1, u_2 > = \| u_1 + u_2 \|^2 - \| u_1 - u_2 \|^2 - i \| u_1 + iu_2 \|^2 + i \| u_1 - iu_2 \|^2 .$$

Exercise 7.3: Prove this.

Using this, we can carry over the induced representation from $\mathcal{H}^\pi$ to $\mathcal{H}_\pi$ and from our Theorem 7.2 come to an equivalent picture as follows.

Theorem 7.4 (*Second Realization*): $\pi = ind_H^G \pi_0$ is realized on $\mathcal{H}_\pi$ by $\tilde{\pi}$ with

$$\tilde{\pi}(g_0)f(x) := A(g_0, x)f(xg_0) \quad \text{for all } f \in \mathcal{H}_\pi$$

with

$$A(g_0, x) := \delta(h(g_0, x))^{1/2}\pi_0(h(g_0, x))$$

for $s(x)g_0 = h(g_0, x)s(xg_0), x \in \mathcal{X}, g_0 \in G.$

Proof: We evaluate the commutative diagram:

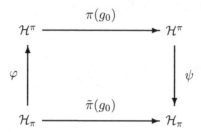

- $f \in \mathcal{H}_\pi$ is mapped by φ onto $\phi \in \mathcal{H}^\pi$ with

$$\phi(g) = \phi(h(x)s(x)) = \delta(h(x))^{1/2}\pi_0(h(x))f(x) \text{ for } g = h(x)s(x) \in G, x \in \mathcal{X}.$$

- $\pi(g_0)$ maps ϕ to ϕ^{g_0} with $\phi^{g_0}(g) = \phi(gg_0).$
- ψ maps ϕ^{g_0} to $\tilde{f}$ with

$$\tilde{f}(x) = \phi^{g_0}(s(x)) = \phi(s(x)g_0).$$

By the Master Equation $s(x)g_0 = h(g_0, x)s(xg_0)$ and the definition of ϕ this can be written as

$$\tilde{f}(x) = \phi(h(x, g_0)s(xg_0)) = \delta(h(g_0, x))^{1/2}\pi_0(h(g_0, x))f(xg_0).$$

Hence, $\tilde{\pi}(g_0)$ transports f to $\tilde{f} = \tilde{\pi}(g_0)f$ with $\tilde{f}$ as given in the Theorem.

Exercise 7.4: Repeat this discussion to get a representation space consisting of functions f living on the space of left cosets $\mathcal{X}' = G/H$ by starting from the induced representation π' given by left translation on the space $\mathcal{H}'^\pi$. Deduce the representation prescription

(7.10) $$\tilde{\pi}(g_0)f(y) := (\delta(h(g_0, y))^{-1/2}\pi_0((h(g_0, y))^{-1})f(g_0^{-1}y)$$

based on the Master equation

$$g_0^{-1}s(y) = s(g_0^{-1}y)h(g_0, y).$$

Remark 7.2: In the special case that π_0 is the trivial one-dimensional representation with $\pi_0(h) = 1$ for all $h \in H$, we fall back to the situation of our Example 1.5 in Section 1.3 where we constructed a representation on a space of functions on a homogeneous space $\mathcal{X} = H \backslash G$. This may be seen as starting point of the construction of induced representations.

Remark 7.3: From the cocycle condition (7.8) for $h = h(g, x)$, one can deduce that the *multiplier system* A in Theorem 7.4 as well fulfills this relation, i.e. we have

$$A(g_1 g_2, x) = A(g_1, x) A(g_2, x g_1) \quad \text{for all } g_1, g_2 \in G, x \in \mathcal{X}.$$

And, vice versa, one can take such a system as tool to construct a representation on a homogeneous space $\mathcal{X}$ with a right G-action (see for instance [Ki1] p.388).

For instance in [BR] p.478 a proof can be found that essentially all this does not depend on the choice of the section s. In most practical cases one has a more or less natural decomposition of the given group into subgroups as at the beginning of this section, so one can avoid to talk about different sections, since one has a natural decomposition. We already discussed the Heisenberg group as an example for this. Here the Schrödinger representation came out as the Second Realization with $\tilde{\pi} = \pi_m$ acting on $\mathcal{H}_\pi = L^2(\mathbf{R})$. The First Realization by functions living on the group is called *Heisenberg representation* .

7.1.3   Final Approach

Before we give more concrete examples, we reproduce some general considerations (which may be skipped at first reading) linked to the report on structure theory in section 6.6 and take over the following situation over from [Kn] p.168: Let G be (not uniquely) decomposed as $G = KH$ and H in *Langlands decomposition* $H = MAN$. We do not give all the details for a proper definition as to be found in [Kn] p.132. But it may suffice to say that one has a corresponding direct sum decomposition with corresponding Lie algebras $\mathfrak{g} = \mathfrak{k} \oplus \mathfrak{m} \oplus \mathfrak{a} \oplus \mathfrak{n}$ where $\mathfrak{a}$ is abelian, $\mathfrak{n}$ nilpotent, and $\mathfrak{m}$ and $\mathfrak{a}$ normalize $\mathfrak{n}$. (Our standard example is $G = \mathrm{SL}(2, \mathbf{R})$. Here we have as in section 6.6.4 the (unique) Iwasawa decomposition $G = KAN$ with $K = \mathrm{SO}(2)$, A the positive diagonal elements, $N = \{n(x); x \in \mathbf{R}\}$, and $H = MAN$ with $M = \{\pm E_2\}$.) Let Δ^+ denote the roots of $(\mathfrak{g}, \mathfrak{a})$ positive for N and $\rho := (1/2) \Sigma_{\alpha \in \Delta^+} (\dim \mathfrak{g}_\alpha) \alpha$. Moreover, let σ be an irreducible unitary representation of M on V^σ, and $\nu \in (\mathfrak{a}^*)^{\mathbf{C}}$, i.e. a complex linear functional on $\mathfrak{a}$. Then in this context, the induced representation is written as

$$\pi = ind_{MAN}^G (\sigma \otimes \exp \nu \otimes 1)$$

and in the *induced picture* given by left translation $\pi(g_0)\phi(g) = \phi(g_0^{-1} g)$ on the representation space, which is given by the dense subspace of continuous functions $\phi : G \longrightarrow V^\sigma$ with functional equation

$$\phi(gman) = e^{-(\nu + \rho) \log a} \sigma(m)^{-1} \phi(g)$$

and the norm

$$\|\phi\|^2 := \int_K |\phi(k)|^2 dk.$$

By restriction to the compact group K one gets the *compact picture*. A dense subspace is given by continuous functions $f : K \longrightarrow V^\sigma$ with

$$f(km) = \sigma(m)^{-1} f(k)$$

and the same norm as above. Here the representation $\tilde{\pi}$ is given by

$$\tilde{\pi}(g_0) f(k) := e^{-(\nu + \rho) H(g_0^{-1} k)} \sigma(\mu(g_0^{-1} k))^{-1} f(\kappa(g_0^{-1} k))$$

where g decomposes under $G = KMAN$ as $g = \kappa(g)\mu(g)e^{H(g)} n$.

In [Kn] p.169 there is a third picture, the *noncompact picture*, given by restriction to $\bar{N} = \Theta N$. Here the representation $\tilde{\pi}$ is given by

$$\tilde{\pi}(g_0)f(\bar{n}) := e^{-(\nu+\rho)\log a(g_0^{-1}\bar{n})}\sigma(m(g_0^{-1}\bar{n}))^{-1}f(\bar{n}(g_0^{-1}\bar{n}))$$

if one takes account of the fact that $\bar{N}MAN$ exhausts G except for a lower dimensional set and that in general one has a decomposition $g = \bar{n}(g)m(g)a(g)n$. The representation space is here $L^2(\bar{N}, e^{2\mathrm{Re}\,\nu H(\bar{n})}d\bar{n})$.

7.1.4   Some Questions and two Easy Examples

Several questions arise immediately:

i) How to construct non-trivial functions in $\mathcal{H}^\pi$ resp. $\mathcal{H}_\pi$?
ii) When is an induced representation π irreducible?
iii) Which representations can be constructed by induction?
iv) If one restricts an induced representation to the subgroup, does one get back the representation of the subgroup?
v) What happens by iteration of the induction procedure?
vi) What about the induced representation if the inducing representation is the sum or tensor product of two representations ?

Here we can not give satisfying answers to these questions (but hope that our examples later on will be sufficiently illuminating). We only mention some remarks.

Ad i) One can prove ([BR] p.477) the following statement.
Proposition 7.1: Let φ be a continuous $\mathcal{H}_0$-valued function on G with compact support. Define ϕ_φ by

$$\phi_\varphi(g) := \int_H \pi_0(h^{-1})\varphi(hg)d_r h.$$

Then ϕ_φ is a continuous function on G whose support goes to a compact set by the projection $p : G \longrightarrow X$ and is an element of $\mathcal{H}^\pi$. The set

$$\mathcal{C}^\pi := \{\phi_\varphi;\ \varphi(g) = \xi(g)v,\ \xi \in \mathcal{C}_c(G), v \in \mathcal{H}_0\}$$

is a dense set in $\mathcal{H}^\pi$.

Ad ii) We have seen that the Schrödinger representation of the Heisenberg group is an irreducible induced representation. In general, this is not the case and in our examples below we will have to discuss this. We shall find conditions to be added to the definition of our space if one wants to come to an irreducible representation (e.g. *holomorphic induction*). In [BR] p.499, there are general criteria to get irreducibility (using the concept of a *canonical system of imprimitivity*).

Ad iii) This concept of canonical system of imprimitivity is also used to give a *criterium of inducibility* stating under which conditions a unitary representation is equivalent to an induced representation. One finds versions of this in [BR] p.495, [Ki] p.191, and in [Ki1] p.389 (Mackey Inducibility Criterion). We do not reproduce these because the statements need more new notions.

Ad iv) Induction and restriction are *adjoint functors*. We do not try to make this more precise. At least for finite groups, one finds statements of the *Frobenius Reciprocity* as in [Se] p.57 or [Ki] p.185, which, using our notation (see section 1.2), for a representation π of G can be written as

$$c(\pi, ind_H^G \pi_0) = c(\pi|_H, \pi_0).$$

Ad v) One has the concept of *induction in stages*, which symbolically can be written as

$$ind_H^G ind_K^H \simeq ind_K^G$$

if we have a subgroup chain $G \supset H \supset K$ ([Ki] p.184, [BR] p.489).

Ad vi) If π_0 and π_0' are representations of the subgroup H of G the operations of taking direct sum and induction are interchangeable

$$ind_H^G (\pi_0 \oplus \pi_0') \simeq ind_H^G \pi_0 \oplus ind_H^G \pi_0'$$

([BR] p.488). And if π_1 and π_2 are unitary representations of the closed subgroups H_1 and H_2 of the separable, locally compact groups G_1 resp. G_2, then one has ([BR] p.491)

$$ind_{H_1 \times H_2}^{G_1 \times G_2} (\pi_1 \otimes \pi_2) \simeq ind_{H_1}^{G_1} \pi_1 \otimes ind_{H_2}^{G_2} \pi_2.$$

Up to now, we treated only the example of the Heisenberg group. Before we do more serious work with $G = \mathrm{SL}(2, \mathbf{R})$, we discuss another example we already looked at, namely $G = \mathrm{SU}(2)$, and the two-dimensional form of the Poincaré group, which later on will be discussed more thoroughly in the four-dimensional form in the context of Mackey's method for semidirect products.

Example 7.2: Let

$$G = \{g = (\begin{smallmatrix} A & n \\ 0 & 1 \end{smallmatrix}) \in \mathrm{SL}(3, \mathbf{R}); \ A \in \mathrm{SO}(1,1), n = (\begin{smallmatrix} a \\ t \end{smallmatrix}) \in \mathbf{R}^2\}$$

be the *two-dimensional Poincaré group*. We abbreviate $g =: (A, n)$, so that matrix multiplication leads to the composition law

$$gg' = (AA', An' + n).$$

In analogy to the case of $\mathrm{SO}(2)$, we write the elements of $\mathrm{SO}(1,1)$ in the form

$$A = (\begin{smallmatrix} \cosh \vartheta & \sinh \vartheta \\ \sinh \vartheta & \cosh \vartheta \end{smallmatrix}) =: rh(\vartheta), \ \vartheta \in \mathbf{R}.$$

From the addition theorem we have $AA' = rh(\vartheta)rh(\vartheta') = rh(\vartheta + \vartheta')$, and $d\vartheta$ is a bi-invariant measure on $\mathrm{SO}(1,1)$. We take the subgroup

$$H := \{h = (\begin{smallmatrix} E_2 & n \\ 0 & 1 \end{smallmatrix}); \ n \in \mathbf{R}^2\}$$

and the one-dimensional representation π_0 defined by

$$\pi_0(h) := e^{\pi i(a\alpha + t\beta)} = e^{\pi i^t n\hat{n}} =: \chi_{\hat{n}}(n), \ \hat{n} := (\begin{smallmatrix} \alpha \\ \beta \end{smallmatrix}) \in \mathbf{R}^2.$$

Moreover, we denote by K the image of the injection of $\mathrm{SO}(1,1)$ into G

$$K := \{k = \begin{pmatrix} A & 0 \\ 0 & 1 \end{pmatrix}; A \in \mathrm{SO}(1,1)\}$$

Then every $g \in G$ has a Mackey decomposition $g = (A, n) = (E, n)(A, 0)$ and we have

$$G \longrightarrow \mathcal{X} = H \backslash G \simeq K \simeq \mathbf{R}, \; g = (A, n) \longmapsto A = rh(\vartheta) \longmapsto \vartheta.$$

The Master Equation $s(x)g_0 = h(g_0, x)s(xg_0)$ in this case simply is

$$(A, 0)(A_0, n_0) = (AA_0, An_0) = (E, An_0)(AA_0, 0).$$

We see that $d\vartheta$ is not only quasi-invariant but invariant and we have a trivial normalizing function $\delta(h) = 1$ for all $h \in H$. Hence, the induced representation $\pi = ind_H^G \chi_{\hat{n}}$ is given by right translation on the space $\mathcal{H}^\pi$, which is the completion of the space of continuous functions

$$\phi : G \longrightarrow \mathbf{C}, \; \phi(hg) = \chi_{\hat{n}}(h)\phi(g), \; \int_{\mathbf{R}} |\phi((rh(\vartheta), 0))|^2 \, d\vartheta < \infty.$$

The induced representation has its Second Realization on the space $\mathcal{H}_\pi = L^2(\mathcal{X}, d\mu(x)) = L^2(\mathbf{R}, d\vartheta)$. We have the isomorphism given by

$$\mathcal{H}^\pi \ni \phi \longmapsto f \in \mathcal{H}_\pi \text{ with } f(\vartheta) := \phi((rh(\vartheta), 0)),$$

and the prescription for the representation from Theorem 7.4, namely

$$\tilde{\pi}(g_0)f(x) = A(g_0, x)f(xg_0),$$

in this case leads to

$$\tilde{\pi}((rh(\vartheta_0), n_0))f(\vartheta) = \chi_{\hat{n}}(rh(\vartheta)n_0)f(\vartheta + \vartheta_0) = e^{\pi i^t n_0 rh(\vartheta)\hat{n}}f(\vartheta + \vartheta_0).$$

The representation is irreducible. This is a byproduct of Mackey's general theory for representations of semidirect products, upon which we will report later in section 7.3. The reader is invited to find a direct proof using the infinitesimal method analogous to the example in 6.5 (**Exercise 7.5**). In the next example the induced representation will come out to be reducible.

Example 7.3: For the group

$$G = \mathrm{SU}(2) = \{g = \begin{pmatrix} a & b \\ -\bar{b} & \bar{a} \end{pmatrix}; a, b \in \mathbf{C}, |a|^2 + |b|^2 = 1\}$$

we obtained in 4.2 the result that every irreducible representation of G is equivalent to a representation $\pi_j, j \in (1/2)\mathbf{N}_0$, given on the space $V^{(j)} = \mathbf{C}[x, y]_{2j}$ of homogeneous polynomials P of degree $2j$ by the prescription

$$\pi_j(g_0)P(x, y) := P(\bar{a}_0 x - b y_0, \bar{b}_0 x + a_0 y).$$

This formula is based on the action on functions produced by

$$\begin{pmatrix} x \\ y \end{pmatrix} \longmapsto g_0^{-1} \begin{pmatrix} x \\ y \end{pmatrix} = {}^t\bar{g}_0 \begin{pmatrix} x \\ y \end{pmatrix} = \begin{pmatrix} \bar{a}_0 x - b_0 y \\ \bar{b}_0 x + a_0 y \end{pmatrix}.$$

We want to see how this fits into our scheme of construction of induced representations: It is very natural to take as subgroup H the group of diagonal matrices

$$H := \{h = \begin{pmatrix} a & \\ & \bar{a} \end{pmatrix}; a = e^{i\varphi}, \varphi \in \mathbf{R}\} \simeq \mathrm{U}(1) \simeq \mathrm{SO}(2)$$

and as its representation π_0 given by $\pi_0(h) = e^{ik\varphi} =: \chi_k(h), k \in \mathbf{N}_0$. Since in π_j we have a left action, we deal here also with an action of G by left translation, i.e.

$$\pi(g_0)\phi(g) = \phi(g_0^{-1}g)$$

for functions ϕ from a representation space $\mathcal{H}'^\pi$, which fulfill the functional equation

$$\phi(gh) = \pi_0(h^{-1})\phi(g) = \lambda^k\phi(g), \; \lambda = e^{-i\varphi}.$$

(We know that G and H are compact and, hence, unimodular so that the normalizing factor δ is trivial.) The homogeneous space $\mathcal{X}' = G/H$ showing up here is one of the most fundamental examples of homogeneous spaces: One has

$$SU(2)/H \simeq \mathbf{P}^1(\mathbf{C}),$$

the *one-dimensional complex projective space*. We already met with projective spaces in 4.3 while introducing the concept of projective representations. Here we use the following notation: $G = SU(2)$ acts on $V = \mathbf{C}^2$ by

$$(g, v) \longmapsto gv, \; v := \left(\begin{smallmatrix} x \\ y \end{smallmatrix} \right).$$

We have $\mathbf{P}^1(\mathbf{C}) = (V\backslash 0)_\sim$ where for $v, v' \in V$ one has $v \sim v'$ iff there exists a $\lambda \in \mathbf{C}^*$ with $v' = \lambda v$. The action of G on V induces an action on $\mathbf{P}^1(\mathbf{C})$ given by $g(v_\sim) := (gv)_\sim$. We write $(x : y) = {}^t(x, y)_\sim$. For $v_0 = {}^t(0, 1)$ we have $gv_0 = {}^t(b, \bar{a})$, i.e. $g(0 : 1) = (b : \bar{a})$, and the stabilizing group of $(v_0)_\sim = (0 : 1)$ is just $H \simeq U(1)$. Hence we have

$$SU(2)/H \longrightarrow \mathbf{P}^1(\mathbf{C}), \; g = \left(\begin{smallmatrix} a & b \\ -\bar{b} & \bar{a} \end{smallmatrix} \right) \longmapsto (b : \bar{a}).$$

A homogeneous function in two variables $F(x, y)$ induces via $f(x : y) := F(x, y)$ a function f on $\mathbf{P}^1(\mathbf{C})$, and this in turn induces a function ϕ_F living on G given by $\phi_F(g) := F({}^t(g\,{}^t(0, 1))) = F((0, 1)\,{}^tg)$. Hence we associate the function ϕ_P to a homogeneous polynomial P of degree $k = 2j$ from our representation π_j, namely

$$\phi_P(g) := P(b, \bar{a}) \text{ for } g = \left(\begin{smallmatrix} a & b \\ -\bar{b} & \bar{a} \end{smallmatrix} \right).$$

The fact that P is homogeneous of degree k translates to the functional equation

$$\phi_P(gh) = P((0, 1)\,{}^t(gh)) = P((0, e^{-i\varphi})\,{}^tg) = e^{-i\varphi k}P((0, 1)\,{}^tg) = \lambda^k\phi_P(g)$$

where $h = \left(\begin{smallmatrix} e^{i\varphi} & \\ & e^{-i\varphi} \end{smallmatrix} \right)$ and $\lambda = e^{-i\varphi}$ as above. One has

$$g_0^{-1}g = \left(\begin{smallmatrix} \bar{a}_0 & -b_0 \\ \bar{b}_0 & a_0 \end{smallmatrix} \right)\left(\begin{smallmatrix} a & b \\ -\bar{b} & \bar{a} \end{smallmatrix} \right) = \left(\begin{smallmatrix} \bar{a}_0 a + b_0\bar{b} & \bar{a}_0 b - b_0\bar{a} \\ \bar{b}_0 a - a_0\bar{b} & \bar{b}_0 b + a_0\bar{a} \end{smallmatrix} \right)$$

and hence

$$\phi_P(g_0^{-1}g) = P(\bar{a}_0 b - b_0\bar{a}, \bar{b}_0 b + a_0\bar{a}) = \phi_{\pi(g_0)P}(g).$$

We see that π_j is a candidate for a subrepresentation of the induced representation $\pi = ind_H^G \chi_k$. We propose to show as **Exercise 7.6** that the finite norm condition is fulfilled and to search for further conditions to be imposed such that we get a subspace of $\mathcal{H}^\pi$, which is isomorphic to the space $V^{(j)}$ of π_j. We shall return to this question and the example in Section 7.6 in the framework of the more sophisticated and elegant concept of an interpretation of the induction process via *line bundles*.

7.2   Unitary Representations of $\mathrm{SL}(2,\mathbf{R})$

Serge Lang dedicated a whole book, namely "$\mathrm{SL}(2,\mathbf{R})$" ([La1]), to this group and its representations. Also, there is a lot of refined material concerning the representations of $\mathrm{SL}(2,\mathbf{R})$ throughout Knapp's book [Kn]. Summaries presenting models for the different representations can be found in many places, for instance in [KT] p.1–24 or in [Do]. The most classical reference is Bargmann's article [Ba] from 1947. We shall need the representations of $G = \mathrm{SL}(2,\mathbf{R})$ later in the discussion of the representations of the Poincaré group and most essentially in the Outlook to Number Theory.

Using the Infinitesimal Method, in Theorem 6.2 in section 6.4, we obtained a list fixing the possible equivalence classes of irreducible unitary representations of $G = \mathrm{SL}(2,\mathbf{R})$. Here we will show that these representations really exist by presenting concrete realizations or, as one also says, *models*. In most occasions it is important to have different models, depending on the occasion where the representation is showing up. The method we follow here is just the application of the induction procedure that we developed in the preceding section. Thus, we have to start by choosing a subgroup H of our G. Already in several occasions in this text, we specified subgroups of $G = \mathrm{SL}(2,\mathbf{R})$. We recall (for instance from 6.6.4) the notation

$$
\begin{aligned}
K &:= \mathrm{SO}(2) = \{r(\vartheta) = \begin{pmatrix} \cos\vartheta & \sin\vartheta \\ -\sin\vartheta & \cos\vartheta \end{pmatrix}; \vartheta \in \mathbf{R}\}, \\
A &:= \{t(y) = \begin{pmatrix} y^{1/2} & \\ & y^{-1/2} \end{pmatrix}; y \in \mathbf{R}, y > 0\}, \\
N &:= \{n(x) = \begin{pmatrix} 1 & x \\ & 1 \end{pmatrix}; x \in \mathbf{R}\}, \\
\bar{N} &:= \{\bar{n}(x) = \begin{pmatrix} 1 & \\ x & 1 \end{pmatrix}; x \in \mathbf{R}\}, \\
M &:= \{\pm E_2\}, \\
B &:= MAN = \{h = \begin{pmatrix} a & b \\ & a^{-1} \end{pmatrix}; a,b \in \mathbf{R}, a \neq 0\}, \\
\bar{B} &:= MA\bar{N}.
\end{aligned}
$$

The subgroup B of upper triangular matrices is a special example of a *parabolic subgroup*. It is also called the standard *Borel group*. One has the disjoint Iwasawa decomposition $G = KAN = ANK$ and $G = KB = BK$ with $B \cap K = \{\pm E_2\}$. It makes sense to start the induction procedure from all these subgroups but the best thing to do is *parabolic induction*:

1. The Principal Series

We take $H := B$ and π_0 the one-dimensional representation, which for fixed $s \in \mathbf{R}$ and $\epsilon \in \{0,1\}$, is given by

$$\pi_0(h) := |\, a\, |^{is} (\operatorname{sgn} a)^\epsilon.$$

We put $\mathcal{X} := H\backslash G$ and take the Mackey decomposition (7.5)

$$
\begin{aligned}
g = \begin{pmatrix} a & b \\ c & d \end{pmatrix} &= \begin{pmatrix} d^{-1} & b \\ & d \end{pmatrix}\begin{pmatrix} 1 & \\ c/d & 1 \end{pmatrix} &= h(x)s(x) \text{ for } d \neq 0, \\
&= \begin{pmatrix} b & -a \\ & b^{-1} \end{pmatrix}\begin{pmatrix} & 1 \\ -1 & \end{pmatrix} &= h(x)s(x) \text{ for } d = 0.
\end{aligned}
$$

Hence $\mathcal{X}$ is in bijection to the set $S := \{(\begin{smallmatrix} 1 & \\ x & 1 \end{smallmatrix}); x \in \mathbf{R}\} \cup \{(\begin{smallmatrix} & 1 \\ -1 & \end{smallmatrix})\}$. As in our

context a point has measure zero, we can disregard the element $(\begin{smallmatrix} & 1 \\ -1 & \end{smallmatrix})$ and work

with

$$G \longrightarrow \mathcal{X} = H\backslash G \supset \mathcal{X}' \simeq \mathbf{R}, \ g \longmapsto x := c/d, \ d \neq 0$$

with section $s : \mathcal{X}' \longrightarrow G, x \longmapsto s(x) = (\begin{smallmatrix} 1 & \\ x & 1 \end{smallmatrix})$. For $g_0 = (\begin{smallmatrix} a_0 & b_0 \\ c_0 & d_0 \end{smallmatrix})$ our Master

Equation (7.7)

$$s(x)g_0 = h(g_0, x)s(xg_0)$$

is realized by

$$h(x, g_0) = (\begin{smallmatrix} (b_0 x + d_0)^{-1} & b_0 \\ 0 & b_0 x + d_0 \end{smallmatrix}), \ s(xg_0) = (\begin{smallmatrix} 1 & \\ xg_0 & 1 \end{smallmatrix}), \ xg_0 = \frac{a_0 x + c_0}{b_0 x + d_0}.$$

We see that $d\mu(x) = dx$ is a quasi-invariant measure on $\mathcal{X}' \simeq \mathbf{R}$ with the quasi-measure relation (7.6)

$$\frac{d\mu(xg_0)}{d\mu(x)} = \frac{1}{(b_0 x + d_0)^2} =: \rho_{g_0}(x) = \delta(h(g_0, x)).$$

Hence, we have $\delta(h) = a^2$ and the induced representation $\pi = ind_H^G \pi_0$ is given in the First Realization by right translation on the space $\mathcal{H}^\pi$, which is the completion of the space of smooth functions $\phi : G \longrightarrow \mathbf{C}$ with

(7.11) $$\phi(hg) = (\delta(h))^{1/2}\pi_0(h)\phi(g) \quad \text{for all } h \in H, g \in G$$

and

$$\|\phi\|^2 := \int_{\mathbf{R}} |\phi(s(x))|^2 \, dx < \infty.$$

As we saw in our discussion of the representations of the Heisenberg group, it is more illuminating to look at the Second Realization. This is given by the representation space $\mathcal{H}_\pi = L^2(\mathcal{X}, d\mu(x)) = L^2(\mathbf{R}, dx)$. We see from Theorem 7.4 (or by a direct calculation) that the representation acts by

$$\tilde{\pi}(g_0)f(x) = |b_0 x + d_0|^{-is-1} (\text{sgn}\,(b_0 x + d_0))^\epsilon f(\frac{a_0 x + c_0}{b_0 x + d_0}) \text{ for } f \in L^2(\mathbf{R}, dx).$$

Exercise 7.7: Determine the derived representation and verify that this induced representation is a model for the up to now only hypothetical representation $\pi_{is,\pm}$ from Theorem 6.2 in 6.4.

Remark 7.4: This exercise provides as byproduct a proof that this induced representation is irreducible. There are more direct ways to see this (see for instance [Kn] p.36).

Exercise 7.8: We propose to verify in a special situation that the construction essentially does not depend on the choice of the quasi-invariant measure: As in [BR] p.483, repeat the former discussion for the quasi-invariant measure $d\mu(x) = \varphi(x)\,dx$ with $\varphi(x) = (1 + x^2)^{-2}$ to get a representation space $\tilde{\mathcal{H}}_\pi$ linked to our $\mathcal{H}_\pi$ by the intertwining operator $\tilde{f} \longmapsto f, \ f(x) = \tilde{f}(x)/(1 + x^2)^{1/2}$.

To get more familiarity with these constructions and because of the general importance of the result, we also discuss the realization of our induced representation by left translation, s.t., for functions ϕ living on G, we have a representation given by $\pi'(g_0)\phi(g) = \phi(g_0^{-1}g)$.

We keep H and π_0 as above and now put $\mathcal{X}' = G/H$. For $g = \begin{pmatrix} a & b \\ c & d \end{pmatrix}$ with $a \neq 0$, we have a Mackey decomposition

$$g = s(y)h(y), \; s(y) = \begin{pmatrix} 1 & \\ y & 1 \end{pmatrix}, \; y = c/a, \; h(y) = \begin{pmatrix} a & b \\ & a^{-1} \end{pmatrix},$$

i.e. a map $\qquad p: G \longrightarrow G/H = \mathcal{X}' \supset \mathring{\mathcal{X}}' \simeq \mathbf{R}, \; g \longmapsto y = c/a,$

with section $y \longmapsto s(y) = \begin{pmatrix} 1 & \\ y & 1 \end{pmatrix}$. The Master Equation is in this case

$$g_0^{-1}s(y) = s(g_0^{-1}y)h(y, g_0)$$

with

$$s(g_0^{-1}y) = \begin{pmatrix} 1 & \\ g_0^{-1}y & 1 \end{pmatrix}, \; g_0^{-1}y = \frac{a_0 y - c_0}{-b_0 y + d_0},$$

$$h(y, g_0) = \begin{pmatrix} d_0 - b_0 y & -b_0 \\ & (d_0 - b_0 y)^{-1} \end{pmatrix} =: h^*.$$

We take $d\mu(y) = dy$ as quasi-invariant measure on $\mathring{\mathcal{X}}' \simeq \mathbf{R}$ and have the quasi-measure relation

$$\frac{d\mu(g_0^{-1}y)}{d\mu(y)} = \frac{1}{(-b_0 y + d_0)^2} = \rho_{g_0}(y).$$

To be on the safe side and to provide for some more exercise in the handling of these notions, we verify that left translation preserves the norm: We take a function ϕ with

$$\phi(gh) = (\delta(h))^{-1/2}\pi_0(h^{-1})\phi(g) \quad \text{for all } h \in H, g \in G,$$

put $\phi^{g_0}(g) := \phi(g_0^{-1}g)$, and calculate using the Master Equation $g_0^{-1}s(y) = s(g_0^{-1}y)h^*$ and the unitarity of π_0

$$
\begin{aligned}
\|\phi^{g_0}\|^2 &= \int_{\mathbf{R}} |\phi^{g_0}(s(y))|^2 \, d\mu(y) \\
&= \int_{\mathbf{R}} |\phi(s(g_0^{-1}y)h^*)|^2 \, d\mu(y) \\
&= \int_{\mathbf{R}} |\phi(s(g_0^{-1}y))|^2 \, \delta(h^*)^{-1}d\mu(y) \\
&= \int_{\mathbf{R}} |\phi(s(\tilde{y}))|^2 \, d\mu(\tilde{y}),
\end{aligned}
$$

since one has by the quasi-measure relation

$$\frac{d\mu(\tilde{y})}{d\mu(y)} = \frac{d\mu(g_0^{-1}y)}{d\mu(y)} = \frac{1}{(-b_0 y + d_0)^2} = \delta(h^*)^{-1}.$$

So we have as the First Realization the induced representation defined by functions ϕ on G with the functional equation

$$\phi(gh) = (\delta(h))^{-1/2}\pi_0(h^{-1})\phi(g) \quad \text{for all } h \in H, g \in G,$$

and the norm condition

$$\|\phi\|^2 = \int_{\mathbf{R}} |\phi(s(y))|^2 \, d\mu(y) < \infty.$$

From here, by the map $\phi \longmapsto f$, $f(y) := \phi((\begin{smallmatrix} 1 & \\ y & 1 \end{smallmatrix}))$, we come to the Second Realization given on the space $\mathcal{H}_\pi = L^2(\mathbf{R}, dy)$ by the prescription

$$
\begin{aligned}
\tilde{\pi}(g_0) f(y) &= (\delta(h(y, g_0)))^{-1/2} \pi(h(y, g_0)^{-1}) f(g_0^{-1} y) \\
&= |d_0 - b_0 y|^{-is-1} (\operatorname{sgn}(d_0 - b_0 y))^\epsilon f(\frac{a_0 y - c_0}{-b_0 y + d_0}).
\end{aligned}
$$

This is the same formula as in [KT] p.17 and [Kn] p.36 and p.167 where this is an example for a *noncompact picture*. We remark that Knapp comes to this formula from his *induced picture*, which is prima facie different from our First Realization as he does not use the definition of the norm we used above but the same definition of the norm as in (7.3)

$$
\|\phi\|^2 = \int_K | \phi(k) |^2 \, dk.
$$

In [Kn] p.168, one finds a proof that this norm and the one we used above are equal. We propose to search here for a direct proof (**Exercise 7.9**).

Moreover, in [Kn] p.37, there is a proof of the irreducibility of these induced representations $\pi_{is,\pm}$ except for $\pi_{0,-}$, which is based on euclidean Fourier Transform.

2. The Discrete Series

We will discuss several models, which realize the discrete series representations.

The Upper Half Plane Model

We take $H = K = \mathrm{SO}(2)$ and $\pi_0(h) = \chi_k(r(\vartheta)) = e^{ik\vartheta}$, $k \in \mathbf{Z}$, for $h = r(\vartheta) \in \mathrm{SO}(2)$. It is a custom to construct the representations $\pi_k^\pm$ from our list in Theorem 6.2 in 6.4 by functions living on the space of left cosets $G/K = \mathrm{SL}(2, \mathbf{R})/\mathrm{SO}(2)$. This space is again an eminent example of a homogeneous space. It is called the *Poincaré upper half plane* and we denote it by $\mathfrak{H} := \{\tau = x + iy \in \mathbf{C}; y > 0\}$. Since this is important for our constructions here but also as a standard example in complex function theory, we assemble some related standard notions and facts.

For $g = (\begin{smallmatrix} a & b \\ c & d \end{smallmatrix}) \in \mathrm{SL}(2, \mathbf{R})$ and $\tau \in \mathfrak{H}$, we put

$$
g(\tau) := \frac{a\tau + b}{c\tau + d}
$$

and call this a *linear fractional transformation*.

Exercise 7.10: Prove that this is well defined and provides a transitive left action of $G = \mathrm{SL}(2, \mathbf{R})$ (even of $\mathrm{GL}(2, \mathbf{R})$) on $\mathfrak{H}$.
Verify that $H = \mathrm{SO}(2)$ is the stabilizing group of $\tau = i$.
Show that the *automorphic factor*

$$
j(g, \tau) := (c\tau + d)^{-1}
$$

fulfills the cocycle relation (7.8) (in the version for the action from the left)

$$
j(gg', \tau) = j(g, g'(\tau)) j(g', \tau).
$$

One has a map

$$p : G = SL(2, \mathbf{R}) \longrightarrow SL(2, \mathbf{R})/SO(2) \simeq \mathfrak{H}, \quad g \longmapsto g(i) =: \tau = x + iy$$

for the Iwasawa decomposition $g = n(x)t(y)r(\vartheta)$, which here can be interpreted as a Mackey decomposition $g = s(\tau)h, s(\tau) = n(x)t(y), h = r(\vartheta)$. Moreover, one easily verifies the following formulae

$$(7.12) \qquad y = \frac{1}{c^2 + d^2}, \quad x = \frac{ac + bd}{c^2 + d^2}, \quad e^{i\vartheta} = \frac{d - ic}{\sqrt{c^2 + d^2}} \quad \text{for } g = \begin{pmatrix} a & b \\ c & d \end{pmatrix}$$

and, for $g = r(\vartheta')t(y')n(x')$,

$$(7.13) \qquad \cos \vartheta' = \frac{a}{\sqrt{a^2 + c^2}}, \quad \sin \vartheta' = \frac{-c}{\sqrt{a^2 + c^2}}, \quad y' = \frac{1}{a^2 + c^2}, \quad x' = \frac{ab + cd}{a^2 + c^2}.$$

Exercise 7.11: Verify

$$\frac{d(g(\tau))}{d\tau} = \frac{1}{(c\tau + d)^2}$$

and show that $d\mu(\tau) = y^{-2}dxdy$ is a G-invariant measure on $\mathfrak{H}$.

This is also an occasion to practise again the calculation of infinitesimal operations initiated in 6.2 and 6.3: For a smooth complex function ϕ on G and $U \in \mathfrak{g}$, we get a right invariant differential operator $\mathcal{R}_U$ defined by

$$\mathcal{R}_U\phi(g) := \frac{d}{dt}\phi((\exp tU)^{-1}g) \mid_{t=0}$$

and a left invariant operator $\mathcal{L}_U$ by

$$\mathcal{L}_U\phi(g) := \frac{d}{dt}\phi(g(\exp tU)) \mid_{t=0} .$$

From 6.4 we recall that we have

$$\mathfrak{g} = \mathfrak{sl}(2, \mathbf{R}) = \{X \in M_2(\mathbf{R}); \operatorname{Tr} X = 0\} = < F, G, H >$$

with

$$F = \begin{pmatrix} & 1 \\ 1 & \end{pmatrix}, \ G = \begin{pmatrix} & \\ 1 & \end{pmatrix}, \ H = \begin{pmatrix} 1 & \\ & -1 \end{pmatrix}$$

and

$$\mathfrak{g}_c := \mathfrak{g} \otimes_{\mathbf{R}} \mathbf{C} = < F, G, H >_{\mathbf{C}} = < X_+, X_-, Z >_{\mathbf{C}}$$

where

$$X_\pm = (1/2)(H \pm i(F + G)) = (1/2)\begin{pmatrix} 1 & \pm i \\ \pm i & -1 \end{pmatrix}, \ Z = -i(F - G) = \begin{pmatrix} & -i \\ i & \end{pmatrix}$$

with

$$[Z, X_\pm] = \pm 2X_\pm, \ [X_+, X_-] = Z.$$

Using the coordinates x, y, ϑ of the Iwasawa decomposition as parameters for g and the commutation rule $t(y)n(x) = n(yx)t(y)$, one obtains

$$
\begin{aligned}
\mathcal{R}_H\phi(g) &= \tfrac{d}{dt}\phi((\exp tH)^{-1}g) \mid_{t=0} \\
&= \tfrac{d}{dt}\phi((\begin{smallmatrix} e^{-t} & \\ & e^t \end{smallmatrix})n(x)t(y)r(\vartheta)) \mid_{t=0} \\
&= \tfrac{d}{dt}\phi(n(e^{-2t}x)t(e^{-2t}y)r(\vartheta)) \mid_{t=0} \\
&= -(2x\partial_x + 2y\partial_y)\phi(g).
\end{aligned}
$$

The same way one has

$$
\begin{aligned}
\mathcal{R}_F\phi(g) &= \tfrac{d}{dt}\phi((\exp tF)^{-1}g)\,|_{t=0} \\
&= \tfrac{d}{dt}\phi((\begin{smallmatrix} 1 & -t \\ & 1 \end{smallmatrix})n(x)t(y)r(\vartheta))\,|_{t=0} \\
&= \tfrac{d}{dt}\phi(n(x-t)t(y)r(\vartheta))\,|_{t=0} \\
&= -\partial_x\phi(g),
\end{aligned}
$$

and

$$
\begin{aligned}
\mathcal{R}_G\phi(g) &= \tfrac{d}{dt}\phi((\exp tG)^{-1}g)\,|_{t=0} \\
&= \phi((\begin{smallmatrix} 1 & \\ -t & 1 \end{smallmatrix})n(x)t(y)r(\vartheta))\,|_{t=0} \\
&= \phi(n(x')t(y')r(\vartheta'))\,|_{t=0} \\
&= ((x^2 - y^2)\partial_x + 2xy\partial_y + y\partial_\vartheta)\phi(g),
\end{aligned}
$$

where in the last formula we used (7.12) to get

$$
\begin{aligned}
x' &= \frac{x - t(x^2 + y^2)}{t^2y^2 + (1 - tx)^2}, \\
y' &= \frac{y}{t^2y^2 + (1 - tx)^2}, \\
e^{i\vartheta'} &= e^{i\vartheta}\,\frac{1 - t(x - iy)}{\sqrt{t^2y^2 + (1 - tx)^2}}.
\end{aligned}
$$

Hence we have the differential operators

$$
\begin{aligned}
\mathcal{R}_F &= -\partial_x, \\
\mathcal{R}_G &= (x^2 - y^2)\partial_x + 2xy\partial_y + y\partial_\vartheta, \\
\mathcal{R}_H &= -2x\partial_x - 2y\partial_y,
\end{aligned}
\tag{7.14}
$$

which combine to

$$
\begin{aligned}
\mathcal{R}_Z &= i((1 + x^2 - y^2)\partial_x + 2xy\partial_y + y\partial_\vartheta), \\
\mathcal{R}_{X_\pm} &= \pm(i/2)(((x \pm i)^2 - y^2)\partial_x + 2(x \pm i)y\partial_y + y\partial_\vartheta).
\end{aligned}
\tag{7.15}
$$

A similar calculation (**Exercise 7.12**) leads to

$$
\begin{aligned}
\mathcal{L}_Z &= -i\partial_\vartheta, \\
\mathcal{L}_{X_\pm} &= \pm(i/2)e^{\pm 2i\vartheta}(2y(\partial_x \mp i\partial_y) - \partial_\vartheta).
\end{aligned}
\tag{7.16}
$$

In 6.4 we already met with the (multiple of the Casimir) element

$$
\Omega := X_+X_- + X_-X_+ + (1/2)Z^2,
$$

which, by the way, is also (**Exercise 7.13**) $\Omega = FG + GF + (1/2)H^2$. By the natural extension of the definition of the differential operators from $\mathfrak{g}$ to $U(\mathfrak{g}_c)$ (via $\mathcal{R}_{UU'} := \mathcal{R}_U\mathcal{R}_{U'}$) and a small calculation we get the SL(2, **R**)-*Laplacian*

$$
\Delta := \mathcal{R}_\Omega = \mathcal{L}_\Omega = 2y^2(\partial_x^2 + \partial_y^2) - 2y\partial_x\partial_\vartheta.
\tag{7.17}
$$

After these general preparations, we finally come to the discrete series representations. There are a lot of different models in the literature. We present the most customary one and then discuss how it fits into the scheme of the induction procedure.

For instance in [La1] p.181, we find the representation π_m with the representation space $\mathcal{H}_m := L^2_{hol}(\mathfrak{H}, \mu_m), m \geq 2$. This space consists of *holomorphic* (i.e. complex differentiable) functions f on the upper half plane $\mathfrak{H}$, which have a finite norm $\|f\|_m < \infty$, i.e. with

$$\|f\|^2_m := \int_{\mathfrak{H}} \mid f(x,y) \mid^2 y^m \frac{dxdy}{y^2} < \infty$$

and one has an scalar product defined by

$$< f_1, f_2 > := \int_{\mathfrak{H}} \overline{f_1(x,y)} f_2(x,y) y^m \frac{dxdy}{y^2}.$$

As will be apparent soon (even for those readers, which are not so familiar with complex function theory), the main point is here reduced to the fact that holomorphic functions f are essentially smooth complex valued functions characterized by the *Cauchy Riemann differential equation*

$$(\partial_x + i\partial_y)f = 0.$$

We write $f(\tau) := f(x,y)$ and put $g^{-1} =: \begin{pmatrix} a & b \\ c & d \end{pmatrix}$ (this is some trick to get nicer formulae but, if not remembered carefully, sometimes leads to errors). Then the representation π_m is in the *half plane model* given by

(7.18) $\pi_m(g)f(\tau) := f(g^{-1}(\tau))j_m(g^{-1}, \tau) = f(g^{-1}(\tau))(c\tau + d)^{-m}.$

Here we have used the important notation of the automorphic factor

$$j_m(g, \tau) := j(g, \tau)^m$$

extending the one introduced in Exercise 7.10.

Remark 7.5: In [Kn] p.35, one finds a representation given on the same space by the prescription

$$\check{\pi}_m(\begin{pmatrix} a & b \\ c & d \end{pmatrix})f(\tau) := (-b\tau + d)^{-m} f(\frac{a\tau - c}{-b\tau + d}).$$

And in [GGP] p.53, the authors define a representation T_m given by

$$T_m(\begin{pmatrix} a & b \\ c & d \end{pmatrix})f(\tau) := (b\tau + d)^{-m} f(\frac{a\tau + c}{b\tau + d}).$$

The reader is invited to compare this to our definition. We stick to it because we want to take over directly some proofs from Lang's presentation in [La1] p.181ff. Namely, one has to verify several things:

- $\mathcal{H}_m$ is complete and non-trivial.
- π fulfills the algebraic relation $\pi(gg') = \pi(g)\pi(g')$.
- π is continuous and unitary.

The algebraic relation is an immediate consequence of the facts stated as Exercise 7.10. We verify the unitarity using Exercise 7.11:

$$\|\pi(g)f\|_2^2 = \int_0^\infty \int_{-\infty}^\infty |f(g^{-1}\tau)|^2 |c\tau + d|^{-2m} y^{m-2} \, dx dy$$
$$= \int_0^\infty \int_{-\infty}^\infty |f(\tau)|^2 y^{m-2} \, dx dy$$
$$= \|f\|_2^2.$$

For the other statements we refer to Lang. The idea is to transfer the situation from the upper half plane to the unit disc where the function theory is more lucid. Since this is an important technique, we report on some facts:

The Unit Disc Model

By the *Cayley transform* c the Poincaré half plane $\mathfrak{H}$ is (biholomorphically) equivalent to the unit disc

$$\mathfrak{D} := \{\zeta \in \mathbf{C}; |\zeta| < 1\} :$$

In all cases, this bijective map sends the point i to 0 but otherwise is not uniquely fixed in the literature. We take the map

$$c : \mathfrak{H} \longrightarrow \mathfrak{D}, \ \tau \longmapsto \zeta := \frac{\tau - i}{\tau + i} = C(\tau) \text{ for } C = \begin{pmatrix} 1 & -i \\ 1 & i \end{pmatrix},$$

which has as its inverse the map given by

$$\zeta \longmapsto \tau = -i\frac{\zeta + 1}{\zeta - 1} = C^{-1}(\zeta) \text{ with } C^{-1} = (1/(2i))\begin{pmatrix} i & i \\ -1 & 1 \end{pmatrix}.$$

It is a standard fact that $G = \mathrm{SL}(2, \mathbf{R})$ is isomorphic to

$$G' := \mathrm{SU}(1,1) = \{g' = \begin{pmatrix} \alpha & \beta \\ \bar{\beta} & \bar{\alpha} \end{pmatrix}; |\alpha|^2 - |\beta|^2 = 1\}.$$

We take

$$\tilde{c} : \mathrm{SL}(2, \mathbf{R}) \longrightarrow \mathrm{SU}(1,1), \ g = \begin{pmatrix} a & b \\ c & d \end{pmatrix} \longmapsto g' := CgC^{-1} = \begin{pmatrix} \alpha & \beta \\ \bar{\beta} & \bar{\alpha} \end{pmatrix}$$

with

(7.19) $\alpha = \alpha(g) = (1/2)(a + d + i(b - c)), \beta = \beta(g) = (1/2)(a - d - i(b + c)).$

Then we have

$$r(\vartheta) := \begin{pmatrix} \cos \vartheta & \sin \vartheta \\ -\sin \vartheta & \cos \vartheta \end{pmatrix} \longmapsto \begin{pmatrix} e^{i\vartheta} & \\ & e^{-i\vartheta} \end{pmatrix} =: s(\vartheta),$$

i.e. $\tilde{c}(\mathrm{SO}(2)) = \{s(\vartheta); \vartheta \in \mathbf{R}\} = K'$, which we identify with U(1). By the linear fractional transformation $G' \times \mathfrak{D} \longrightarrow \mathfrak{D}, (g', \zeta) \longmapsto g'(\zeta)$, G' acts (transitively) on $\mathfrak{D}$. The stabilizing group of $\zeta = 0$ is U(1) $= K'$. Hence in analogy to the map

$$p : G \longrightarrow G/\mathrm{SO}(2), g \longmapsto g(i) = \tau = x + iy$$

with the Iwasawa decomposition $g = n(x)t(y)r(\vartheta)$, we have the map

$$p' : G' \longrightarrow G'/K' = \mathfrak{D}, \ g' \longmapsto g'(0) = \zeta = u + iv$$

and the commutative diagram

$$G = \mathrm{SL}(2, \mathbf{R}) \xrightarrow{\ \tilde{c}\ } G' = \mathrm{SU}(1,1)$$

$$p \downarrow \qquad\qquad\qquad\qquad \downarrow p'$$

$$\mathfrak{H} = \mathrm{SL}(2, \mathbf{R})/\mathrm{SO}(2) \xrightarrow{\ c\ } \mathfrak{D} = \mathrm{SU}(1,1)/\mathrm{U}(1)$$

with

$$\tilde{c}(n(x)t(y)) = (1/2) y^{-1/2} \begin{pmatrix} 1 + i\bar{\tau} & -1 - i\tau \\ -1 + i\bar{\tau} & 1 - i\tau \end{pmatrix}$$

and $c(\tau) = \zeta = \frac{\tau - i}{\tau + i}$.

Now, it is a rather easily acceptable fact that the space of holomorphic functions on $\mathfrak{D}$ is spanned by the monomials $\zeta^n, n \in \mathbf{N}_0$. And some calculation shows that

$$d\nu(\zeta) := \frac{dudv}{1 - |\zeta|^2}, \quad d\nu_m(\zeta) := 4^{1-m}(1 - |\zeta|^2)^m \frac{dudv}{1 - |\zeta|^2},$$

is a G'–invariant measure resp. a quasi-invariant measure on $\mathfrak{D}$ and one has an isomorphism between the corresponding spaces of holomorphic functions on $\mathfrak{H}$ and $\mathfrak{D}$

$$C_m : \mathcal{H}_m = L^2_{hol}(\mathfrak{H}, d\mu_m) \longrightarrow L^2_{hol}(\mathfrak{D}, d\nu_m), \ f \longmapsto C_m f := \tilde{f}$$

given by

$$\tilde{f}(\zeta) = f(-i\frac{\zeta + 1}{\zeta - 1})(\frac{-2i}{\zeta - 1})^m.$$

Using a convenient extension of the group action given by (7.18), this can also be written as

$$\tilde{f}(\zeta) = f(C^{-1}(\zeta)) j_m(C^{-1}, \zeta).$$

The map C_m intertwines the representation π_m from the half plane model given by (7.18) with the equivalent representation $\tilde{\pi}_m$ on the space of functions $\tilde{f} \in L^2_{hol}(\mathfrak{D}, d\nu_m)$ given by the prescription

$$\tilde{\pi}_m(g)\tilde{f}(\zeta) := \tilde{f}(\tilde{c}(g)^{-1}(\zeta)) j_m(\tilde{c}(g)^{-1}, \zeta).$$

We call this the *unit disc model*. By the way, if one replaces $\tilde{c}(g)$ by $g' \in G'$, one also has a representation of $G' = \mathrm{SU}(1,1)$.

Already above, we stated the (plausible) fact that $L^2_{hol}(\mathfrak{D}, d\nu_m)$ is spanned by functions $\tilde{f}_n$ with $\tilde{f}_n(\zeta) = \zeta^n, n \in \mathbf{N}_0$. As to be seen easily, via C_m, these functions correspond to functions f_n living on $\mathfrak{H}$ with

$$f_n(\tau) := (\frac{\tau - i}{\tau + i})^n (\tau + i)^{-m}.$$

The Group Model

Intertwined with these two equivalent models, there is a third one, which we will call the *group model* because its representation space consists of functions living on the group $G = \mathrm{SL}(2, \mathbf{R})$. We go back to the map $\tilde{c} : G \longrightarrow \tilde{G}$. In the formulae (7.19) one has the elements of $\tilde{G}$ as functions of $g \in G$. We introduce the parameters x, y, ϑ describing g in the Iwasawa decomposition as in (7.12) and get by a small calculation

$$
\begin{aligned}
\alpha(g) &= (1/2)(a + d + i(b - c)) = (1/2)y^{-1/2}e^{i\vartheta}i(\bar{\tau} - i), \\
\beta(g) &= (1/2)(a - d - i(b + c)) = (1/2)y^{-1/2}e^{-i\vartheta}(-i)(\tau - i),
\end{aligned}
$$

and by another one

$$
\phi_n(g) := \beta(g)^n \bar{\alpha}(g)^{-(n+m)}(1/(2i))^m = y^{m/2}e^{im\vartheta}f_n(\tau).
$$

Inspired by this formula, we define a *lift* φ_m, which associates functions ϕ living on G to functions f on $\mathfrak{H}$ via

$$
f(\tau) \longmapsto \phi_f(g) = (\varphi_m f)(g) := f(g(i))j_m(g, i) = f(\tau)y^{m/2}e^{im\vartheta}.
$$

Here we use the fact that one has $j_m(g, i) = (c\tau + d)^{-m} = y^{m/2}e^{im\vartheta}$ if g is fixed by our Iwasawa parametrization (7.12). With the help of the cocycle condition for the automorphic factor (or multiplier) j_m, we deduce for these lifted functions

$$
\begin{aligned}
\phi(g_0^{-1}g) &= f(g_0^{-1}g(i))j_m(g_0^{-1}g, i) \\
&= f(g_0^{-1}\tau)j_m(g_0^{-1}, g(i))j_m(g, i) \\
&= (\pi_m(g_0)f)(g(i))j_m(g, i).
\end{aligned}
$$

This shows that the lift φ_m intertwines (algebraically) the representation π_m on the space $L^2_{hol}(\mathfrak{H}, d\mu_m)$ with the left regular representation λ on the space of functions ϕ living on G and coming out by the lift φ_m. In [La1] p.180 one finds a proof that the functions ϕ_n are square integrable with respect to the (biinvariant) measure dg on G given (again in our Iwasawa parameters) by

$$
dg = y^{-2}dx\,dy\,d\vartheta.
$$

We call this representation, given by left translation on the space spanned by the functions ϕ_n living on the group, the *group model*.

Already earlier, while introducing the other two models, we could have identified the class of our representation with the one of the representation π_m^- of highest weight $-m$ from the list of (possible) irreducible unitary representations obtained by infinitesimal considerations in 6.4. We will do this now. The reader is invited to watch very closely because Lang in [La1] p.184 (erroneously if our reasoning is correct) identifies it as a lowest weight representation.

There are several ways to proceed: For the isomorphism $\tilde{c} : G = \mathrm{SL}(2, \mathbf{R}) \longrightarrow \mathrm{SU}(1, 1)$ one has $\tilde{c}(r(\vartheta)) = s(\vartheta), \tilde{c}(r(\vartheta)^{-1}g)) = s(-\vartheta)\tilde{c}(g)$. Hence we deduce

$$
\alpha(r(\vartheta)^{-1}g) = e^{-i\vartheta}\alpha(g), \quad \beta(r(\vartheta)^{-1}g) = e^{-i\vartheta}\beta(g),
$$

and

$$
\phi_n(r(\vartheta)^{-1}g) = e^{-(2n+m)i\vartheta}\phi_n,
$$

i.e. our representation has the weights $-m, -(m+2), -(m+4), \ldots$. We compare this with our results from 6.4 and conclude that we have a representation equivalent to the highest weight representation π_m^-. But we can do more: In (7.15) we determined the operators $\mathcal{R}_F, \mathcal{R}_G, \mathcal{R}_H$ and $\mathcal{R}_Z, \mathcal{R}_{X_\pm}$. These operators are just the operators producing the derived representation of the left regular representation λ. Thus, by a small calculation (**Exercise 7.14**), one can verify the formulae

$$\mathcal{R}_Z \phi_n = -(2n+m)\phi_n, \ \mathcal{R}_{X_+}\phi_n = -n\phi_{n+1}, \ \mathcal{R}_{X_-}\phi_n = (n+m)\phi_{n+1}.$$

We can get corresponding formulae by determination of the derived representations of the other models. We propose the following

Exercise 7.15: Verify the formulae

$$\begin{array}{rcl}
\mathcal{R}_Z^m & = & \hat{\pi}_m(Z) & = & i((1+\tau^2)\partial_\tau + m\tau), \\
\mathcal{R}_{X_\pm}^m & = & \hat{\pi}_m(X_\pm) & = & \pm(i/2)((i\pm\tau)^2\partial_\tau + m(\tau\pm i)),
\end{array}$$

and

$$\begin{array}{rcl}
\mathcal{R}_Z^m f_n & = & -(m+2n)f_n, \\
\mathcal{R}_{X_+}^m f_n & = & -nf_{n-1}, \\
\mathcal{R}_{X_-}^m f_n & = & (n+m)f_{n+1}.
\end{array}$$

Similarly the representations π_m^+ of lowest weight can be realized by holomorphic functions on the *lower half plane* $\bar{\mathfrak{H}} = \{\tau = x+iy \in \mathbf{C}, y < 0\}$ or by antiholomorphic functions f, i.e. with $(\partial_x - i\partial_y)f = 0$.

Exercise 7.16: Determine the class of the representation $\check{\pi}_m$ taken over from Knapp in the Remark 7.5 above.

The Holomorphically Induced Model

Finally, we want to show how these realizations of the discrete series representations fit into the discussion of the induction procedure. Namely one can obtain them as subrepresentations of suitable induced representations:

We start with the subgroup $H = K = SO(2)$ and its representation π_0 given by

$$\pi_0(r(\vartheta)) = \chi_k(r(\vartheta)) = e^{ik\vartheta}, \ k \in \mathbf{Z}.$$

The Iwasawa decomposition can be understood as a Mackey decomposition

$$g = s(\tau)r(\vartheta), \ s(\tau) := n(x)t(y),$$

and we have our standard projection $p : G \longrightarrow G/SO(2) = \mathfrak{H}, \ g \longmapsto g(i) = x + iy = \tau$ with section $s : \mathfrak{H} \longrightarrow G, \ \tau \longmapsto s(\tau)$. Then the induced representation $\pi_k^{ind} := ind_{SO(2)}^G \chi_k$ is given by (inverse) left translation λ on the completion $\mathcal{H}^{\chi_k}$ of the space of smooth functions ϕ on G with functional equation

$$\phi(gr(\vartheta)) = e^{-ik\vartheta}\phi(g), \quad \text{for all } g \in G, r(\vartheta) \in SO(2)$$

and the norm condition

$$\|\phi\|_2^2 = \int_{\mathfrak{H}} |\phi(s(\tau))|^2 \ y^{-2}dxdy < \infty.$$

From the functional equation we learn that the functions ϕ in the representation space are all of the type

$$\phi(g) = e^{-ik\vartheta} F(x, y)$$

with a function F depending only on x, y if g is given in our Iwasawa parametrization. Comparing this to the functions establishing the group model above, we see that we have to find a condition, which further restricts these functions F. Here we remember from the infinitesimal considerations in 6.4 that functions spanning the space of a discrete series representation $\pi_k^{\pm}$ have to be eigenfunctions of the Casimir element

$$\Omega = X_+ X_- + X_- X_+ + (1/2)Z^2.$$

In (7.17) we already determined that Ω acts on smooth functions ϕ living on G as the differential operator

$$\Delta = \mathcal{R}_\Omega = \mathcal{L}_\Omega = 4y^2(\partial_x^2 + \partial_y^2) - 4y\partial_x\partial_\vartheta.$$

The preceding discussion in mind, we try the ansatz

$$\phi(g) = e^{-ik\vartheta} y^{-k/2} f(x, y).$$

We get

$$\Delta\phi = [(4y^2(\partial_x^2 + \partial_y^2) + 4yik(\partial_x + i\partial_y) + (k/2)((k/2) + 1))f]y^{-k/2}e^{-ik\vartheta},$$

i.e. we have

$$\Delta\phi = (k/2)((k/2) + 1)\phi$$

if

$$\Delta_k f := 4y^2(\partial_x^2 + \partial_y^2) + 4yik(\partial_x + i\partial_y)f = 0,$$

and this is obviously the case if f fulfills the equation

$$\partial_{\bar{z}} f := (1/2)(\partial_x + i\partial_y)f = 0,$$

i.e. if f is holomorphic. Hence for $k = -m$, we have recovered the lift of the representation π_m as a subrepresentation of our induced representation π_k^{ind}.
There is a shorter way to single out in $\mathcal{H}^{\chi_k}$ a subspace carrying π_m: To the two conditions characterizing the space of the induced representation, namely the functional equation and the norm condition, we add simply the condition

$$\mathcal{L}_{X_-}\phi = 0.$$

This comes out as follows: From (7.16) we have

$$\mathcal{L}_{X_-} = -(i/2)e^{-2i\vartheta}(2y(\partial_x + i\partial_y) - \partial_\vartheta).$$

For $\phi(g) = e^{-ik\vartheta} y^{-k/2} f(x, y)$ this again leads to the holomorphicity condition $\partial_{\bar{z}} f = 0$.

Remark 7.6: This is a special example of a general procedure propagated in particular by W. Schmid and called *holomorphic induction*.

Remark 7.7: Here we began by stating the prescription for the representation π_m in a space of functions living on the homogeneous space $\mathfrak{H} = G/SO(2)$. It is a nice exercise to rediscover this formula by our general procedure to change from the First to the Second Realization of an induced representation:

We start with Mackey's decomposition as above, $g = s(\tau)r(\vartheta)$, and for $g_0^{-1} = \begin{pmatrix} a & b \\ c & d \end{pmatrix}$ the Master Equation, using $s(\tau) = n(x)t(y)$, states

$$g_0^{-1}s(\tau) = s(g_0^{-1}(\tau))r(\vartheta^*)$$

with

$$e^{i\vartheta^*} = \sqrt{\frac{c\tau + d}{c\bar{\tau} + d}}.$$

Moreover one has the standard formulae

$$\frac{d\mu(g_0^{-1}(\tau))}{d\mu(\tau)} = \frac{1}{(c\tau + d)^2}, \ \mathrm{Im}\,(g_0^{-1}(\tau)) = \frac{y}{\mid c\tau + d \mid^2},$$

such that the left regular representation $\pi(g_0)\phi(g) = \phi(g_0^{-1}g)$ in the Second Realization with $F(\tau) = \phi(s(\tau))$ translates to the prescription

$$\pi(g_0)F(\tau) = (\frac{d\mu(g_0^{-1}(\tau))}{d\mu(\tau)})^{1/2}\pi_0^{-1}(r(\vartheta^*))F(g_0^{-1}(\tau)).$$

With $\pi_0 = \chi_k$ and $F(\tau) = y^{-k/2}f(\tau)$ for $k = -m$ by a small calculation this leads to the prescription (7.18)

$$\pi_m(g_0)f(\tau) = f(g_0^{-1}(\tau))(c\tau + d)^{-m}.$$

Remark 7.8: The functions ϕ in the representation space of the Ffirst Realization and the *matrix coefficients* $< \pi(g)\phi, \phi' >$ are square-integrable for $m \geq 2$. This is a general criterium to distinguish discrete series representations from the other representations.

The representations $\pi_1^{\pm}$ are called *mock discrete series representations* or *limits of discrete series*. They are realized by the same prescriptions as the $\pi_m^{\pm}$ with $m \geq 0$ but here the space of holomorphic functions f on $\mathfrak{H}$ for π_1^{-} gets the new norm

$$\|f\|^2 = \sup_{y>0} \int_{\mathbf{R}} \mid f(x + iy) \mid^2 dx,$$

and analoguously for the other case.

The courageous reader may be tempted to apply this principle of holomorphic induction to reconstruct as **Exercise 7.17** the irreducible unitary representations of $G = \mathrm{SU}(2)$.

3. The Complementary Series

For the sake of completeness, we indicate here also a model, but only briefly, as done in most sources. (The best idea is to go back to Bargmann's original paper [Ba].) We take up again the prescriptions as in the construction of the principal series but to get a model for $\pi_s, 0 < s < 1$ we change the norm on the space of complex valued functions on $\mathbf{R}$ and introduce

$$\|f\|_s^2 := \int_{\mathbf{R}}\int_{\mathbf{R}} \frac{\overline{f(x)}f(y)}{\mid x - y \mid^{1-s}}dxdy.$$

For $g^{-1} = \begin{pmatrix} a & b \\ c & d \end{pmatrix}$ the prescription is here

$$\pi_s(g)f(x) = |\, cx + d \,|^{-1-s} \, f(g^{-1}(x)).$$

Some more models can be found in [vD] p.465 and [GGP] p.36.

A Final General Remark

We close this section by pointing out to the *Subrepresentation Theorem*, which goes back to Harish–Chandra and roughly states (for a precise formulation and proof see [Kn] p.238) that a certain class of representations, the *admissible representations*, may be realized as subrepresentations of parabolically induced representations starting eventually by non-unitary representations of the subgroup. Admissible representations are very important, in particular in number theory. An admissible representation (π, V) of G is (in first approximation) characterized as follows ([Ge] p.55): Let K be a maximal compact subgroup of G. For an irreducible representation χ of K, denote by $V(\chi)$ the subspace of vectors in V, which transform according some multiple of χ. Then one has a decomposition

$$V = \oplus_\chi V(\chi)$$

where each $V(\chi)$ is finite dimensional.

Irreducible unitary representations of reductive groups are admissible. So all representations in this text have this property (with exception of the representations of the (non-reductive) Jacobi group appearing later).

The interested reader can exemplify the statement of the subrepresentation theorem in the case of $G = SL(2, \mathbf{R})$ by constructing as **Exercise 7.18** another model of the discrete series representation by starting the induction procedure from

$$H = \{h = n(x)t(y); x, y \in \mathbf{R}, y > 0\}, \ \pi_0(h) = y^s, \ s \in \mathbf{C}.$$

7.3 Unitary Representations of SL(2,**C**) and of the Lorentz Group

The representations of the group $G = SL(2, \mathbf{C})$ are of particular importance for physicists because this group is the double cover of the identity component of the complete homogeneous Lorentz group in four dimensions (to which we will restrict ourselves in this section). At first, we will discuss the representations of $SL(2, \mathbf{C})$ and then the homomorphism into the Lorentz group.

As central fact we have:

Theorem 7.5: For $G = SL(2, \mathbf{C})$ there are (up to equivalence) only two series of irreducible unitary representations:
– the *unitary principal series* $\pi_{k,iv}$ depending on two parameters $k \in \mathbf{Z}, v \in \mathbf{R}$, and
– the *complementary series* $\pi_{0,u}$ depending on a real parameter u with $0 < u < 2$.
Among these the only equivalences are $\pi_{k,iv} \simeq \pi_{-k,-iv}$.

Thus, in contrast to the smaller group $SL(2, \mathbf{R})$, we have no discrete series representations. We reproduce models for these representations from [Kn] p. 33/4 and discuss their construction via the induction procedure afterwards.

As representation space we take the space $\mathcal{H} = L^2(\mathbf{C})$ of square-integrable complex functions f on $\mathbf{C}$ with measure $dzd\bar{z}$. For $z \in \mathbf{C}$, $f \in \mathcal{H}$, the prescription for the unitary principal series representation is given by

$$(7.20) \qquad \pi_{k,iv}((\begin{smallmatrix} a & b \\ c & d \end{smallmatrix}))f(z) := | -bz + d |^{-2-iv} \left(\frac{-bz + d}{| -bz + d |} \right)^{-k} f\left(\frac{az - c}{-bz + d} \right).$$

Here, the notion of the fractional linear transformation used in the last section as a map $SL(2, \mathbf{R}) \times \mathfrak{H} \longrightarrow \mathfrak{H}$, $(g, \tau) \longmapsto g(\tau)$, is naturally extended to a map

$$SL(2, \mathbf{C}) \times \{\mathbf{C} \cup \infty\} \longrightarrow \{\mathbf{C} \cup \infty\}$$

with

$$\begin{aligned} (\begin{smallmatrix} a & b \\ c & d \end{smallmatrix})(z) &:= \tfrac{az+b}{cz+d} && \text{for } z \in \mathbf{C},\ cz + d \neq 0, \\ &:= \infty && \text{for } z \in \mathbf{C},\ cz + d = 0, \\ &:= \tfrac{a}{c} && \text{for } z = \infty. \end{aligned}$$

In a similar way, for $k \in \mathbf{Z}$, $w = u + iv \in \mathbf{C}$, there is a *non–unitary principal series* $\pi_{k,w}$ given by the prescription

$$\pi_{k,w}((\begin{smallmatrix} a & b \\ c & d \end{smallmatrix}))f(z) := | -bz + d |^{-2-w} \left(\frac{-bz + d}{| -bz + d |} \right)^{-k} f\left(\frac{az - c}{-bz + d} \right)$$

for functions f in the space $L^2(\mathbf{C}, d\mu_w)$ with $d\mu_w(z) = (1 + | z |^2)^u dzd\bar{z}$. However, in the special case $k = 0$, $w = u \in \mathbf{R}$, $0 < u < 2$, the representation $\pi_u := \pi_{0,u}$ is unitary if the representation space is provided with the inner product

$$< f, \tilde{f} > := \int_{\mathbf{C}} \int_{\mathbf{C}} \frac{\overline{f(z)} \tilde{f}(\zeta)}{| z - \zeta |^{2-u}} dzd\zeta.$$

This is our model for the *complementary series*.

As to the construction, these representations are induced representations where one takes the subgroup

$$H := \{h = (\begin{smallmatrix} \alpha & \beta \\ & \alpha^{-1} \end{smallmatrix});\ \alpha, \beta \in \mathbf{C}, \alpha \neq 0\}$$

and the representation π_0 with $\pi_0(h) := | \alpha |^w \left(\frac{\alpha}{|\alpha|} \right)^k$. We have the Mackey decomposition

$$g = (\begin{smallmatrix} a & b \\ c & d \end{smallmatrix}) = s(\tau)h$$

where (as in our discussion of the principal series for $SL(2, \mathbf{R})$ in the previous section) it suffices to treat the case $a \neq 0$, such that one has

$$s(\tau) = (\begin{smallmatrix} 1 & \\ z & 1 \end{smallmatrix}),\ z := b/a \text{ and } h = (\begin{smallmatrix} a & b \\ & a^{-1} \end{smallmatrix}),$$

and the Master Equation $g^{-1}s(z) = s(z^*)h^*$ with

$$z^* = \frac{az - c}{-bz + d}, \quad h^* = \begin{pmatrix} -bz + d & -b \\ & (-bz + d)^{-1} \end{pmatrix}.$$

Moreover, we have

$$dz^* d\bar{z}^* = |-bz + d|^{-4} dz d\bar{z}.$$

Hence, since the induced representation in left regular version in its Second Realization on the space $\mathcal{H} = L^2(\mathbf{C})$ has the general prescription

$$\pi(g)f(z) = \sqrt{\frac{d\mu(z^*)}{d\mu(z)}} \pi_0(h^*)f(z^*), \quad \text{for all } f \in \mathcal{H},$$

with $w = iv$ this leads to the formula (7.20).

In [Kn] p.33/4 Knapp proves the irreducibility of $\pi_{k,iv}$ using Schur's Lemma in the unitary version cited in 3.2. One could also do this (and verify that really we only have the above mentioned irreducible unitary representations) by infinitesimal considerations as in our discussion of SL(2, **R**). But since $\mathfrak{sl}(2, \mathbf{C})$ has real dimension 6, this is leads to considerable non-trivial computations (to be found for instance in Naimark [Na] or Cornwell [Co] p.668ff in the framework of the discusssion of the Lorentz group).

The interest in the representations of $G = \text{SL}(2, \mathbf{C})$ is enlarged by the fact that they immediately lead to projective representations of the *proper orthochronous Lorentz group*, resp. those, which are trivial on $-E_2$, to genuine representations, the same way as in our discussion of SU(2) and SO(3) in 4.3. Before we describe the relation between SL(2, **C**) and the Lorentz group we have to fix some notation.

We write

$$G^L := \{A \in \text{GL}(4, \mathbf{R}); {}^t A D_{3,1} A = D_{3,1} = \begin{pmatrix} 1 & & & \\ & 1 & & \\ & & 1 & \\ & & & -1 \end{pmatrix}\}$$

for the *complete homogeneous Lorentz group*, i.e. the group of linear transformations, which leave invariant the *Minkowski metric* from the theory of special relativity

$$ds^2 = dx_1^2 + dx_2^2 + dx_3^2 - dx_4^2.$$

One may be tempted to replace (x_1, x_2, x_3, x_4) by (x, y, z, ct), (c the velocity of light) and put

$$ds^2 = c^2 dt^2 - dx^2 - dy^2 - dz^2,$$

as often done in the physics literature, but we will not do this here. In our former notation from 0.1 we also have $G^L = \text{O}(3, 1)$ and $G^L_+ = \text{SO}(3, 1)$ for the *proper homogeneous Lorentz group*, i.e. the invariant subgroup consisting of matrices A with $\det A = 1$. We write the matrices in block form

$$A = \begin{pmatrix} B & x \\ {}^t y & a \end{pmatrix}, \quad B \in \text{GL}(3, \mathbf{R}), x, y \in \mathbf{R}^3, a \in \mathbf{R}.$$

G^L_+ has two connected components distinguished by $a > 0$ and $a < 0$ since one has

$$\begin{pmatrix} {}^tB & y \\ {}^tx & a \end{pmatrix}\begin{pmatrix} E_3 & \\ & -1 \end{pmatrix}\begin{pmatrix} B & x \\ {}^ty & a \end{pmatrix} = \begin{pmatrix} {}^tBB - y^ty & {}^tBx - ya \\ {}^txB - a^ty & {}^txx - a^2 \end{pmatrix},$$

and the condition that this should be equal to $\begin{pmatrix} E_3 & \\ & -1 \end{pmatrix}$ asks for

(7.21) $\qquad\qquad {}^tBB - y^ty = E_3,\ {}^tBx - ya = 0,\ {}^txx - a^2 = -1,$

and, hence, in particular $a^2 \geq 1$.

The subgroup $G^{L^+}_+$ with $a > 0$ is called the *proper orthochronous homogeneous Lorentz group*. It is the connected component of the identity in $SO(3, 1)$. We will abbreviate this here to *Lorentz group* and write $G_0 := SO(3,1)^0$ for it. As easily to be seen, one can imbed $SO(3)$ into G_0 by putting

$$SO(3) \ni B \longmapsto \tilde{B} := \begin{pmatrix} B & \\ & 1 \end{pmatrix}.$$

These elements are called *Lorentz rotations*. Moreover, we introduce the *Lorentz boosts*, which are of the type

$$B(t) = \begin{pmatrix} \cosh t & & & \sinh t \\ & 1 & & \\ & & 1 & \\ \sinh t & & & \cosh t \end{pmatrix}, \quad t \in \mathbf{R}.$$

The structure of G_0 is clarified by the following remark.

Proposition 7.2: For every $A \in G_0$ there are $R, S \in SO(3)$ and $t \in \mathbf{R}$ such that

$$A = \tilde{R}B(t)\tilde{S},$$

i.e. G_0 is generated by Lorentz rotations and Lorentz boosts.

A proof can be found in any source treating the Lorentz group, for instance in [He] p.75. Our main fact in this context is the next theorem.

Theorem 7.6: There is a surjective homomorphism

$$\rho : G = SL(2, \mathbf{C}) \longrightarrow G_0 = SO(3, 1)^0$$

with $\ker \rho = \{\pm E_2\}$.

Proof (sketch): We put

$$V := \{X \in M_2(\mathbf{C}); X = {}^t\bar{X} =: X^*\} = \left\{\begin{pmatrix} \alpha & z \\ \bar{z} & \beta \end{pmatrix}; \alpha, \beta \in \mathbf{R}, z \in \mathbf{C}\right\}$$

and look at the map

$$\varphi : \mathbf{R}^4 \longrightarrow V,\ v = {}^t(a, b, c, d) \longmapsto \begin{pmatrix} -a+d & b+ic \\ b-ic & a+d \end{pmatrix}.$$

For $X = \varphi(v), Y = \varphi(w)$ this map transforms the Minkowski (quasi-)inner product

$$< v, w > = aa' + bb' + cc' - dd'$$

into the corresponding one defined on V by

$$\sigma : V \times V \longrightarrow \mathbf{R}, \ (X, Y) \longmapsto -(1/2)(\det(X + Y) - \det X - \det Y)$$

as easily to be verified by a small calculation. Hence our φ induces a bijection between $O(3, 1)$ and $O(V, \sigma)$, which can be recognized as an isomorphism and we take this to identify both sets.
We define a map

$$\rho : \mathrm{SL}(2, \mathbf{C}) \longrightarrow \mathrm{GL}(V), \ A \longmapsto \rho(A) \text{ with } \rho(A)X := AXA^* \text{ for } X \in V,$$

which obviously is a homomorphism. We have

$$\ker \rho = \{A; \ AXA^* = X \quad \text{for all } X \in V\}$$

and deduce $\ker \rho = \{\pm E_2\}$ by evaluating the defining condition for $X = E_2, \begin{pmatrix} 1 & \\ & -1 \end{pmatrix}$,

and $\begin{pmatrix} & 1 \\ 1 & \end{pmatrix}$. The main task is to prove the surjectivity of ρ. We go back to the proof of Theorem 4.7 in 4.3 and use the fact that G_0 is generated in $\mathrm{GL}(V)$ by elements $\tilde{S}_2(\alpha), \tilde{S}_3(\alpha)$ and $B(t)$. By some computation, one verifies that these are images under ρ of respectively $r(\alpha/2), s(\alpha/2)$ and $s(-it/2)$.

This theorem justifies the statement made above that the representations π of $\mathrm{SL}(2, \mathbf{C})$ with $\pi(-E_2) = id$ create the representations of $\mathrm{SO}(3, 1)^0$. E.g., in [Co] p.671ff one finds a discussion of the finite-dimensional (non-unitary) representations of the Lorentz group.

7.4 Unitary Representations of Semidirect Products

The semidirect products form a class of groups whose importance is also enhanced by its applications in physics as the Euclidean and Poincaré groups are prominent examples, which fit into the following general concept:

We take two groups G_0 and N and a (left) action of G_0 on N given by

$$\rho : G_0 \times N \longrightarrow N, \ (a, n) \longmapsto \rho(a)n$$

with the additional condition that every $\rho(a)$ is an automorphim of N, i.e. $\rho(a) \in \mathrm{Aut}\, N$ for all $a \in G_0$. Then we define the associated *semidirect product* $G := G_0 \ltimes N$ as the set of pairs $g = (a, n), a \in G_0, n \in N$ provided with the composition law

(7.22) $$gg' := (aa', n\rho(a)n').$$

G comes out as a group with neutral element e consisting of the pair of neutral elements in G_0 and in N. The inverse to $g = (a, n)$ is $g^{-1} := (a^{-1}, \rho(a^{-1})n^{-1})$. (Verification of the associative law as **Exercise 7.19.**)

Examples:

Example 7.4: The *Euclidean group* $G^E(n)$ in n dimensions: We take $G_0 = SO(n)$ and $N = \mathbf{R}^n$. For $A \in SO(n), x \in \mathbf{R}^n$, the action ρ is given by matrix multiplication $(A, x) \longmapsto Ax$. We denote the semidirect product $SO(n) \ltimes \mathbf{R}^n$ by $G^E(n)$ and write this again as a matrix group

$$G^E(n) = \{g = (A, a) = \begin{pmatrix} A & a \\ & 1 \end{pmatrix}; \; A \in SO(n), a \in \mathbf{R}^n\}.$$

Here the matrix multiplication in $M_{n+1}(\mathbf{R})$

$$gg' = \begin{pmatrix} A & a \\ & 1 \end{pmatrix}\begin{pmatrix} A' & a' \\ & 1 \end{pmatrix} = \begin{pmatrix} AA' & a + Aa' \\ & 1 \end{pmatrix}$$

reflects exactly the composition law (7.22) leading in this case to $gg' = (AA', a + Aa')$. This law comes out very naturally if one asks for the composition of the affine transformations $x \longmapsto Ax + a$, which preserve the euclidean norm $\|x\|^2 = x_1^2 + \cdots + x_n^2$.

Example 7.5: The *Poincaré group* $G^P(n)$ in n dimensions: Paralell to the case above we take $G_0 = SO(n-1, 1)$ and $N = \mathbf{R}^n$ and get

$$G^P(n) = \{g = (A, a) = \begin{pmatrix} A & a \\ & 1 \end{pmatrix}, \; A \in SO(n-1, 1), a \in \mathbf{R}^n\}$$

with the same composition law. This is the invariance group of the affine transformations preserving the Minkowski (pseudo-)norm $\|x\|^2 = x_1^2 + \cdots + x_{n-1}^2 - x_n^2$.

Example 7.6: Interprete the group $B := \{g = \begin{pmatrix} a & b \\ & a^{-1} \end{pmatrix}; \; a, b \in \mathbf{R}, a > 0\}$ as a semidirect product (**Exercise 7.20**).

Example 7.7: Interprete the Heisenberg group

$$\text{Heis}'(\mathbf{R}) := \{g = \begin{pmatrix} 1 & x & z \\ & 1 & y \\ & & 1 \end{pmatrix}; \; x, y, z \in \mathbf{R}\}$$

as semidirect product of $G_0 = \mathbf{R}$ acting on $N = \mathbf{R}^2$. Can you do the same with $Heis(\mathbf{R})$ as realized in 0.3? (**Exercise 7.21**)

Example 7.8: The *Jacobi group* G^J is the semidirect product of $G_0 = SL(2, \mathbf{R})$ acting on the Heisenberg group $N = \text{Heis}(\mathbf{R})$ via

$$(\begin{pmatrix} a & b \\ c & d \end{pmatrix}, (\lambda, \mu, \kappa)) \longmapsto (d\lambda - c\mu, -b\lambda + a\mu, \kappa).$$

This group can be presented in several ways as a matrix group (see our Section 8.5 or for instance [EZ] or [BeS] p.1).

With exception of the last one, in all these examples the group N is abelian. In the sequel we will restrict our treatment to this case as things become easier under this hypothesis and use additive notation in the second item of the pair $g = (a, n)$ (as we already did in the examples). Whoever is interested in more general cases is refered to Mackey's original papers or the treatment of the representation theory of the Jacobi group in [BeS].

Remark 7.9: We use the following embeddings

$$G_0 \longrightarrow G = G_0 \ltimes N, \ a \longmapsto (a, 0) =: a$$

and

$$N \longrightarrow G = G_0 \ltimes N, \ n \longmapsto (e, n) =: n.$$

Here one has to be a bit careful because we have $n \cdot a = (e, n)(a, 0) = (a, n) = g$ but $a \cdot n = (a, 0)(e, n) = (a, \rho(a)n) \neq (a, n) = g$ in general. There is the useful relation

$$(a, 0)(e, n)(a^{-1}, 0) = (e, \rho(a)n),$$

which, using our embeddings, can be stated as $ana^{-1} = \rho(a)n$.

Example 7.9: We take

$$
\begin{aligned}
G &:= \text{Heis}'(\mathbf{R}) = \{g = (x, y, z); x, y, z \in \mathbf{R}\}, \\
G_0 &:= \{a = (x, 0, 0); x \in \mathbf{R}\},
\end{aligned}
$$

and

$$N := \{n = (0, y, z); \ y, z \in \mathbf{R}\}. \qquad \models$$

Using the multiplication law $gg' = (x + x', y + y', z + z' + xy')$, one calculates

$$ana^{-1} = (0, y, z + xy) = \rho(a)n$$

and this is consistent with the usual semidirect product composition law (7.22) given by $gg' = (a, n)(a', n') = (aa', n\rho(a)n')$.

Example 7.10: For a presentation as semidirect product of the group

$$B := \{g = \begin{pmatrix} a & b \\ & a^{-1} \end{pmatrix} =: (a, b); \ a, b \in \mathbf{R}, a > 0\}$$

one may be tempted to try $G_0 := \{a = (a, 0); a > 0\}$ and $N := \{n = (1, b); b \in \mathbf{R}\}$. From matrix multiplication we have $gg' = (aa', ab' + ba'^{-1})$ and (using the embedding introduced above) $ana^{-1} = (1, a^2 b) = \rho(a)n$. But then the composition law would give $gg' = (a, b)(a', b') = (aa', n\rho(a)n') = (aa', b + a^2 b')$ which is **not** the multiplication law in this group. To make things consistent (and thus solve the problem from Exercise 7.20), one constructs the semidirect product $G = \mathbf{R}_{>0} \ltimes \mathbf{R} = \{[a, x], a, x \in \mathbf{R}, x > 0\}$ with $\rho(a)x = a^2 x$ and defines an isomorphism

$$\varphi : B \longrightarrow G, \ (a, b) \longmapsto [a, ab].$$

The agreable thing about these semidirect products $G = G_0 \ltimes N$, N abelian, is the fact that Mackey's Theory provides an explicit way to construct their irreducible unitary representations as induced representations. We describe this method without going everywhere into the intricacies of the proofs, which rely on some serious functional analysis:

We start by an easy but important fact relating the representations of $G = G_0 \ltimes N$ with G_0 acting on N by ρ to those of its constituents G_0 and N.

Proposition 7.3: Let π be a representation of G and π', π'' the restrictions of π to N resp. G_0, i.e $\pi'(n) = \pi((e, n))$ for $n \in N$ and $\pi''(a) = \pi((a, 0))$ for $a \in G_0$. Then π' and π'' completely determine π and are related by the condition

$$(7.23) \qquad \pi'(\rho(a)n) = \pi''(a)\pi'(n)\pi'(a^{-1}) \quad \text{for all } a \in G_0, n \in N.$$

Proof: π is determined by π' and π'' as one has

$$\pi((a, n)) = \pi((e, n)(a, 0)) = \pi((e, n))\pi((a, e)) = \pi'(n)\pi''(a).$$

And from the composition law in G

$$(a_1, n_1)(a_2, n_2) = (a_1 a_2, n_1 + \rho(a_1)n_2)$$

we deduce

$$\pi'(n_1)\pi''(a_1)\pi'(n_2)\pi''(a_2) = \pi'(n_1)\pi'(\rho(a_1)n_2)\pi''(a_1)\pi''(a_2)$$

and with $a_1 = a, n_2 = n$ this leads immediately to (7.23).

Hence to find representations of G, one has to find representations of G_0 and N fulfilling (7.23). This condition is important in the context of Mackey's *system of imprimitivity*, which is the main tool in the proof that the construction we describe in the sequel leads to a way to construct *all irreducible* unitary representations of semidirect products. We essentially follow [BR] p.503ff and [Ki] p.195ff (where the more general case of representations of *group extensions* is treated). A decisive element in the construction comes from the fact that the action ρ of G_0 on N leads also to an action on the unitary dual $\hat{N}$ of N, which in our abelian case consists of (classes of) characters $\hat{n}$ of N. Namely, we have

$$G_0 \times \hat{N} \longrightarrow \hat{N}, \quad (a, \hat{n}) \longmapsto a\hat{n}$$

given by $a\hat{n}(n) := \hat{n}(\rho(a^{-1})n)$. Since the (abelian) group N acts (trivially) on $\hat{N}$ by conjugation, i.e. via $n_0\hat{n}(n) := \hat{n}(n_0^{-1}nn_0)$, one has an action of the semidirect product G on $\hat{N}$. The G–orbit of $\hat{n} \in \hat{N}$ is written as

$$\hat{O}_{\hat{n}} := G\hat{n}.$$

Mackey's construction works under the assumption that the semidirect product is *regular*. All the groups we treat as examples are of this type. Hence, since we will not give complete proofs anyway, a not so ambitious reader at first sight may skip the following rather technical condition (from [BR] p.506).

Definition 7.1: We say that G is a *regular* semidirect product of N and G_0 if $\hat{N}$ contains a countable family $Z_1, Z_2, \ldots$ of Borel subsets, each a union of G–orbits, such that every orbit in $\hat{N}$ is the intersection of the members of a subfamily $Z_{n_1}, Z_{n_2}, \ldots$ containing that orbit. Without loss of generality, we can suppose that the intersection of a finite number of Z_i is an element of the family (Z_i). This is equivalent to the assumption that any orbit is the limit of a deceasing subsequence of (Z_i).

One also has a slightly different looking condition for regularity used by Mackey ([Ma] p.77):
There is a subset $J \subset \hat{N}$, which meets each $G-$orbit $\hat{O}$ exactly once and is an analytic subset.

Now as our central prescription for the construction of representations in this section, we take over from [BR] Theorem 4 on p.508:

Theorem 7.7: Let $G = G_0 \ltimes N$ be a regular semidirect product of separable, locally compact groups G_0 and N, N abelian. Then every irreducible unitary representation of G is unitarily equivalent to an induced representation $\pi = ind_H^G \pi_0$, which can be constructed as follows.

– **Step 1:** Determine the set $\hat{N}$ of characters $\hat{n}$ of N, i.e. the dual group of N.
– **Step 2:** Classify and determine the G_0-orbits $\hat{O}$ in $\hat{N}$.
– **Step 3:** Choose an element $\hat{n}_0$ in each orbit $\hat{O}$ and determine the stabilizing group $G_{0\hat{n}_0} =: H_0$. (This group is called *the little group* in the physics literature.)
– **Step 4:** Determine the unitary dual $\hat{H}_0$ of H_0.
– **Step 5:** From every element of $\hat{H}_0$ take a representation π_1 of H_0 and extend it to a representation π_0 of $H := H_0 \ltimes N$ by

$$\pi_0((a, n)) := \hat{n}_0(n)\pi_1(a) \quad \text{for all } a \in H_0, n \in N.$$

– **Step 6:** Take H and π_0 from the last step and proceed to (normalized) induction $\pi = ind_H^G \pi_0$.

It is quite evident that this recipe produces unitary representations. But it is far from evident that these are irreducible and that we really get the unitary dual of the semidirect product and, hence, have a classification scheme at hand. As already said, we can not go into the proofs here (which essentially use Mackey's imprimitivity theorem) and refer to the Chapters 16 and 17 of [BR] or §13 of [Ki] and Section 10 of [Ma] where slightly different approaches are adopted. Now, we discuss some examples to show how the method works. In these examples one can verify directly the validity of the procedure encapsulated in Theorem 7.7 for instance with the help of the infinitesimal method from our Chapter 6.

Examples:

Example 7.11: $G := \text{Heis}'(\mathbf{R}) = \{g = (x, y, z); x, y, z \in \mathbf{R}\}$ with

$$gg' = (x + x', y + y', z + z' + xy')$$

is isomorophic to the semidirect product of $G_0 := \{a = (x, 0, 0); x \in \mathbf{R}\}$ acting on $N := \{n = (0, y, z); y, z \in \mathbf{R}\}$ via $\rho(a)n := (0, y, z + xy)$ as seen in the **Example** 7.9 above.
– **Step 1:** The dual $\hat{N}$ of $N \simeq \mathbf{R}^2$ is again to be identified with $\mathbf{R}^2$. With $(l, m) \in \mathbf{R}^2$ we write

$$\hat{n}(n) = e^{i(ly+mz)} =: \chi_{(l,m)}(y, z).$$

– **Step 2:** $a \in G_0 \simeq \mathbf{R}$ acts on $\hat{n} \in \hat{N} \simeq \mathbf{R}^2$ by

$$a\hat{n}(n) := \hat{n}(\rho(a^{-1})n) = e^{i(ly+m(z-xy))} = \chi_{(l-xm,m)}(y, z).$$

Hence $\hat{N}$ is disjointly decomposed into exactly two types of G_0−orbits:

- **Type 1:** For $m \neq 0$ and $\hat{n}_0 = (0, m)$ one has the line $\hat{O}_{n_0} = \{(-xm, m), x \in \mathbf{R}\}$,
- **Type 2:** For $m = 0$ and $\hat{n}_0 = (l, 0)$ there is the point $\hat{O}_{n_0} = \{(l, 0)\}$.

We get the following picture

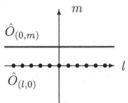

One can see that G is a regular semidirect product: The set $\{(l, m) \in \mathbf{R}^2; \, lm = 0\}$ is an analytic set, which meets each orbit exactly once.

– **Step 3:** The stabilizing group of $\hat{n}_0 = (0, m), m \neq 0$ is $H_0 = G_{0(0,m)} = \{0\}$ and that of $\hat{n}_0 = (l, 0), l \in \mathbf{R}$ is $H_0 = G_{0(l,0)} = G_0 \simeq \mathbf{R}$.

– **Step 4:** In the first case the unitary dual of H_0 simply consists of the trivial identity representation and in the second case we have $\hat{H}_0 \simeq \mathbf{R}$ with representatives given by $\pi_1(a) = e^{ijx} =: \chi_j(x), j \in \mathbf{R}$.

– **Step 5** and **6:** In the second case, the type 2 orbits, the product of the stabilizing group $H_0 = G_0$ with N already exhausts the whole group G. Hence there is no induction necessary and one has for $(j, l) \in \mathbf{R}^2$ a one-dimensional representation π of $G = G_0 \ltimes N$ given by

$$\pi((x, y, z)) = e^{ijx} e^{ily}.$$

In the first case, belonging to the type 1 orbits, one has $H = H_0 \ltimes N \simeq N \simeq \mathbf{R}^2$ with a representation π_0 given by $\pi_0(h) = e^{imz}$ for all $h = (0, y, z)$. Here the induction makes sense and we have a representation $\pi = \text{ind}_H^G \pi_0$ (for instance) given by right translation on the completion $\mathcal{H}$ of the space of smooth functions $\phi : G \longrightarrow \mathbf{C}$ with functional equation

$$\phi(hg) = e^{imz} \phi(g) \quad \text{for all } h \in H, g \in G,$$

and $\|\phi\|^2 = \int_{\mathbf{R}} |\phi((x, 0, 0)|^2 \, dx < \infty$. As the composition law says

$$gg_0 = (x + x_0, y + y_0, z + z_0 + xy_0) \text{ and } (0, y, z)(x, 0, 0) = (x, y, z),$$

an element ϕ in $\mathcal{H}$ is of the form $\phi(g) = e^{imz} f(x)$ with $f \in L^2(\mathbf{R})$ and one has

$$\phi(gg_0) = e^{imz} e^{im(z_0 + xy_0)} f(x + x_0).$$

So we have a second realization of this representation given by

$$\tilde{\pi}(g_0) f(x) = e^{im(z_0 + xy_0)} f(x + x_0) \text{ for } f \in L^2(\mathbf{R}).$$

This is the Schrödinger representation π_m from our Example 1.6 in 1.3 translated to Heis$'(\mathbf{R})$ by the isomorphism

$$\varphi : \text{Heis}'(\mathbf{R}) \longrightarrow \text{Heis}(\mathbf{R}), \, g = (x, y, z) \longmapsto (\lambda, \mu, \kappa) = (x, y, 2z - xy),$$

from Exercise 0.1. Hence the unitary dual of the Heisenberg group G is

$$\hat{G} = \mathbf{R}^2 \sqcup (\mathbf{R} \setminus \{0\}).$$

Example 7.12: $G := G^E(2) = SO(2) \ltimes \mathbf{R}^2$ has $G_0 = SO(2)$ and $N = \mathbf{R}^2$. We write $g = (a, n)$ with $a = r(\vartheta)$ and $n = x := {}^t(x_1, x_2)$. One has $\rho(a)n := r(\vartheta)x$

– **Step 1:** As in example 7.11 we have $\hat{N} = \mathbf{R}^2$. This time, for $y = (y_1, y_2) \in \mathbf{R}^2$ we write

$$\hat{n}(n) = e^{i(y_1 x_1 + y_2 x_2)} = \chi_{(y_1, y_2)}(x_1, x_2).$$

– **Step 2:** $a = r(\vartheta) \in G_0 = SO(2)$ acts on $\hat{n} \in \hat{N}$ by

$$a\hat{n}(n) := \hat{n}(\rho(a^{-1})n) = e^{iyr(-\vartheta)x} = \chi_{yr(-\vartheta)}(x).$$

Hence $\hat{N}$ is again decomposed into two types of orbits:
• Type 1: For $\hat{n}_0 = (r, 0), r > 0$, the circle

$$\hat{O}_{n_0} = \{y \in \mathbf{R}^2, \|y\|^2 = r^2\}.$$

• Type 2: For $\hat{n} = (0, 0)$, the point $\hat{O}_{n_0} = \{(0, 0)\}$.

We can verify the regularity of the semidirect product: One has a countable family of Borel subsets of $\hat{N}$ consisting of the following sets: $Z_{0,0} :=$ the point $\hat{O}_{n_0} = \{(0, 0)\}$ and for any two positive rational numbers $r_1 < r_2$ $Z_{r_1, r_2} :=$ the union of all orbits of type 2 with $r_1 < r < r_2$. Then each orbit is the intersection of the members of the subfamily, which contain the orbit.

– **Step 3:** For the type 1 orbits the stabilizing group is $G_{0(r,0)} = \{E_2\}$ and for the type 2 one has $G_{0(0,0)} = G_0$.

– **Step 4:** The unitary dual of $G_0 = SO(2)$ is $\mathbf{Z}$.

– **Step 5 and 6:** In the second case, the type 2 orbit, the product of N and the stabilizing group is G. Hence for every $k \in \mathbf{Z}$, we have a one-dimensional representation π of G given by

$$\pi((r(\vartheta), x)) = e^{ik\vartheta}.$$

In the first case, the type 1 orbit, one has for $r > 0$ the genuine subgroup $H \simeq N$ and its one-dimensional representation π_0 given by

$$\pi_0((E_2, x)) = \hat{n}_0(n) = e^{irx_1}.$$

Hence, for every $r > 0$, we have an induced representation $\pi = ind_H^G \pi_0$. In consequence, the unitary dual of $G = G^E(2)$ is

$$\hat{G} = \mathbf{R}_{>0} \sqcup \mathbf{Z}.$$

Example 7.13: As **Exercise 7.22** apply our construction recipe to construct representations of

$$G = B := \{g = \begin{pmatrix} a & b \\ & a^{-1} \end{pmatrix} =: (a, b); \ a, b \in \mathbf{R}, a > 0\}$$

and

$$G = B' := \{g = \begin{pmatrix} a & b \\ & a^{-1} \end{pmatrix} =: (a, b); \ a, b \in \mathbf{R}, a \neq 0\}$$

and determine the unitary dual.

Exercise 7.23: Do the same for the Euclidean group in three dimensions $G^E(3)$.

7.5   Unitary Representations of the Poincaré Group

The Poincaré or inhomogeneous Lorentz group is generally known as the *symmetry group of space-time*. In continuation to our discussion of the Lorentz group in 7.3, we put

$$G = G^P(4) := G_0 \ltimes N \text{ with } G_0 := SO(3,1)^0, \; N := \mathbf{R}^4.$$

Sometimes it is necessary to take more generally $G_0 = SO(3,1)$ or even $G_0 = O(3,1)$, which also preserve the Minkowski (pseudo)norm $| \cdot |$ given by

$$| x |^2 := x_1^2 + x_2^2 + x_3^2 - x_4^2 = {}^t x \left(\begin{matrix} E_3 & \\ & -1 \end{matrix} \right) x, \quad \text{for all } x = {}^t(x_1, x_2, x_3, x_4) \in \mathbf{R}^4.$$

We restrict our treatment to a discussion of the simplest case (it should not be too difficult to extend our results to the more general cases). It should also be not difficult to translate our procedure to the situation apparently prefered by the physics literature ([Co] p.677ff, [BR] p.513ff]), namely that the coordinates are x_0, x_1, x_2, x_3 and the norm is given by $| x |^2 = x_0^2 - x_1^2 - x_2^2 - x_3^2$. Moreover, there are reasons from physics (see the hints in 4.3 while discussing the relation of the representations of $SO(3)$ and $SU(2)$) and mathematics to define and treat the semidirect product

$$\tilde{G} := SL(2, \mathbf{C}) \ltimes \mathbf{R}^4.$$

For the definition of this product, one goes back to the proof of Theorem 7.6 in 7.3 where we fixed the surjection $\rho : SL(2, \mathbf{C}) \longrightarrow SO(3,1)$: There we had an identification $\varphi : \mathbf{R}^4 \longrightarrow V = \mathrm{Herm}_2(\mathbf{C})$ and defined an action of $SL(2, \mathbf{C})$ on V (and hence on $\mathbf{R}^4$) by

$$\rho(\tilde{A})X = \tilde{A} X \tilde{A}^* \quad \text{for all } X \in V, \; \tilde{A} \in SL(2, \mathbf{C}).$$

We start with the discussion of $G = G_0 \ltimes N, G_0 = SO(3,1)^0, \; N = \mathbf{R}^4$ (say, for aesthetical reasons) but soon shall be lead to take in also $\tilde{G}$. We follow the construction procedure from Theorem 7.7 in the last section.

Step 1: The dual $\hat{N}$ of $N = \mathbf{R}^4$ is again to be identified with $\mathbf{R}^4$. We treat N as consisting of (coordinate) columns $x \in \mathbf{R}^4$ and $\hat{N}$ as consisting of (*momentum*) rows ${}^t p, p = (p_1, p_2, p_3, p_4) \in \mathbf{R}^4$. Hence we write

$$\hat{n}(n) = e^{i \Sigma p_j x_j} = e^{i {}^t p x} =: \chi_p(x).$$

(In the physics literature often minus signs appear in this identfication of N and its dual, but we want to avoid the discussion of co- and contravariant vectors at this stage.)

Step 2: $a \in G_0$, here given by $A \in SO(3,1)^0$, acts on $\hat{n} \in \hat{N}$, given by $p \in \mathbf{R}^4$, by

$$a\hat{n}(n) := \hat{n}(\rho(A^{-1})n) = \chi_{pA^{-1}}(x),$$

i.e. A transforms p into pA^{-1}. It is elementary calculus that the orbits for this action are subsets in the three types of quadrics

- Type 1: the two-sheeted hyperboloid given by

$$- | p |^2 = p_4^2 - p_3^2 - p_2^2 - p_1^2 = m^2 \text{ for } m \in \mathbf{R}, m \neq 0,$$

- Type 2: the one-sheeted hyperboloid given by

$$-\mid p \mid^2 = p_4^2 - p_3^2 - p_2^2 - p_1^2 = (im)^2 \text{ for } m \in \mathbf{R}, m \neq 0,$$

- Type 3: the *light cone* given by

$$-\mid p \mid^2 = p_4^2 - p_3^2 - p_2^2 - p_1^2 = 0.$$

More precisely, one verifies using the explicit description of $SO(3,1)$ in 7.3 that $\mathbf{R}^4$ decomposes exactly into the following orbits:

- Type 1^+: $\hat{O}_m^+ := \hat{O}_{\hat{n}_0} = G_0\hat{n}_0$ for $\hat{n}_0 = (0,0,0,m)$,
- Type 1^-: $\hat{O}_m^- := \hat{O}_{\hat{n}_0} = G_0\hat{n}_0$ for $\hat{n}_0 = (0,0,0,-m)$,
- Type 2: $\hat{O}_{im} := \hat{O}_{\hat{n}_0} = G_0\hat{n}_0$ for $\hat{n}_0 = (m,0,0,0)$,
- Type 3^+: $\hat{O}_0^+ = \hat{O}_{\hat{n}_0} = G_0\hat{n}_0$ for $\hat{n}_0 = (-1,0,0,1)$,
- Type 3^-: $\hat{O}_0^- := \hat{O}_{\hat{n}_0} = G_0\hat{n}_0$ for $\hat{n}_0 = (-1,0,0,-1)$,
- Type 3^0: $\hat{O}_0^0 := \hat{O}_{\hat{n}_0} = G_0\hat{n}_0 = \{(0,0,0,0)\}$ for $\hat{n}_0 = (0,0,0,0)$,

The type 1^+ orbit $\hat{O}_m^+$ is called the *positive mass shell*, $\hat{O}_m^-$ the *negative mass shell* and $\hat{O}_0^\pm$ the *forward* resp. *backward light cone*. The situation is sketched in the picture:

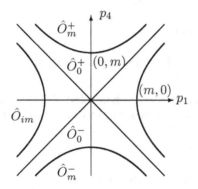

The regularity of the semidirect product can be verified analogously as in our example 2 in the last section (see [BR] p.517/8).

Now, for each orbit $\hat{O}$, we have to determine its *little group*, i.e. the stabilizing group $H_0 := G_{0\hat{n}_0}$.

Step 3: We get the following four cases.

$$
\begin{aligned}
H_0 := G_{0\hat{n}_0} &\simeq SO(3) && \text{for the type } 1^\pm, && \text{i.e. } \hat{n}_0 = (0,0,0,\pm m), \\
&\simeq SO(2,1) && \text{for the type } 2, && \text{i.e. } \hat{n}_0 = (m,0,0,0), \\
&\simeq G^E(2) && \text{for the type } 3^\pm, && \text{i.e. } \hat{n}_0 = (-1,0,0,\pm 1), \\
&\simeq SO(3,1)^0 && \text{for the type } 3^0, && \text{i.e. } \hat{n}_0 = (0,0,0,0).
\end{aligned}
$$

Proof: i) As in 7.3 we write the elements of $G_0 = \mathrm{SO}(3,1)$ as

$$A = \begin{pmatrix} B & x \\ {}^t y & a \end{pmatrix}, \; B \in \mathrm{SO}(3), x, y \in \mathbf{R}^3, a \in \mathbf{R}.$$

For $m \neq 0$, the condition

$$(0,0,0,m)A = (0,0,0,m)$$

enforces $y = 0$ and $a = 1$. Moreover, from $A \in \mathrm{SO}(3,1)$ one deduces (via (7.21)) $x = 0$ and, hence, the first assertion $G_{0\hat{n}_0} \simeq \mathrm{SO}(3)$.

ii) The other assertions can be proved similarly by direct computations. But as announced above, it is more agreable to use the fact from 7.3 that one has $\mathrm{SL}(2,\mathbf{C}) =: \tilde{G}_0$ via the surjection $\rho : \tilde{G}_0 \longrightarrow G_0$ as a double cover of the Lorentz group. In the proof of Theorem 7.6 in 7.3 we identified $\mathbf{R}^4$ via a map φ with the space V of hermitian 2×2-matrices

$$^t(a,b,c,d) \longmapsto \begin{pmatrix} -a+d & b+ic \\ b-ic & a+d \end{pmatrix},$$

and defined the map

$$\rho : \mathrm{SL}(2,\mathbf{C}) \longrightarrow \mathrm{GL}(V), \; \tilde{A} \longmapsto \rho(\tilde{A}) \text{ with } \rho(\tilde{A})X = \tilde{A}X\tilde{A}^* \text{ for } X \in V.$$

Hence for $m \neq 0$, the condition

$$(m,0,0,0)A = (m,0,0,0), \; A \in G_0$$

can be translated into the condition

$$\rho(\tilde{A})X_0 = \tilde{A}X_0\tilde{A} = X_0 \text{ for } X_0 := \varphi(m,0,0,0) = \begin{pmatrix} -m & \\ & m \end{pmatrix},$$

i.e. $\tilde{A} \in \mathrm{SU}(1,1)$. It is not difficult to see (**Exercise 7.24**) that the homomorphism $\rho : \mathrm{SL}(2,\mathbf{C}) \longrightarrow \mathrm{SO}(3,1)^0$ restricts to a surjection $\mathrm{SU}(1,1) \longrightarrow \mathrm{SO}(2,1)^0$.

It will be useful later to take also

$$X_0 := \varphi(0,0,m,0) = \begin{pmatrix} & mi \\ -mi & \end{pmatrix}.$$

By a small calculation, one verifies that the stabilizing group $\{\tilde{A}; \tilde{A}X_0\tilde{A} = X_0\}$ here is $\mathrm{SL}(2,\mathbf{R})$: For $\sigma = \begin{pmatrix} & i \\ -i & \end{pmatrix}$ one has $\sigma = \sigma^{-1}$ and $^tA^{-1} = \sigma A\sigma^{-1}$. Hence from $A\sigma A^* = \sigma$ we deduce $\sigma^{-1}A\sigma = A^{*-1} = {}^tA^{-1}$, i.e. $A = \bar{A}$.

iii) One has $\varphi(-1,0,0,1) = \begin{pmatrix} 2 & \\ & \end{pmatrix} =: X_0$, and $\tilde{A}X_0\tilde{A} = X_0$ calls for

$$\tilde{A} = \begin{pmatrix} a & b \\ & a^{-1} \end{pmatrix}, \text{ with } a,b \in \mathbf{C}, |a| = 1.$$

We put $a =: e^{i\vartheta/2}$, $b =: e^{-i\vartheta/2}z$, $\vartheta \in [0,4\pi]$, $z \in \mathbf{C}$, and get the multiplication law

$$\tilde{A}\tilde{A}' = \begin{pmatrix} a'' & b'' \\ & a''^{-1} \end{pmatrix}, \text{ with } a'' = e^{i(\vartheta+\vartheta')/2}, \; b'' = e^{-i(\vartheta+\vartheta')/2}(z + e^{i\vartheta}z').$$

This shows that in this case the stabilizing group $\tilde{H}_0 \subset \tilde{G}$ is isomorphic to the semidirect product of a rotation group $\{e^{i\vartheta}\}$ and a translation group $\mathbf{R}^2 \simeq \mathbf{C}$ with the composition law

$$(e^{i\vartheta/2}, z)(e^{i\vartheta'/2}, z') = (e^{i(\vartheta+\vartheta')/2}, z + e^{i\vartheta}z').$$

The homomorphism $(e^{i\vartheta/2} \longmapsto (r(\vartheta))$ makes $\tilde{S}^1 := \{e^{i\vartheta/2}; 0 < \vartheta < 4\pi\}$ a double cover of $SO(2)$ and this extends to a surjection of $\tilde{H}_0$ to the Euclidean group $G^E(2)$.

iv) In the last case there is nothing to prove.

Step 4: We already determined the unitary dual of covering groups $\tilde{H}_0$ of all the little groups coming in here:

i) In the first case (type $1^\pm$), for $\tilde{H}_0 = SU(2)$, the elements of the unitary dual are represented by the representations $\pi_j, j \in (1/2)\mathbf{N}_0$ discussed in 4.2.

ii) In the second case (type 2), one has $\tilde{H}_0 = SU(1,1)$ and from 6.2 we know that $SU(1,1)$ is isomorphic to $SL(2,\mathbf{R})$ and that the representations appear in three series $\pi_k^\pm, k \in \mathbf{N}; \pi_{is,\pm}, s \in \mathbf{R}; \pi_s, 0 < s < 1$.

iii) For the third case (type $3^\pm$), as Example 7.12 to our construction procedure in 7.4 we elaborated that for the two-dimensional Euclidean group one has one-dimensional representations $\chi_k, k \in \mathbf{Z}$ (hence $\chi_j, j \in (1/2)\mathbf{Z}$ for the double cover) and infinite-dimensional representations $\pi_r, r > 0$.

iv) For the type 3^0, one has the representations of the Lorentz group resp. its covering $SL(2,\mathbf{C})$ discussed in 7.3, which come in two series $\pi_{k,iv}, k \in \mathbf{Z}, v \in \mathbf{R}; \pi_{0,u}, 0 < u < 2$.

Step 5 and 6: Finally, we have to extend the representations of the little group to its product with $N = \mathbf{R}^4$ and, if this product is not the whole group, proceed to an induced representation. This way we get the following representations (as usual, we list those of the covering group $\tilde{G}$ of the Poincaré group):

The Unitary Dual of the Poincaré Group

i) In the first case, one has $\tilde{H} = SU(2) \ltimes \mathbf{R}^4$. For the type 1^+ with $m > 0$, we have for this group the representation π_0 given by

$$\pi_0((B,x)) = \pi_j(B)e^{imx_4} \text{ for } B \in SU(2), \ x = {}^t(x_1, x_2, x_3, x_4) \in \mathbf{R}^4.$$

We write for the induced representation

$$ind_{\tilde{H}}^{\tilde{G}}\pi_0 =: \pi_{m,+;j}.$$

Similarly, the type 1^- leads to

$$\pi_0((B,x)) = \pi_j(B)e^{-imx_4} \text{ for } B \in SU(2), \ x = {}^t(x_1, x_2, x_3, x_4) \in \mathbf{R}^4$$

and we denote the induced representation by

$$\pi_{m,-;j}, \ m > 0, j \in (1/2)\mathbf{N}_0.$$

ii) In the second case, one has

$$\tilde{H} = \mathrm{SL}(2, \mathbf{R}) \ltimes \mathbf{R}^4$$

and its representation

$$\pi_0((A, x)) = \pi(A)e^{imx_3} \text{ for } A \in \mathrm{SL}(2, \mathbf{R}), \ x = {}^t(x_1, x_2, x_3, x_4) \in \mathbf{R}^4$$

where π here is one of the representations of $\mathrm{SL}(2, \mathbf{R})$ listed above in part ii) of Step 4 resp. in Section 7.2. For the corresponding induced representation we write

$$
\begin{aligned}
ind_{\tilde{H}}^{\tilde{G}} \pi_0 \ &=: \ \pi_{im;k,\pm} \quad && \text{for } \pi = \pi_k^\pm, \ k \in \mathbf{N}; m > 0, \\
&=: \ \pi_{im;is,\pm} \quad && \text{for } \pi = \pi_{is,\pm}, \ s \in \mathbf{R}; m > 0, \\
&=: \ \pi_{im;s} \quad && \text{for } \pi = \pi_s, \ s \in \mathbf{R}, \ 0 < s < 1; m > 0.
\end{aligned}
$$

iii) In the third case, for type $3^\pm$, one has

$$\tilde{H} = (\tilde{S}^1 \ltimes \mathbf{R}^2) \ltimes \mathbf{R}^4$$

and its representation

$$\pi_0((\tilde{A}, x)) = \pi(\tilde{A})e^{i(x_1 \pm x_4)} \text{ for } \tilde{A} \in (\tilde{S}^1 \ltimes \mathbf{R}^2), \ x = {}^t(x_1, x_2, x_3, x_4) \in \mathbf{R}^4$$

where π here is one of the representations of $(\tilde{S}^1 \ltimes \mathbf{R}^2)$ indicated in Step 4. For the corresponding induced representation we write

$$
\begin{aligned}
ind_{\tilde{H}}^{\tilde{G}} \pi_0 \ &=: \ \pi_{0,\pm;j} \quad && \text{for } \pi = \chi_j, j \in (1/2)\mathbf{Z}, \\
&=: \ \pi_{0,\pm;r} \quad && \text{for } \pi = \pi_r, \ r \in \mathbf{R}, r > 0.
\end{aligned}
$$

iv) In the forth case (for type 3^0) the little group is the whole group and we get the representations of $\tilde{G}$ by the trivial extension of the representations π of $\mathrm{SL}(2, \mathbf{C})$, i.e. one has $\pi((\tilde{A}, x)) = \pi(\tilde{A})$ for $(\tilde{A}, x) \in \tilde{G}_0$. For the corresponding representation we write

$$
\begin{aligned}
\pi \ &=: \ \pi_{0;j,iv,\pm} \quad && \text{for } \pi = \pi_{j,iv,\pm}, v \in \mathbf{R} \\
&=: \ \pi_{0;0,u} \quad && \text{for } \pi = \pi_{0,u}, \ u \in \mathbf{R}, 0 < u < 2.
\end{aligned}
$$

Among these, the representations $\pi_{m,+;j}, \ m > 0, j \in (1/2)\mathbf{N}_0$, coming from the type 1^+ orbit and $\pi_{0,\pm;j}, j \in (1/2)\mathbf{Z}$, coming from the type $3^\pm$ orbit have a meaning for physics since they are interpreted as describing elementary particles of spin j and mass m resp. of zero mass.

Example 7.14: As an illustration we discuss an explicit model for the representation $\pi_{m,+;j}$ following [BR] p.522:

From the preceding discussion we take over

$$\tilde{G} = \tilde{G}_0 \ltimes N, \ \tilde{G}_0 = \mathrm{SL}(2, \mathbf{C}), \ N = \mathbf{R}^4$$

and write $\tilde{G} \ni g = (\tilde{A}, x), \tilde{A} \in \mathrm{SL}(2, \mathbf{C}), x \in \mathbf{R}^4$ (a column). A bit more explicitely than at the beginning of this section, the action of $\tilde{G}_0$ on N is described as follows. We have the identification

$$\varphi : \mathbf{R}^4 \longrightarrow V = \mathrm{Herm}_2(\mathbf{C}),$$

$$x = {}^t(x_1, x_2, x_3, x_4) \longmapsto X = \begin{pmatrix} x_4 - x_1 & x_2 + ix_3 \\ x_2 - ix_3 & x_4 + x_1 \end{pmatrix}.$$

Via φ the action of $\tilde{G}_0$ on V given by $(\tilde{A}, X) \longmapsto \rho(\tilde{A})X = \tilde{A}X\tilde{A}^*$ is translated into the action of $\tilde{G}_0$ on $\mathbf{R}^4$ given by

$$\rho(\tilde{A})x := \varphi^{-1}(\rho(\tilde{A})\varphi(x)).$$

By a slight abuse of notation, as we already did above, we also take $\rho(\tilde{A})$ to be a matrix from $G_0 = \mathrm{SO}(3,1)^0$ and understand $\rho(\tilde{A})x$ as multiplication of the column x by the matrix $\rho(\tilde{A})$. We identify the unitary dual $\hat{N}$ with $\mathbf{R}^4$ by writing $\hat{n}(n) = e^{i^t px}$, i.e. $\hat{n}$ is coordinatized by (the row) $^t p$, $p \in \mathbf{R}^4$. Then the action of $\tilde{G}_0$ on $\hat{N}$ is simply given by the multiplication of the row $^t p$ by the matrix $\rho(\tilde{A}^{-1})$, i.e.

$$^t p \longmapsto {}^t p \rho(\tilde{A}^{-1}) =: \tilde{A} \cdot p.$$

The construction of the representation $\pi_{m,+;j}$ is based on the type 1^+ orbit

$$\hat{O}_m = \tilde{G}_0 \cdot p_0, \ p_0 = (0, 0, 0, m), m > 0.$$

As we remarked already, p_0 corresponds to $X_0 = mE_2$ and has the stabilizing group $\tilde{H}_0 = \mathrm{SU}(2)$. We take the representation π_j of $\mathrm{SU}(2)$ and extend it to $\tilde{H} = \tilde{H}_0 \ltimes N$ to get π_0 given by

$$\pi_0((R, x)) = e^{imx_4}\pi(R) = e^{i^t p_0 x}\pi_j(R) \quad \text{for all } R \in \mathrm{SU}(2), x \in \mathbf{R}^4.$$

To get some more practice and since it is important for application in physics, we work out the "Second Realization" of the induced representation $ind_{\tilde{H}}^{\tilde{G}}\pi_0$, i.e. with a representation space consisting of functions u living on the homogeneous space

$$\tilde{G}/\tilde{H} \simeq \tilde{G}_0/\tilde{H}_0 \simeq \hat{O}_m^+.$$

(At this moment, one usually does not bother with any finiteness or integrability condition and simply takes smooth functions. We shall have occasion to do better in the next chapter.)

The general theory of the Iwasawa decomposition or a direct calculation say that we have a decomposition

$$\tilde{G}_0 = \mathrm{SL}(2, \mathbf{C}) \ni \tilde{A} = A_p R_p, R_p \in \mathrm{SU}(2), A_p = \begin{pmatrix} \lambda & z \\ & \lambda^{-1} \end{pmatrix}, \lambda \in \mathbf{R}^*, z \in \mathbf{C}.$$

Here it is customary in physics to coordinatize A_p by coordinates $^t p = (p_1, p_2, p_3, p_4)$ such that $p = A_p \cdot p_0$. Then one has the Mackey decomposition

$$\tilde{G} \ni g = s(p)h, \ s(p) = (A_p, 0), h = (R_p, x) \in \tilde{H}$$

and the Master Equation

$$g_0^{-1}s(p) = s(g_0^{-1}(p))h^*, h^* = (R^*, x^*) \in \tilde{H}$$

with

$$
\begin{aligned}
g_0^{-1}s(p) &= (A_0^{-1}, -\rho(A_0^{-1})x_0)(A_p, 0) \\
&= (A_0^{-1}A_p, -\rho(A_0^{-1})x_0) \\
&= (A_{g_0^{-1}(p)}, 0)((A_{g_0^{-1}(p)})^{-1}A_0^{-1}A_p, -\rho((A_{g_0^{-1}(p)})^{-1}A_0^{-1})x_0).
\end{aligned}
$$

This shows that one has

$$R^* = (A_{g_0^{-1}(p)})^{-1} A_0^{-1} A_p, \quad x^* = -\rho((A_{g_0^{-1}(p)})^{-1} A_0^{-1}) x_0.$$

One can verify that $d\mu(p) = p_4^{-1} dp_1 dp_2 dp_3$ is an invariant measure on $\hat{O}_m^+$ and that the Radon-Nikodym derivative is $d\mu(g_0^{-1}(p))/d\mu(p) = 1$. Hence by (7.10), in this case the representation is given by the prescription

$$\tilde{\pi}(g_0) u(p) = \pi_0(h^{*-1}) u(g_0^{-1} p)$$

with

$$\pi_0(h^{*-1}) = e^{i\,{}^t p x_0} \pi_j(R^{*-1})$$

as one can verify by a small calculation starting from $\pi_0((R, x)) = e^{i\,{}^t p_0 x} \pi_j(R)$: Namely, one has

$$h^{*-1} = (R^{*-1}, \tilde{x}) \text{ with } \tilde{x} = -\rho(R^{*-1}) x^*$$

and hence

$$^t p_0 \tilde{x} = -{}^t p_0 \rho(R^{*-1}) x^* = {}^t p_0 \rho(R^{*-1}(A_{g_0^{-1}(p)})^{-1} A_0) x_0$$

i.e.

$$^t p_0 \tilde{x} = {}^t p_0 \rho(A_p^{-1}) x_0 = {}^t p x_0$$

since we have $R^* = (A_{g_0^{-1}(p)})^{-1} A_0^{-1} A_p$.

The functions u we are treating here are vector valued functions

$$u : \hat{O}_m^+ \longrightarrow V_j \simeq \mathbf{C}^{2j+1},$$

i.e. can be written as $u(p) = (u_n(p))_{n=-j,-j+1,...,j}$. Here we write $D_{nn'}^j$ for the matrix elements of the representation π_j from our discussion in 4.2 and then come to

(7.24) $$(\pi((\tilde{A}_0, x_0)) u)_n (p) = \sum_{n'} e^{i\,{}^t p x_0} D_{nn'}^j (R^{*-1}) u_{n'}(\tilde{A}_0^{-1} \cdot p).$$

From here one derives the following interpretation that we have a free particle of spin j and mass m :

In the *rest system* $p = p_0$ and under rotations $(R, 0), R \in \mathrm{SU}(2)$ (7.24) restricts to

$$(\tilde{\pi}((R, 0)) u_n)(p_0) = \sum_{n'} D_{nn'}^j (R) u_{n'}(p_0),$$

i.e. the elementary expression of SU(2)-symmetry belonging to the representation π_j. And for the *infinitesimal generators* $P_\mu = \partial_\mu, \mu = 1, .., 4$ of the translations in the μ-th direction $x \longmapsto x_\mu(t) := x + (t\delta_{\mu\mu'})_{\mu'}$ one obtains from (7.24)

$$P_\mu u_n(p) = p_\mu u_n(p).$$

Hence we have $M u_n = m u_n$, i.e. u_n is an eigenfunction with eigenvalue m for the *mass operator*

$$M := \sqrt{P_4^2 - P_3^2 - P_2^2 - P_1^2}.$$

7.6   Induced Representations and Vector Bundles

The classic approach to the inducing mechanism we followed up to now, in the meantime has found a more geometric version using the notion of *bundles*. As Warner remarks in [Wa] p.376, this does not simplify in any essential way the underlying analysis but has the advantage that important generalizations of the entire process immediately suggest themselves (e.g. *cohomological induction*). Bundles play an important rôle in most parts of contemporary mathematics and we shall have to use them also in our next chapter concerning the beautiful orbit method. Hence, we present here some rudiments of the theory, which nowadays appears in most books on differential geometry, analytic and/or algebraic geometry etc. For instance, for a thourough study we recommend the book [AM] by Abraham and Marsden or 5.4 in [Ki] or, more briefly, the appendices in [Ki1] or [Be]. Whoever is in a hurry to get to the applications in Number Theory may skip this at first lecture. Before we try to shed some light on the notion of line, vector and Hilbert bundles, we look again, but from a different angle, at the construction of the representations of SU(2) from 4.2 and Example 7.3 in 7.1.5. We do this in the hope to create some motivation for the bundle approach.

We take $G := \mathrm{SU}(2) \supset H \simeq \mathrm{U}(1)$ and $\pi_0(h) := \chi_k(\zeta) = \zeta^k$ for $h = \begin{pmatrix} \zeta & \\ & \bar\zeta \end{pmatrix} \in H, k \in \mathbf{N}_0$.

As we discussed in Example 7.3 in 7.1.5, we have the homogeneous space

$$\mathcal{X} = G/H \simeq \mathbf{P}^1(\mathbf{C}), \quad g = \begin{pmatrix} a & b \\ -\bar b & \bar a \end{pmatrix} \longmapsto (b : \bar a)$$

and the *unnormalized induction* $\pi_k^u = ind^u{}_H^G \chi_k$ given by left (inverse) translation on the space $\mathcal{H}^\pi$ of smooth functions $\phi : G \longrightarrow \mathbf{C}$ with functional equation

$$\phi(gh) = \chi_k(h^{-1})\phi(g) \quad \text{for all } g \in G, h \in H.$$

(In this example everything is unimodular, hence it does not matter whether we treat un- or normalized induction, but in general several authors stick to unnormalized induction though one can also treat normalized induction as we shall see below.) For $V = \mathbf{C}$ we introduce as our first example of a *line bundle*

$$\mathcal{B}_k := G \times_H V := \{[g,t] := (g,t)_\sim; \; g \in G, t \in V\}$$

where the equivalence $\sim$ of the pairs is defined by

$$(gh, t) \sim (g, \pi_0(h)t) \text{ for } g \in G, h \in H, t \in V.$$

G acts on $\mathcal{B}_k$ by

$$g_0[g,t] := [g_0 g, t] \quad \text{for all } g_0 \in G, [g,t] \in \mathcal{B}_k.$$

One has a projection

$$pr : \mathcal{B}_k \longrightarrow G/H = \mathcal{X}, \; [g,t] \longmapsto x := gH.$$

Obviously the projection and the action are well defined and one has the *fiber over* x

$$(\mathcal{B}_k)_x := pr^{-1}(x) = \{[g,t]; gH = x\} \simeq V \simeq \mathbf{C}.$$

This leads to the picture that $\mathcal{B}_k$ is a bundle standing over the base space $\mathcal{X}$ and consisting of the fibers $(\mathcal{B}_k)_x$ over all base points $x \in \mathcal{X}$.

Our main object is the space $\Gamma(\mathcal{B}_k)$ of *sections* of $\mathcal{B}_k$, i.e. maps

$$\varphi : \mathcal{X} \longrightarrow \mathcal{B}_k \text{ with } pr \circ \varphi = id_{\mathcal{X}}.$$

These sections are of the form $\varphi(x) = [g, \phi(g)]$ for $x = gH$ with a function $\phi : G \longrightarrow V$. Since the coset gH may be represented also by $gh, h \in H$, the welldefinedness of the section asks for

$$[g, \phi(g)] = [gh, \phi(gh)] = [gh, \pi_0(h^{-1})\phi(g)]$$

where the last equation is expression of the equivalence introduced in the definition of the bundle $\mathcal{B}_k$. This leads to the fact that the function ϕ has to fulfill the functional equation

$$\phi(gh) = \pi_0(h^{-1})\phi(g) \quad \text{for all } h \in H, g \in G,$$

which we know from the construction of the representation space for the induced representation. We define an action of G on the space $\Gamma(\mathcal{B}_k)$ of sections by

$$(g_0 \cdot \varphi)(gH) := g_0\varphi(g_0^{-1}g) = [g, \phi(g_0^{-1}g)] \quad \text{for all } g_0 \in G, \varphi \in \Gamma(\mathcal{B}_k),$$

and this leads to an action of G on the space $\mathcal{H}$ of functions $\phi : G \longrightarrow V$ (satisfying the functional equation) defined by

$$g_0 \cdot \phi(g) := \phi(g_0^{-1}g).$$

Hence at least formally, we recover the prescription for the definition of the induced representation π. This can be completed to a precise statement if we provide our bundle with an appropriate topology. And this appears to be largely independent of the special example $G = \mathrm{SU}(2)$, which we proposed at the beginning. In the following, we will try to show how this works by giving some rudiments of the theory of bundles. We start by following the treatment of Warner [Wa] p.377, which is adapted to the application in representation theory, and proceed to another more general approach afterwards.

Bundles, First Approach

Definition 7.2: Let $\mathcal{X}$ be a fixed Hausdorff topological space. A *bundle* $\mathcal{B}$ *over* $\mathcal{X}$ is a pair (B, p) where B is a Hausdorff topological space and p is a continuous open map of B onto $\mathcal{X}$.

One calls $\mathcal{X}$ the *base space*, B the *bundle space*, and p the *bundle projection*. By abuse of notation the pair (B, p) is often abbreviated by the letter B to indicate the bundle.

For $x \in \mathcal{X}$, the set $p^{-1}(x)$ is called the *fiber over x* and often denoted by B_x. A *cross section* or simply *section* of the bundle is a map $s : \mathcal{X} \longrightarrow B$ with $p \circ s = id_{\mathcal{X}}$.

If $b \in B$ and $s(p(b)) = b$, then we say that s *passes through* b. If to every $b \in B$ there is a continuous cross section s passing through b, then the bundle is said to have *enough cross sections*.

A *Hilbert bundle* $\mathcal{B}$ over $\mathcal{X}$ is a bundle (B, p) over X together with operations and a norm making each fiber $B_x, x \in \mathcal{X}$ into a complex Hilbert space and satisfying the following conditions:

1.) The map $B \longrightarrow \mathbf{R}, b \longmapsto \|b\|$ is continuous,
2.) The operation of addition is a continuous map of $\{(b_1, b_2) \in B \times B, p(b_1) = p(b_2)\}$ into B,
3.) The operation of scalar multiplication is a continuous map of $\mathbf{C} \times B$ into B,
4.) The topology of B is such that its restriction to each fiber B_x is the norm topology of the Hilbert space B_x. (This condition is formulated more precisely in [Wa] p.377.)

Example 7.15: Let E be any Hilbert space. *The trivial bundle* with constant fiber E is given by $B := \mathcal{X} \times E$ equipped with the product topology and the projection p given by $p(x, a) = x$ for $x \in \mathcal{X}, a \in E$. The cross sections are given by the functions $s : \mathcal{X} \longrightarrow E$.

Let $\mathcal{B} = (B, p)$ be a Hilbert bundle over $\mathcal{X}$ and $\Gamma(\mathcal{B})$ the set of continuous cross sections of $\mathcal{B}$. Then it is clear that $\Gamma(\mathcal{B})$ is a complex vector space under pointwise addition and scalar multiplication. In the application to representation theory this space shall provide the representation space of a representation. So we need a topology on it. And here things start to be a bit technical. Following [Wa], we assume that $\mathcal{X}$ is locally compact and for $\Gamma(\mathcal{B})$ define the *uniform on compacta topology*. For our purposes it is sufficient that in particular this leads to the consequence that each sequence (s_n) in $\Gamma(\mathcal{B})$ converges to a continuous cross section φ iff for each compact subset $C \subset \mathcal{X}$ one has $\sup_{x \in C} \|s_n(x) - s(x)\| \longrightarrow 0$ for $n \longrightarrow \infty$. (In [Wa] the sequence is replaced by the more general notion of *net* but we will not go into this here.)

The integration theory in this context is still more delicate. So for details, we also refer to [Wa]. For our purpose it will suffice to fix a measure μ on $\mathcal{X}$ and think of $L^q(\mathcal{B}, \mu)$ as the space of sections s of $\mathcal{B}$ with $\|s\|_q := (\int_{\mathcal{X}} \|s(x)\|^q d\mu(x))^{1/q} < \infty$. Then the inner product on the space of sections is given by

$$< s, t > := \int_{\mathcal{X}} < s(x), t(x) > d\mu(x) \quad \text{for all } s, t \in L^2(\mathcal{B}, \mu).$$

Let G be a linear group, H a closed subgroup, δ the quotient of the modular functions of G and H treated in 7.1 (normalized by the condition $\delta(1) = 1$) and π_0 a unitary representation of H on the Hilbert space V.

We define a continuous right action of H on the product space $G \times V$ as follows:

$$(g, a)h := (gh, \delta(h)^{1/2} \pi_0(h^{-1})a) \quad \text{for all } g \in G, a \in V.$$

The orbit under H of a pair $(g, a) \in G \times E$ will be denoted by $[g, a] := (g, a)H$. The space of all such orbits equipped with the quotient topology will be called $E(G, H, \pi_0)$ or simply E. This space is Hausdorff and the natural projection map

$$p : E \longrightarrow G/H =: \mathcal{X}, \quad [g, a] = (g, a)H \longmapsto gH =: x$$

is a continuous and open surjection. Therefore $\mathcal{E} := (E, p)$ is a bundle over $\mathcal{X} = G/H$, which becomes a Hilbert bundle by the following natural assignments:

We define addition and scalar multiplication in each fiber $E_x := \{[g, a]; a \in V\}$ by

$$c_1[g, a_1] + c_2[g, a_2] := [g, c_1 a_1 + c_2 a_2] \quad \text{for all } c_1, c_2 \in \mathbf{C}$$

and the scalar product by

$$< [g, a_1], [g, a_2] >_{E_x} := \delta(g)^{-1} < a_1, a_2 > .$$

Obviously, there are some verifications to be done. Whoever does not want to do these on his own can look into [Wa] p.380/1. The bundle is called the *Hilbert bundle induced by* π_0 (*and* δ) and denoted by $\mathcal{E}(\pi_0)$ or again $\mathcal{E}$. G acts continuously from the left on $\mathcal{E}$ by

$$g_0[g, a] := [g_0 g, a].$$

Finally, we rediscover the normalized induced representation $\pi = ind_H^G \pi_0$ of G as a representation by unitary operators on the space of square integrable sections of $\mathcal{E}$: To this end, one first observes that there exists a one-to-one correspondence between cross sections φ for $\mathcal{E} = \mathcal{E}(\pi_0)$ and functions $\phi : G \longrightarrow V$ fulfilling the functional equation

$$\phi(gh) = \delta(h)^{1/2} \pi_0(h^{-1})\phi(g) \quad \text{for all } g \in G, h \in H.$$

The correspondence is given by

$$\varphi(gH) = (g, \phi(g))H.$$

If ϕ and φ are connected in this way, one says that ϕ is the *unitary form* of φ and φ is the *bundle form* of ϕ. Evidently a cross section φ is continuous on $\mathcal{X} = G/H$ iff its unitary form ϕ is continuous on G. It is a bit less evident but true (see [Wa] p.381) that $\mathcal{E}(\pi_0)$ admits enough cross sections. Therefore it makes sense to consider the Hilbert space

$$\mathcal{H}_{\pi_0} := L^2(\mathcal{E}, \mu)$$

where μ is the quasi-invariant measure on $\mathcal{X}$ associated with δ and the ρ–function on $\mathcal{X}$ as in 7.1. (It can be shown that the construction is independent of the choice of the function δ.) By a (small) abuse of notation, one denotes the space of bundle forms ϕ of elements φ in the space of sections as well by $\mathcal{H}_{\pi_0}$. This space inherits the Hilbert space structure and has as scalar product

$$< \phi, \tilde{\phi} > := \int_{\mathcal{X}} \delta(s(x))^{-1} < \phi(s(x)), \tilde{\phi}(s(x)) > d\mu(x)$$

where s denotes the Mackey section we introduced in 7.1.

The bundle theoretic definition of the unitarily induced representation π is as follows: For $g_0 \in G$ and $\varphi \in \mathcal{H}_{\pi_0}$, put

$$\pi(g_0)\varphi(gH) := g_0 \varphi(g_0^{-1} gH)$$

and correspondingly for the unitary form ϕ of φ

$$\pi(g_0)\phi(g) = \phi(g_0^{-1} g).$$

Some routine verifications parallel to those we did in 7.1 (see [Wa] p.382) show that we have got a unitary representation equivalent to the one constructed in 7.1. And one should get the feeling that all this is realized by the example of the (complex!) bundle $\mathcal{E} = \mathcal{B}_k$ over $\mathcal{X} = \mathbf{P}^1(\mathbf{C})$ from the beginning of this section. We will say more when we come back to this at the end of the section.

As Warner remarks ([Wa] p.382), the preceding bundle-theoretic construction makes no use of the local triviality of the constructed bundle. Moreover, "the definition of Hilbert bundle is, strictly speaking, not in accord with the customaty version - however, it has the advantage of allowing to place Hilbert space structures on the fibers without having to worry about transition functions". As we shall need this later, we present some elements of this other way to treat bundles.

Bundles, Second Approach

Here we need the notion of a differentiable manifold, which already was alluded to in 6.2 and is to be found in most text books on advanced calculus. We recommend in particular the Appendix II in [Ki1] (or the Appendix A in [Be]).

Definition 7.3 : A $p-$*manifold* M (or a p-*dimensional manifold*) is a Hausdorff topological space that admits a covering by open sets $U_i, i \in I$, endowed with one-to-one continuous maps $\varphi_i : U_i \longrightarrow V_i \subset \mathbf{R}^p$.
M is called *separable* if it can be covered by a countable system of such open sets.
A *smooth* resp. *analytic* p-manifold is a separable p-manifold so that all maps

$$\varphi_{i,j} := \varphi_i \circ \varphi_j^{-1} \mid_{\varphi_j(U_i \cap U_j)}$$

are infinitely differentiable (resp. analytic).

We use the following terminology:

– the pairs (U_i, φ_i) (and often also the sets U_i) are called *charts*;

– for $m \in U_i =: U$ and $\varphi_i =: \varphi$ the members $x_1, \dots, x_p$ of the $p-$tuple $\varphi(m) =: x \in \mathbf{R}^p$ are called *local coordinates*;

– the functions $\varphi_{i,j}$ are called *transition functions*;

– the collection $(U_i, \varphi_i)_{i \in I}$ is called an *atlas on M*.

It has to be clarified which atlases are really different: We say that two atlases $(U_i)_{i \in I}$ and $(U'_j)_{j \in J}$ are called *equivalent* if the transition functions from any chart of the first atlas to any chart of the second one are smooth (resp. analytic). The *structure of a smooth* (resp. *analytic*) *manifold* on M is an equivalence class of atlases.

When dealing with a smooth manifold, we use only one atlas keeping in mind that we can always replace it by an equivalent one.

In parallel to the just given notions for real manifolds, one defines a smooth (or holomorphic) complex p-manifold where $\mathbf{R}^p$ is replaced by $\mathbf{C}^p$ and the transition functions are required to be holomorphic, i.e. complex differentiable.

Examples:

Example 7.16: $M = \mathbf{R}^p$ and open subsets are real p–manifolds.

Example 7.17: $M \subset \mathbf{R}^q$ is a p-dimensional submanifold if for any $m \in M$ there are an open set $V \subset \mathbf{R}^q$ and differentiable functions $f_i : V \longrightarrow \mathbf{R}, i = 1, \ldots, q - p$, such that

– a) $M \cap V = \{x \in V; \ f_1(x) = \cdots = f_{q-p}(x) = 0\}$,
– b) the functional matrix $\left(\frac{\partial f}{\partial x}\right)$ has rank $q - p$ for all x.

The hyperboloids appearing as orbits in our discussion of the representations of the Poincaré group in 7.5 are three-dimensional smooth submanifolds of $\mathbf{R}^4$.

Example 7.18: $M = \mathbf{P}^1(\mathbf{C})$ (as in Example 7.3 in 7.1.5 and the beginning of this section) is a smooth one-dimensional complex manifold:
We cover $\mathbf{P}^1(\mathbf{C})$ by two open subsets $U := \{(u : v); \ u, v \in \mathbf{C}, v \neq 0\}$ and $U' := \{(u : v); \ u, v \in \mathbf{C}, u \neq 0\}$. We have the homeomorphisms

$$\varphi : U \longrightarrow \mathbf{C}, \ (u : v) \longmapsto u/v =: w \in \mathbf{C},$$
(7.25)
$$\varphi' : U' \longrightarrow \mathbf{C}, \ (u : v) \longmapsto v/u =: w' \in \mathbf{C}.$$

And the map $\varphi' \circ \varphi \mid_{\{w \neq 0\}} (w) = 1/w = w'$ is complex differentiable. So here one has an atlas consisting of two charts U and U'.

Certainly, at the same time, $\mathbf{P}^1(\mathbf{C})$ is an example for a two-dimensional real manifold.

Example 7.19: Show as **Exercise 7.25** that M is a smooth p–dimensional (real) manifold where

$$M = \mathbf{P}^p(\mathbf{R}) := \{[u_1 : \cdots : u_{p+1}] = (u_1, \ldots, u_{p+1})_\sim; \ (u_1, \ldots, u_{p+1}) \in \mathbf{R}^{p+1} \backslash \{0\}$$

with $(u_1, \ldots, u_{p+1}) \sim (u'_1, \ldots, u'_{p+1})$ iff there exists a $\lambda \in \mathbf{R}^*$ with $u'_i = \lambda u_i$ for all i.

If M is a smooth p–manifold and M' a smooth p'–manifold, there is a natural notion of a *smooth map* $F : M \longrightarrow M'$. Namely, for any point $m \in M$ with $F(m) = m'$ and any charts $U \ni m$ and $U' \ni F(m) = m'$, the local coordinates of m' must be smooth functions of local coordinates of $m \in U$.
A smooth map from one p–manifold to another, which has a smooth inverse map is called a *diffeomorphism*. Two manifolds are called *diffeomorphic* if there is a diffeomorphism from one to another.

A manifold is called *oriented* iff it has an *oriented* atlas, i.e. an atlas where for any two charts with coordinates x and y the determinant of the functional matrix $\left(\frac{\partial x}{\partial y}\right)$ is positive whenever it is defined.

As already said, there is much more to be discussed in this context (and we refer again to e.g. [Ki1]) but, we hope, this rudimentary presentation suffices to grab an understanding of our second approach to bundles:

Definition 7.4: $\mathcal{E}$ is called a (fiber) *bundle* over the base $\mathcal{X}$ with fiber F iff $\mathcal{E}, F$ and $\mathcal{X}$ are smooth manifolds and there is given a smooth surjective map $p : \mathcal{E} \longrightarrow \mathcal{X}$ such that locally $\mathcal{E}$ is a direct product of a part of the base and the fiber F.

More precisely, we require that any point $x \in \mathcal{X}$ has a neighbourhood U_x such that $E_{U_x} := p^{-1}(U_x)$ can be identified with $U_x \times F$ via a smooth map h_x so that, for $m \in E_{U_x}$, one has $h_x(m) = (x', v)$ if $x' = p(m)$, i.e. the following diagram (where p_1 denotes the projection to the first factor) is commutative

$$
\begin{array}{ccc}
p^{-1}(U_x) & \xrightarrow{\;h_x\;} & U_x \times F \\[4pt]
\downarrow p & & \downarrow p_1 \\[4pt]
U_x & \xrightarrow{\;\sim\;} & U_x.
\end{array}
$$

From the definition it follows that all sets $E_x := p^{-1}(x)$, $x \in \mathcal{X}$, called *fibers*, are smooth submanifolds diffeomorphic to F.

One often writes $\mathcal{E} \xrightarrow{F} \mathcal{X}$ to denote a bundle with fiber F and base $\mathcal{X}$.

There is the natural notion of a map between bundles: Let $\mathcal{E} \xrightarrow{F} \mathcal{X}$ and $\mathcal{E}' \xrightarrow{F'} \mathcal{X}'$ be two bundles. Then a *bundle map* is given by the pair of smooth maps

$$
\Phi : \mathcal{E} \longrightarrow \mathcal{E}', f : \mathcal{X} \longrightarrow \mathcal{X}'
$$

such that the following diagram is commutative

$$
\begin{array}{ccc}
\mathcal{E} & \xrightarrow{\;\Phi\;} & \mathcal{E}' \\[4pt]
p \downarrow & & \downarrow p' \\[4pt]
\mathcal{X} & \xrightarrow{\;f\;} & \mathcal{X}'.
\end{array}
$$

Two bundles over the same base $\mathcal{X}$ are called *equivalent* if there is a bundle map with $f = id_{\mathcal{X}}$ and Φ a diffeomorphism.

A bundle $\mathcal{E} \xrightarrow{F} \mathcal{X}$ is called a *trivial bundle* if it is equivalent to the bundle $\mathcal{X} \times F \xrightarrow{F} \mathcal{X}$.

We are mainly interested in *vector* or *line bundles* where the fiber is $F := V$ a vector space (with $\dim V = n = 1$ for the case of a line bundle) and the maps h_x are linear in each fiber $p^{-1}(x'), x' \in U_x$.

Primarily, we think of real vector spaces and real manifolds but we also will treat complex cases.

In any case, the dimension of the fiber as a vector space is called the *rank* of the vector bundle.

To make this more explicit and show a way to construct bundles, we take an atlas of the real manifold $\mathcal{X}$ given by open sets U_i ($i \in I$), and $V = \mathbf{R}^n$. Then we have homeomorphisms

$$h := h_i : E_{U_i} = p^{-1}(U_i) \longrightarrow U_i \times \mathbf{R}^n,$$

which a) respect the fibers, i.e. with

$$h(m) = (x, v) \quad \text{for all } m \in E_x = p^{-1}(x), x \in U_i,$$

and b) restrict to vector space isomorphisms

$$h\mid_{E_x} : E_x \longrightarrow \mathbf{R}^n.$$

We denote by

$$h_{i,j} := h_i \circ h_j^{-1}\mid_{(U_i \cap U_j) \times \mathbf{R}^n} : (U_i \cap U_j) \times \mathbf{R}^n \longrightarrow (U_i \cap U_j) \times \mathbf{R}^n$$

the *bundle transition maps* and introduce (differentiable) functions, which are called *bundle transition functions*

$$\psi_{i,j} : U_i \cap U_j \longrightarrow \mathrm{GL}(n, \mathbf{R}) \simeq \mathrm{Aut}\,(V)$$

by

$$h_{i,j}(x, v) = (x, \psi_{i,j}(x)v) \quad \text{for all } (x, v) \in (U_i \cap U_j) \times \mathbf{R}^n.$$

It is not difficult to see that these $\psi_{i,j}$ are uniquely determined and for i, j, k with

$$U_i \cap U_j \cap U_k \neq \emptyset$$

fulfill the following *cocycle condition*

$$(7.26) \qquad\qquad\qquad \psi_{i,j} \circ \psi_{j,k} = \psi_{i,k}.$$

Here, our main point is that we can construct bundles with fiber $\mathbf{R}^n$ over a manifold with atlas $(U_i)_{i \in I}$ by *pasting together* columns $U_i \times \mathbf{R}^n$ via such a given cocycle system $\psi_{i,j}$. We have the following construction principle, whose validity is not too difficult to prove.

Proposition 7.4: Let $\mathcal{X}$ be a manifold with atlas $(U_i)_{i \in I}$ and $\psi_{i,j} : U_i \cap U_j \longrightarrow \mathrm{GL}(n, \mathbf{R})$ be a system of smooth functions with

$$\psi_{i,i} = 1 \quad \text{in} \quad U_i,$$
$$\psi_{i,j} \circ \psi_{j,i} = 1 \quad \text{in} \quad U_i \cap U_j,$$
$$\psi_{i,j} \circ \psi_{j,k} \circ \psi_{k,i} = 1 \quad \text{in} \quad U_i \cap U_j \cap U_k.$$

Let $\tilde{\mathcal{E}}$ be the union of the manifolds $U_i \times \mathbf{R}^n$. One says that points $(x_i, v_i) \in U_i \times \mathbf{R}^n$ and $(x_j, v_j) \in U_j \times \mathbf{R}^n$ are equivalent if $x_i = x_j$ and $v_i = \psi_{i,j}(x_j)v_j$. Then the factor space $\mathcal{E}$ of $\tilde{\mathcal{E}}$ defined by this equivalence relation is a vector bundle with fiber $\mathbf{R}^n$ over $\mathcal{X}$ with projection p induced by the natural projections $(x_i, v_i) \longmapsto x_i$. And each such fiber bundle is equivalent to a bundle of this type.

Proposition 7.5: Let $\mathcal{E}$ and $\mathcal{E}'$ be vector bundles of the same rank n over $\mathcal{X}$ defined by the systems of bundle transition functions $\psi_{i,j}$ and $\psi'_{i,j}$ (without loss of generality, by a refinement argument, one can assume that both bundles are defined by the same atlas $(U_i)_{i\in I}$). Then both bundles are equivalent iff there is a family of matrix functions $f_i : U_i \longrightarrow GL(n, \mathbf{R})$ such that one has

$$\psi'_{i,j} = f_i^{-1}\psi_{i,j}f_j \quad \text{for all } i,j \in I.$$

As in our first approach, for open sets $U \subset \mathcal{X}$ we introduce (vector) spaces $\Gamma(U, \mathcal{E})$ of sections s over U, i.e. smooth maps $s : U \longrightarrow \mathcal{E}$ with $p \circ s = id_U$. For $U = \mathcal{X}$ we talk of *global sections*.

Parallel to our treatment of real bundles, one can look at the complex case by replacing appropriately $\mathbf{R}$ by $\mathbf{C}$.

Examples:

Example 7.20: The most prominent example, which led to the introduction of the notion of bundles, is the *tangent bundle* TM, which comes about if for an n-dimensional manifold M with charts $(U_i, \varphi_i), i \in I$, as matrix of bundle transition functions one takes the Jacobian J of the manifold transition functions $\varphi_{i,j}$

$$\psi_{i,j}(m) := J_{\varphi_{i,j}}(\varphi_j(m)).$$

Then TM can be thought of as consisting of classes of pairs $(m, a)_\sim, m \in M, a \in \mathbf{R}^n$, with $(m, a) \sim (m, b)$ iff a and b are related by the *contravariant* transformation property

$$b_i = \sum_{j=1}^n \frac{\partial y_i}{\partial x_j}(y)a_j$$

when m appears in the chart (U, φ) with coordinates $x = (x_1, \ldots, x_n)$ and simultaneously in the chart (U', φ') with coordinates $y = (y_1, \ldots, y_n)$. This has a geometrical background: in first approximation, we take the manifold M and to each point $m \in M$ associate as a fiber E_m the *tangent space* $T_m M$ to M at this point m. From higher calculus one knows that there are different ways to describe the elements $X_m \in T_m M$ i.e. the *tangent vectors* to a manifold. We follow the one already prepared in 6.2 while discussing the definition of the Lie algebra associated to a linear group and think of the tangent space of M at m as the space of all tangent vectors of all smooth curves γ in M passing through m. More precisely, a *tangent vector* X_m to M at the point m is defined as the equivalence class of smooth parametrized curves $\gamma = \gamma(t)$ passing through m. Two curves γ and $\tilde{\gamma}$ in M with $\gamma(0) = \tilde{\gamma}(0) = m$ are equivalent iff their tangent vectors in m coincide in the coordinates of a chart (U, φ) with $m \in U$, i.e. one has $\gamma \sim \tilde{\gamma}$ if there is a chart (U, φ) with $m \in U$ and coordinates $\varphi(m) = x$ such that

$$\frac{d}{dt}\varphi(\gamma(t))\,|_{t=0} = \frac{d}{dt}\varphi(\tilde{\gamma}(t))\,|_{t=0}.$$

With respect to this chart, a tangent vector can be symbolized by the n-tupel of real numbers a_i where a_i is just the i-th component of $\frac{d}{dt}\varphi(\gamma(t))|_{t=0}$ or by the differential operators $X_m = \sum_{i=1}^n a_i\partial_{x_i}$ acting on smooth functions f in the coordinates x. Addition and scalar multiplication of these operators resp. tuples define a vector space structure, which can and will be carried over to the first given description of the space $T_m M$.

If (U', φ') with coordinates y is another chart for m and $(b_1, \ldots, b_n)$ are the components of the tangent vector to γ expressed in this coordinate system, by the chain rule we have with $\varphi'(\varphi^{-1}(x)) = y(x)$

$$b_i = (\frac{d}{dt} \varphi'(\gamma(t)) \mid_{t=0})_i = (\frac{d}{dt} (\varphi'(\varphi^{-1}(\varphi(\gamma(t))))) \mid_{t=0})_i = \sum_j \frac{\partial y_i}{\partial x_j} a_j,$$

i.e. the contravariant transformation property from our general construction above.

Definition 7.5: Global sections $X : M \longrightarrow TM$ of the tangent bundle TM are called *vector fields* on M.

Let (U, φ) be a chart with coordinates $x = (x_1, \ldots, x_n)$. With a (now usual) slight misuse of notation we write

$$X \mid_U =: \sum_{i=1}^n a_i(x) \partial_{x_i}$$

where the coefficients a_i are smooth functions defined in U. For a chart (U', φ') with coordinates $y = (y_1, \ldots, y_n)$, one has as well

$$X \mid_{U'} = \sum_{j=1}^n b_j(y) \partial_{y_j}$$

and on $U \cap U' \neq \emptyset$ we have the contravariant transformation property

$$b_j(y) = \sum_i \frac{\partial y_j}{\partial x_i}(y) \, a_i(x(y)).$$

$V(M)$ denotes the vector space of smooth vector fields X on M.

Example 7.21: The *cotangent bundle* to M has as its fibers the cotangent spaces $T_m^* M$, i.e. the duals to the tangent spaces $T_m M$. It turns up if, in the same situation as above in Example 7.20, as matrix of bundle transition functions we take the transpose of the Jacobian of the map $\varphi_j \circ \varphi_i^{-1} =: \varphi_{j,i}$, i.e.

$$\psi_{i,j}(m) = {}^t J_{\varphi_{j,i}}(\varphi_i(m)).$$

The global sections $\alpha : M \longrightarrow T^* M$ of the cotangent bundle $T^* M$ are called *differential forms of first degree* or simply *one-forms* on M. As a parallel to the notation for vector fields above, with functions a_i^* and b_j^* defined in U resp. U' we write

$$\alpha \mid_U = \sum_i a_i^*(x) dx_i, \quad \text{resp.} \quad \alpha \mid_{U'} = \sum_j b_j^*(y) dy_j,$$

and here have the *covariant* transformation property

(7.27)
$$b_j^*(y) = \sum_j \frac{\partial x_i}{\partial y_j}(x(y)) a_i^*(x(y)).$$

$\Omega^1(M)$ denotes the vector space of one-forms α on M.

Example 7.22: The first two examples concerned real bundles but similarly one can treat the complex case. Here we take the one-dimensional complex manifold $M = \mathbf{P}^1(\mathbf{C})$ covered by the two charts (U, φ) and (U', φ') fixed above in (8.10) while discussing M as a manifold. For $k \in \mathbf{Z}$ we get a line bundle $\mathcal{L}_k$ by choosing $\psi(w) = w^{-k}$ as bundle transition function between these two charts. Let us work out the construction of this bundle along the line of the general construction principle given above: We have two "columns", one over U and one over U', namely

$$L := \{(w, t), w, t \in \mathbf{C}\}, \quad L' := \{(w', t'), w', t' \in \mathbf{C}\}.$$

Using ψ we paste them together over $U \cap U' = \mathbf{C}^*$ by identifying $(w, t) \sim (w', t')$ iff

$$w' = 1/w, \quad t' = w^{-k}t.$$

Remark 7.9: It can be shown that up to (holomorphic) equivalence there are no other holomorphic line bundles over $\mathbf{P}^1(\mathbf{C})$ than these $\mathcal{L}_k$.

Proposition 7.6: For the space of holomorphic sections one has

$$\dim_\mathbf{C} \Gamma(\mathcal{L}_k) = k + 1 \qquad \text{for } k \geq 0,$$
$$= 0 \qquad \text{for } k < 0.$$

Proof: Let s be a section of $\mathcal{L}_k$ with

$$s : (u : v) \longmapsto \ell \in \mathcal{L}_k,$$

where $h(\ell) = (w, t) \in L$ for $w = u/v, v \neq 0$, and $h'(\ell) = (w', t') \in L'$ for $w' = v/u, u \neq 0$. The section is holomorphic iff t is given by a holomorphic function in w and t' by a holomorphic function in w' and we have the compatibility condition $t' = w^{-k}t$. As examples and prototypes for such a holomorphic function we take the monomials $t = w^p$ and $t' = w'^{p'}$ with $p, p' \in \mathbf{N}_0$. The compatibilty condition says for $w \neq 0$

$$t'(w') = w'^{p'} = w^k t(w) = w^{-k+p}$$

and, since one has $w' = 1/w$, we get $w^{-p'} = w^{p-k}$ and see that everything is consistent for $p = 0, 1, \ldots, k$ if k is not negative. And the holomorphicity of the relation breaks down for negative k. Hence we have $k + 1$ essentially different holomorphic sections for non-negative k and none for negative k. One needs a bit more complex function theory to be sure that there are no more (see for instance [Ki] p.80).

Bundles and Representations

After this short digression into the world of bundles, which will be useful later, we return to the problem of the construction of group representations. As we know from our first approach to bundles, the space of sections is meant to provide a representation space and, hence, should get the structure of a Hilbert space. In general this needs some work, which we shall attack in the next chapter. But if the space of sections is a finite-dimensional vector space, this is no problem. To get an idea we return to our example $G = \mathrm{SU}(2) \supset H \simeq \mathrm{U}(1)$ and the unitary representation $\pi_j = ind_H^G \pi_0$ induced from the representation $\pi_0 = \chi_k, k = 2j \geq 0$, of $H \simeq \mathrm{U}(1)$. In our first approach we constructed the bundle space $\mathcal{B}_k = \{[g,t]; g \in G, t \in \mathbf{C}\}$ with projection pr onto $\mathcal{X} = G/H = \mathbf{P}^1(\mathbf{C})$ given by

$$pr([g,t] := gH = (u : v) = (b : \bar{a}) \quad \text{for } g = \begin{pmatrix} a & b \\ -\bar{b} & \bar{a} \end{pmatrix}.$$

We define a map of $\mathcal{B}_k$ to the line bundle $\mathcal{L}_k$ constructed by pasting together the sets $L = \{(w,t); w, t \in \mathbf{C}\}$ and $L' = \{(w'.t'); w', t' \in \mathbf{C}\}$ via the bundle transition function $\psi(w) = w^k$ as follows: For g with $a \neq 0$ we put

$$\mathcal{B}_k \ni [g,t] \longmapsto (w = b/\bar{a}; \bar{b}^k t) \in L$$

and for g with $b \neq 0$

$$\mathcal{B}_k \ni [g,t] \longmapsto (w' := \bar{a}/b, a^k t) \in L'.$$

Both maps are well defined since for $h = \begin{pmatrix} \zeta & \\ & \zeta^{-1} \end{pmatrix}$ one has

$$gh^{-1} = \begin{pmatrix} a\zeta^{-1} & b\zeta \\ -\bar{b}\zeta^{-1} & \bar{a}\zeta \end{pmatrix}$$

and

$$\pi_0(h) = \zeta^k.$$

Hence application of the mapping prescription to the element $[gh^{-1}, \pi_0(h)t]$, which as well represents $[g,t]$, leads to the same images. And the images in L and L' are consistent with the pasting condition $(w',t') = (1/w, w^{-k}t)$ for $w \neq 0$, since with $w = b/\bar{a}$ one has $w^{-k}(\bar{b}^k t) = \bar{a}^k t$. Thus both maps induce a map of $\mathcal{B}_k$ to $\mathcal{L}_k$, which obviously is a bijection.

Here we have a complex line bundle over the one-dimensional complex manifold $\mathbf{P}^1(\mathbf{C})$ whose space of holomorphic sections is a complex vector space and the representation space for the representation $\pi_j, 2j = k$ of $G = SU(2)$. In principle, the base spaces for the bundles in our first approach are real manifolds and the section spaces real vector spaces. This example shows that we may get the wanted irreducible representation by some more restrictions, namely we put more structure on the real construction to get the more restrictive complex case (as in holomorphic induction). We shall see more of this under an angle provided from the necessities of physics in the next chapter.

Chapter 8

Geometric Quantization and the Orbit Method

The *orbit method* is an object of pure mathematics, but it got a lot of impetus from physics under the heading of *geometric quantization*. Historically it was proposed by Kirillov in [Ki2] for the description of the unitary dual of nilpotent Lie groups. By further work of Kirillov and many others, in particular B. Kostant, M. Duflo, M. Vergne and D. Vogan, it grew into an important tool for explicit constructions of representations of various types of groups. The construction mechanism uses elements that also appear in theoretical physics when describing the transition from classical mechanics to quantum mechanics. We therefore start by recalling some of this background, even though we cannot give all the necessary definitions.

8.1 The Hamiltonian Formalism and its Quantization

The purpose of theoretical mechanics is to describe the time development of the state of a physical system. In classical mechanics such a state is identfied with a point in an n-dimensional manifold Q, the *configuration space*. The point is usually described by local coordinates $q_1, \ldots, q_n$, called *position variables*. The time development of the system then is described by a curve γ given in local coordinates as $t \longmapsto (q_i(t))$, which is fixed by an ordinary differential equation of second order. In the Hamiltonian formalism one adds to the position variables $q = (q_1, \ldots, q_n)$ the *momentum variables* $p = (p_1, \ldots, p_n)$ (related to the velocity of the particle or system, we do not need to be more exact here) and the time development $t \longmapsto (q(t), p(t))$ is described by the hypothesis that the curve $t \longmapsto (q(t), p(t)) =: \gamma(t)$ has to fulfill *Hamilton's equations*, i.e. the first order system of partial differential equations

$$(8.1) \qquad \dot{q}_i := \frac{dq_i}{dt} = \frac{\partial H}{\partial p_i}, \quad \dot{p}_i := \frac{dp_i}{dt} = -\frac{\partial H}{\partial q_i}$$

where the function $H = H(q, p, t)$ is the *Hamiltonian* of the system, i.e. encodes all the relevant information about the system. For instance, if our system is simply a mass one (point) particle in $\mathbf{R}^3$ moving under the influence of the force with a radial symmetric potential $V(q) = 1/r, r = \sqrt{q_1^2 + q_2^2 + q_3^2}$, one has $H = p_1^2 + p_2^2 + p_3^2 + V(q)$.

This *Hamilton formalism* has several interpretations:

1. The Geometric Picture

A curve $\gamma = \gamma(t)$ is called an *integral curve* for the *vector field*

$$X = a(q,p)\partial_q + b(q,p)\partial_p := a_1(q,p)\,\partial_{q_1} + \ldots a_n(q,p)\,\partial_{q_n} + b_1(q,p)\,\partial_{p_1} + \cdots + b_n(q,p)\,\partial_{p_n}$$

iff at each point of the curve its tangent vector is a vector from the vector field. We write this as

$$\dot{\gamma}(t) = X(\gamma(t)).$$

The statement that γ is an integral curve for the *Hamiltonian vector field*

$$X = X_H \text{ with } a_i = \frac{\partial H}{\partial p_i},\ b_i = -\frac{\partial H}{\partial q_i},$$

is an expression of the assertion that γ fulfills Hamiltons equations (8.1).

2. The Symplectic Interpretation

To state this, we have to provide some more general notions about differential forms, which we will need anyway in the following sections. Hence we enlarge our considerations about the tangent and cotangent bundle and their sections from the last section: We look again at a (real) smooth n-manifold M.

– A *vector field* X is defined as a section of the tangent bundle TM and in local coordinates $x = (x_1, \ldots, x_n)$ written as

$$X = a(x)\partial_x := a_1(x)\partial_{x_1} + \cdots + a_n(x)\partial_{x_n}.$$

– A *differential form of degree one* or *one-form* α is defined as a section of the cotangent bundle T^*M and in local coordinates written as

$$\alpha = a(x)dx := a_1(x)dx_1 + \cdots + a_n(x)dx_n.$$

As one knows from calculus, there also are higher order differential forms of different types. We shall need the symmetric and the antisymmetric or exterior forms.
– A *symmetric r-form* β is written in the local coordinates from above as

$$\beta := \sum b_{i_1\ldots i_r}(x)dx_{i_1} \cdots \cdots dx_{i_r}$$

where the coefficients b are (smooth) functions of the coordinates x and the differentials dx_i are commuting symbols, i.e. with $dx_i \cdot dx_j = dx_j \cdot dx_i$ for all i, j, the sum is meant over all r-tuples $i_1, \ldots, i_r \in \{1, \ldots, n\}$ and the coefficients b are invariant under all permutations of the indices. For instance, if $n = 2$, one has the two-form

$$\beta = b_{11}dx_1^2 + b_{12}dx_1dx_2 + b_{21}dx_2dx_1 + b_{22}dx_2^2,\ b_{12} = b_{21}.$$

Certainly, there is the possibility to introduce a *reduced form* where one restricts the index-tuples by taking exactly one representative from each orbit of the symmetric group $\mathfrak{S}_r$ acting on the index-tuples. Hence, as reduced form one has in the example above

$$\beta = b_{11}dx_1^2 + \tilde{b}_{12}dx_1dx_2 + b_{22}dx_2^2,\ \tilde{b}_{12} = 2b_{12}.$$

If one wants to be more sophisticated (and precise), one says that a symmetric $r-$form is a (smooth) section of the $r-$th symmetric power of the cotangent bundle. But for our treatment here and for the antisymmetric case below the naive description above should be sufficient.

An *antisymmetric* or *exterior $r-$form* α or *form of degree r* is given in local coordinates by

$$\alpha = \sum a_i(x)dx_{i_1} \wedge \cdots \wedge dx_{i_r}$$

where here the differentials are anticommuting, i.e. with $dx_i \wedge dx_j = -dx_j \wedge dx_i$ for all i, j. The summation is again over all $r-$tuples $i = (i_1, \ldots, i_r)$ and the coefficients have the special property of antisymmetry

$$a_i = \operatorname{sgn}\sigma \, a_{\sigma i} \quad \text{for all } i = (i_1, \ldots, i_r), \ \sigma \in \mathfrak{S}_r,$$

hence, in particular one has $a_i = 0$ if two of the indices coincide. If we take in our example $n = r = 2$, we get

$$\alpha = a_{12}dx_1 \wedge dx_2 + a_{21}dx_2 \wedge dx_1, \ a_{12} = -a_{21}.$$

It is very convenient to introduce here the *reduced form* where the sum is taken only over all $r-$tuples $(i) = (i_1, \ldots, i_r)$ with $1 \leq i_1 < \cdots < i_r \leq n$. Hence the reduced form of our example is

$$\alpha = \tilde{a}_{12}dx_1 \wedge dx_2, \ \tilde{a}_{12} = 2a_{12}.$$

After changing coordinates the differential forms are transformed according to an appropriate generalization of the transformation property (7.27) in Example 7.21 in 7.6 and the complete definition is again given by characterizing exterior $r-$forms as smooth sections of the $r-$th exterior power of the cotangent bundle.

The set of these exterior $r-$forms on $U \subset M$ is denoted by $\Omega^r(U)$. By the operations on the coefficients one has a natural manner to define a vector space structure on $\Omega^r(U)$ and one has the possibility to do scalar multiplication with functions. Moreover one has an associative multiplication if one multiplies a $p-$form with a $q-$form distributively regarding the anticommutativity of the dx_i. It is decisive for the application of differential forms that we have still two more operations, the *differentiation d* augmenting the degree by one and the *inner multiplication $i(X)$* with a vector field X lowering the degree by one (moreover there is the *Lie derivative*, which we will not consider for the moment). We will have to be content again with the description in local coordinates.

For $\alpha = \sum a_{(i)}dx_{i_1} \wedge \cdots \wedge dx_{i_r}$ we define the *exterior differentiation $d\alpha$* by

$$d\alpha := \sum da_{(i)} \wedge dx_{i_1} \wedge \cdots \wedge dx_{i_r}$$

where da denotes the usual *total differential* of a function a, i.e. is given by

$$da := \sum_{i=1}^{n} \frac{\partial a}{\partial x_i} dx_i.$$

To handle the exterior differential, one usually reorders it to its reduced form.

A simple but most important example appears if we look at the coordinates given by $x = (q_1, \ldots, q_n, p_1, \ldots, p_n)$ and take the one-form $\alpha = \sum_{i=1}^n q_i dp_i$. We get

$$(8.2) \qquad\qquad \omega_0 := d\alpha = \sum_{i=1}^n dq_i \wedge dp_i$$

and, obviously, have $d\omega_0 = 0$.
We note that the relation $dd\alpha = 0$ holds in general for all α.

A form α is called *closed* iff one has $d\alpha = 0$ and it is called *exact* iff there is a form β such that $\alpha = d\beta$.

Moreover, we point to the fact that exterior differentiation commutes with substitution of the variables and, if $\Phi : M \longrightarrow M'$ is a smooth map of manifolds and α' a form on M', then one can *draw back* α' to a form $\Phi^*\alpha'$ on M. Without being more pecise here, for practical purposes this comes out by substitution of the local coordinates $x \longmapsto y = y(x)$ and using the well known formula $dy_i = \sum_j \frac{\partial y_i}{\partial x_j} dx_j$. We do not have enough space to go into this more deeply but later on shall see in examples how this works.

For $\alpha = \sum a_{(i)}(x) dx_{i_1} \wedge \cdots \wedge dx_{i_r}$ and the vector field $X = \sum b_j(x) \partial_j$, we have an *inner product* $i(X)\alpha$, which comes out (this certainly is not the most elegant definition, but avoids more generalities) by applying successively the prescriptions

$$i(X)dx_i = b_i(x)$$

and

$$i(X)(\alpha \wedge \beta) = (i(X)\alpha) \wedge \beta + (-1)^r \alpha \wedge (i(X)\beta) \text{ for } \alpha \in \Omega^r, \beta \in \Omega^q.$$

To show how this works, we give an example, which is the most interesting for us: As above, we take coordinates $x = (q_1, \ldots, q_n, p_1, \ldots, p_n)$ and $\alpha = \omega = \sum_{i=1}^n dq_i \wedge dp_i$ and $X = \sum_{i=1}^n (a_i(q,p) \partial_{q_i} + b_i(q,p) \partial_{p_i})$. Then we have

$$i(X)\omega = \sum_{i=1}^n (a_i dp_i - b_i dq_i).$$

Finally, we use this for the symplectic formulation of the Hamilton formalism: We go from the configuration space Q with coordinates q to its cotangent bundle $M = T^*Q$, called the *phase space*, with coordinates (q, p). We provide this space with the two-form $\omega = \sum_{i=1}^n dq_i \wedge dp_i$ (and hence get the protype of a *symplectic manifold*, which will be analysed later more thoroughly). Now we can say that a curve γ provides a solution to Hamilton's equations (8.1) if it is an integral curve of the vector field X, which fulfills the equation

$$(8.3) \qquad\qquad dH = i(X)\omega.$$

Namely, in this case one has

$$\frac{\partial H}{\partial q_i} = -b_i, \quad \frac{\partial H}{\partial p_i} = a_i,$$

i.e. X is a Hamiltonian vector field X_H.

3. The Poisson Interpretation

In 6.1 we introduced the *Poisson bracket* for two smooth functions f and g with coordinates (q, p)

$$\{f, g\} := \sum_{i=1}^{n} \frac{\partial f}{\partial q_i} \frac{\partial g}{\partial p_i} - \frac{\partial f}{\partial p_i} \frac{\partial g}{\partial q_i}.$$

Hence Hamilton's equations (8.1) can also be expressed as

$$\dot{q} = \{q, H\}, \ \dot{p} = \{p, H\}.$$

And more generally, if one replaces the special *observables* positions q_i and momenta p_i by a more general *observable* f, i.e. a smooth function of (q, p), as equation of time development one has

(8.4)
$$\dot{f} = \{f, H\}.$$

Here we find an appropriate starting point to do quantization, i.e. to go over to a quantum mechanical description of the system. A quantum mechanical system is modeled by a complex Hilbert space $\mathcal{H}$. The state of the system is represented by a one-dimensional subspace $L = <\psi>$ of $\mathcal{H}, \psi \in \mathcal{H} \setminus \{0\}$, which can be seen as a point $\psi_\sim$ in the projective space

$$\mathbf{P}(\mathcal{H}) := \{\psi_\sim; \psi \in \mathcal{H} \setminus \{0\}\}, \ \psi \sim \psi' \ \text{iff} \ \psi' = \lambda\psi \ \text{for a} \ \lambda \in \mathbf{C}^*.$$

Then the time development reflects in a curve $\gamma(t) = \psi(t)_\sim$ in $\mathbf{P}(\mathcal{H})$ and the physical law determining this development is encoded into a one-parameter group G of unitary operators $U(t), t \in \mathbf{R}$, such that γ is a G−orbit in $\mathbf{P}(\mathcal{H})$. A physical observable f should correspond to an operator $B = B_f$ acting on or in $\mathcal{H}$ (as there are subtle problems, we will not be more precise here) and $< B\psi, \psi >$ provides the information about the probability to measure the observable in the state $< \psi >$. Similar to our treatment of infinitesimal generators of one-parameter subgroups in section 6.2, one can look for a skew-adjoint operator A such that (in some precise sense) one has $U(t) = \exp tA$ and, hence, the counterpart of the classical equation of motion will be

$$\frac{d\psi}{dt} = A\psi.$$

In the physics literature, usually one introduces the (self-adjoint) *Hamilton operator* $\hat{H}$ by another normalization, namely

$$U(t) =: e^{it\hat{H}}$$

(we use units such that Planck's constant is one) and then has the *Schrödinger equation*

$$\frac{d\psi}{dt} = i\hat{H}\psi.$$

In our standard example of a system consisting of a particle with mass m moving in $\mathbf{R}^3$ in a central force field with potential V, one translates the position variables q_j from the classical world into operators B_{q_j} acting by multiplication with iq_j, the momentum variables p_j go to the differential operators $B_{p_j} = \partial_{q_j}$ acting on (a subspace of smooth functions in) the Hilbert space $\mathcal{H} = L^2(\mathbf{R}^3)$ and one has the Schrödinger equation

$$\frac{d\psi}{dt} = i\hat{H}\psi = i((1/2m) \sum_{j=1}^{3} \partial_{q_j}^2 - V(q))\psi.$$

In general, it can be a serious problem to find a Hilbert space $\mathcal{H}$ and operators corresponding to the symplectic space and the observables from the classical description. For some discussion of this we refer to Chapter 5 in [Be]. For our purposes here, we retain the following observation even though it is somewhat nebulous:

In important cases the behaviour of a physical system is guided by its *symmetry group G*, i.e. a group of transformations, which do not change the physical content of the system. If we determine unitary representations of G we will be provided with a Hilbert space and a lot of unitary operators.

In the next sections we shall see that important groups, like the Heisenberg group, the Euclidean and the Poincaré group, $SU(2)$ and $SL(2, \mathbf{R})$, give rise to symplectic manifolds and line bundles living on these such that the spaces of sections of these bundles are representation spaces of unitary representation of G.

8.2   Coadjoint Orbits and Representations

We start by analysing the construction of bundles on symplectic manifolds. Beside the already mentioned sources [Ki] and [Ki1], for further studies we recommend the book by Woodhouse [Wo]. We return to the realm of mathematics though we still use some terms from physics, which should help to give some motivation.

8.2.1   Prequantization

In Section 8.1 we already established the cotangent bundle T^*Q to a manifold Q and in particular the space $\mathbf{R}^n$ with coordinates $(q_1, \ldots, q_n, p_1, \ldots, p_n)$ and provided with the *standard two-form* (8.2)

$$\omega_0 := \sum_{j=1}^n dq_j \wedge dp_j$$

as prototypes of a symplectic manifold. In general, one has the fundamental notion:

Definition 8.1: A smooth (real) p-manifold M is called a *symplectic manifold* if it is provided with a closed nondegenerate *symplectic* two-form ω defined on M, that is an $\omega \in \Omega^2(M)$, such that

– i) $d\omega = 0$,
and
– ii) on each tangent space $T_m M, m \in M$, the condition holds: if for $X \in T_m M$ one has

$$\omega_m(X, Y) = 0 \quad \text{for all } Y \in T_m M,$$

then $X = 0$.

Slightly paraphrasing the condition ii), we remark that M neccessarily has to have even dimension $p = 2n$ and that for ω in $m \in M$ with local coordinates $(q_1, \ldots, q_n, p_1, \ldots, p_n)$ given by (8.2)

$$\omega_0 = \sum_{j=1}^{n} dq_j \wedge dp_j$$

and for vector fields X, Y given by

$$X = \sum_{j=1}^{n}(a'_j(q,p)\,\partial_{q_j} + a''_j(q,p)\,\partial_{p_j}); \ Y = \sum_{j=1}^{n}(b'_j(q,p)\,\partial_{q_j} + b''_j(q,p)\,\partial_{p_j}),$$

from the general duality formula $dx_i(\partial_{x_j}) = \delta_{ij}$ and the antisymmetry of the forms one has the relation

$$\omega_0(X,Y) = \sum_{j=1}^{n}\left(a'_j(q,p)\,b''_j(q,p) - a''_j(q,p)\,b'_j(q,p)\right).$$

At first sight, this form ω_0 may look very special. But there is a beautiful theorem stating that this really a general situation:

Theorem 8.1 (*Darboux*): For every point m of a symplectic manifold (M, ω) of dimension $2n$, there is an open neighbourhood U of m and a smooth map

$$\varphi : U \longrightarrow \mathbf{R}^{2n} \text{ with } \varphi^*\omega_0 = \omega|_U,$$

where ω_0 is the standard symplectic form (8.2) on $\mathbf{R}^{2n}$.

The examples of symplectic manifolds, which are of interest for us, are the coadjoint orbits belonging to the action of a group G on the dual $\mathfrak{g}^*$ of its Lie algebra $\mathfrak{g}$. For instance the projective space $\mathbf{P}^1(\mathbf{C})$ will come out as such an orbit. Before we go into this, we elaborate a bit on the way to construct bundles on symplectic manifolds. As we see it, this construction essentially goes back to Weil's book *Variétés Kählériennes* [We].

Definition 8.2: A symplectic manifold (M, ω) is called *quantizable* if ω is *integral*.

The integrality of ω can be defined in several equivalent ways. Perhaps the most elementary one is to require (as in [Wo] p.159):

(I.) The integral of ω over any closed oriented surface in M is an integral multiple of 2π.

More elegant is the requirement

(II.) ω determines a class $[\omega]$ in $H^2(M, \mathbf{Z})$.

To understand **(II.)**, one has to have at hand the notions of the *second cohomology group* $H^2(M, \mathbf{R})$ and its integral subgroup $H^2(M, \mathbf{Z})$. Though we are not too far from the possibility to define these, we renounce to do it here but discuss the construction of the *prequantum line bundle* where we can see how this condition works (later we shall be led to a more practical criterion to decide the quantizability of the manifolds showing up as coadjoint orbits):

Recalling our definitions from 7.6, this bundle can be constructed as follows. We take a contractible open cover $\mathfrak{U} = (U_j)$ of M. Then Poincaré's Lemma assures that on each U_j, there exists a real one–form β_j such that

$$\omega = d\beta_j.$$

The same way, one knows that on each contractible $U_j \cap U_k \neq \emptyset$ a smooth real function $f_{jk} = -f_{kj}$ exists such that

$$df_{jk} = \beta_j - \beta_k$$

and, for each contractible $U_j \cap U_k \cap U_\ell \neq \emptyset$, there is a constant $a_{jk\ell}$ such that

$$2\pi a_{jk\ell} = f_{jk} + f_{kl} + f_{\ell j} \text{ on } U_j \cap U_k \cap U_\ell.$$

It is the decisive point that the integrality of ω says that the $a_{jk\ell}$ may be assumed to be integers (this is really the essence of the definition above). If we put

$$c_{jk} := \exp i f_{jk},$$

we have

$$dc_{jk} = ic_{jk}(\beta_j - \beta_k)$$

and for $a_{jkl} \in \mathbf{Z}$ on each non–empty $U_j \cap U_k \cap U_\ell$

$$c_{jk}c_{kl}c_{lj} = \exp(2\pi i a_{jk\ell}) = 1.$$

Hence Proposition 7.4 in 7.6 shows that the c_{jk} are transition functions for a line bundle B over M. This bundle has a *compatible Hermitian structure* $<,>$ (see [Wo] p.264), which we define below in 8.2.3. Since in Poincaré's Lemma a form β with $d\beta = \alpha$ is not unique (it can changed by an exact form), this construction is not unique. The general statement is that these constructions of prequantum bundles B are parametrized by $H^1(M,\mathbf{R})/H^1(M,\mathbf{Z})$ ([Wo] p.286).

As it is our goal to construct unitary group representations, we need a Hilbert space. The Hilbert space $\mathcal{H}$ in our prequantization construction (the *prequantum Hilbert space*) is the space of all $s \in \Gamma(M, B)$, i.e. smooth global sections of B for which in some sense an integral of $(s,s)\varepsilon_\omega$ over M exists and is finite. Here the measure ε_ω is associated to the *Liouville form*

$$(-1)^{n(n-1)/2}\frac{1}{n!}\omega^n$$

known from calculus. For $s_1, s_2 \in \mathcal{H}$ the inner product on $\mathcal{H}$ is

$$< s_1, s_2 > := \frac{1}{(2\pi)^n} \int_M (s_1, s_2)\varepsilon_\omega.$$

8.2.2 Example: Construction of Line Bundles over $M = \mathbf{P}^1(\mathbf{C})$

The topic of line bundles on projective space is treated in each text book on algebraic geometry (for instance, see Griffiths-Harris [GH] p.144) and we already constructed holomorphic line bundles over $\mathbf{P}^1(\mathbf{C})$ at the end of section 7.6. But to illustrate the procedure proposed above, for those who – like the author – enjoy explicit descriptions, we add here the following very elementary constructions (to be skipped if one is in a hurry).

1. Real Analytic Bundles $B \longrightarrow \mathbf{P}^1(\mathbf{C})$

We look at $\mathbf{P}^1(\mathbf{C})$ as a real two-dimensional manifold and take the covering $\mathfrak{U} = (U_j)_{j=1,\dots,6}$ with

$$
\begin{aligned}
U_1 &= \{(z:1); \, z = x + iy \in \mathbf{C}; \, |z| < 1\}, \\
U_2 &= \{(1:w); \, w = u + iv \in \mathbf{C}; \, |w| < 1\}, \, w = 1/z = (x - iy)/(x^2 + y^2), \\
U_{3,4} &= \{(z:1); \, z \in \mathbf{C}; \, \operatorname{Re} z = x \gtrless 0\}, \\
U_{5,6} &= \{(z:1); \, z \in \mathbf{C}; \, \operatorname{Im} z = y \gtrless 0\}.
\end{aligned}
$$

This covering is chosen such that all intersections of the covering neighborhoods are simply connected (this is not the case for the covering by U and U', which we used in Example 7.18 in 7.6 as one has $U \cap U' \simeq \mathbf{C}^*$). With $c = \ell/\pi \in \mathbf{R}$ we take as a symplectic form on M

$$
\begin{aligned}
\omega &:= c\frac{dx \wedge dy}{(1 + x^2 + y^2)^2} = c\frac{i}{2}\frac{dz \wedge d\bar{z}}{(1 + |z|^2)^2} \quad \text{for } (z:1) \in U_j, j \neq 2, \\
&:= c\frac{du \wedge dv}{(1 + u^2 + v^2)^2} = c\frac{i}{2}\frac{dw \wedge d\bar{w}}{(1 + |w|^2)^2} \quad \text{for } (1:w) \in U_2.
\end{aligned}
$$

Exercise 8.1: Verify that this two–form ω in the appropriate coordinates has *potential forms* θ_z and θ_w with $d\theta_z = d\theta_w = \omega$ given by

$$
\begin{aligned}
\theta_z &= \frac{c}{2}\frac{xdy - ydx}{1 + x^2 + y^2} = -c\frac{i}{4}\left(\frac{\bar{z}dz - zd\bar{z}}{1 + |z|^2}\right) \quad \text{for } (z:1) \in U_j, \, j \neq 2, \\
\theta_w &= \frac{c}{2}\frac{udv - vdu}{1 + u^2 + v^2} = -c\frac{i}{4}\left(\frac{\bar{w}dw - wd\bar{w}}{1 + |w|^2}\right) \quad \text{for } (1:w) \in U_2.
\end{aligned}
$$

On $U_j \cap U_2 (j \neq 2)$ we have

$$
\begin{aligned}
\theta_z - \theta_w &= -c\frac{i}{4}\left(\frac{\bar{z}dz - zd\bar{z}}{1 + |z|^2} - \frac{1}{1 + |z|^2}\left(\frac{d\bar{z}}{\bar{z}} - \frac{dz}{z}\right)\right) \\
&= -c\frac{i}{4}\left(\frac{\bar{z}dz - zd\bar{z}}{|z|^2}\right) \\
&= \frac{c}{2}\frac{xdy - ydx}{x^2 + y^2} \\
&= \frac{c}{2}d\tilde{f}_{2j}
\end{aligned}
$$

with a potential function $\tilde{f}_{2j}$. If we fix the tangens as a function on $(-\pi/2, \, \pi/2)$ and $f = \arctan$ as its inverse, we can take

$$
-\tilde{f}_{32}(x, y) = f(y/x) = \arctan(y/x)
$$

and moreover, for instance,

$$
-\tilde{f}_{52}(x, y) = -f(x/y) + \pi/2 = -\arctan(x/y) + \pi/2 \quad \text{and} \quad -\tilde{f}_{42}(x, y) = \pi + \arctan(y/x).
$$

Then, on $U_2 \cap U_4 \cap U_5$, we have

$$\tilde{f}_{24} + \tilde{f}_{45} + \tilde{f}_{52} = -\pi - \arctan(y/x) + \pi/2 - \arctan(x/y)$$
$$= -\pi,$$

and similarly on the other intersections. Thus, we see that ω is *integral* in the sense

i) $\omega = d\theta_j$ on U_j

ii) $\theta_j - \theta_k = df_{jk}$ on $U_j \cap U_k$

iii) $f_{jk} + f_{k\ell} + f_{\ell j} = a_{jk\ell} \in 2\pi\mathbf{Z}$ in $U_j \cap U_k \cap U_\ell$ exactly for $c \in 4\mathbf{Z}$.

This result coincides with the usual volume computation

$$\int_{\mathbf{P}^1(\mathbf{C})} \omega = c \int \frac{dx \wedge dy}{(1+x^2+y^2)^2} = c \int_0^\infty \int_0^{2\pi} \frac{rdrd\delta}{(1+r^2)^2}$$

$$= 2\pi c \int_0^\infty \frac{rdr}{(1+r^2)^2} "s = \pi c \int_1^\infty \frac{du}{u^2} = \pi c$$

telling that the volume is an integral multiple of $4\pi = \mathrm{vol}(S^2)$ exactly for $c = 4n$, $n \in \mathbf{Z}$.

Moreoverer, we see that we can construct the $\mathcal{C}^\infty$–prequantum bundle

$$B \xrightarrow{\pi} M = \mathbf{P}^1(\mathbf{C})$$

in the standard way as

$$B = \{(m, v, j)_\sim; \ m \in M, \ v \in \mathbf{C}, \ j \in I\},$$

where here $I = \{1, \ldots, 6\}$ and

$$(m', v', j') \sim (m, v, j)$$

exactly for

$$m' = m, \ v' = \psi_{j'j}(m)v.$$

The bundle transition functions ψ_{jk} are fixed (in the charts with coordinates x, y) by

$$\psi_{jk}(x, y) = c_{jk} e^{if_{jk}(x,y)},$$

i.e. the different trivializations are glued together by

$$e^{i \arg z^\ell}.$$

By the way, the transition function z^ℓ comes out if one does not take real potential forms as above but complex forms like $\theta'_z = (1 + |z|^2)^{-1}\bar{z}dz$.

2. Holomorphic Bundles $E \longrightarrow \mathbf{P}^1(\mathbf{C})$

Following Weil ([We] p. 90), the same construction can be repeated by looking at $\mathbf{P}^1(\mathbf{C})$ as a one-dimensional holomorphic manifold and searching for holomorphic transition functions

$$F_{jk} : U_j \cap U_k \longrightarrow \mathbf{C}^*.$$

Here we take the same covering $\mathfrak{U} = (U_j)_{j=1,\ldots,6}$ as above and the symplectic form

$$\omega := \frac{\ell}{2\pi i} \frac{dz \wedge d\bar{z}}{(1+|z|^2)^2} = \frac{\ell}{2\pi i} \partial\bar\partial \Phi_j \text{ for } j \neq 2,$$

with

$$\Phi_j(z) := \ell \log(1+|z|^2)$$

resp. for $j = 2$

$$\omega := \frac{\ell}{2\pi i} \frac{dw \wedge d\bar{w}}{(1+|w|^2)^2} = \frac{\ell}{2\pi i} \partial\bar\partial \Phi_2$$

with

$$\Phi_2(w) := \ell \log(1+|w|^2).$$

On $U_j \cap U_2, j \neq 2$, we have

$$\Phi_j(z) - \Phi_2(1/z) = \ell \log(|z|^2).$$

Fixing appropriate branches of the complex log-function, we take for $j \neq 2$

$$f_{2j} := \frac{\ell}{2\pi i} \log z.$$

Then we have

$$\Phi_j(z) - \Phi_2(1/z) = -2\pi i(f_{2j} - \bar{f}_{2j})$$

and

$$\eta_j = -\frac{\ell}{2\pi i} d\Phi_j = \frac{\ell}{2\pi i} d\log(1+|z|^2) = \frac{\ell}{2\pi i} \frac{\bar{z}\,dz}{1+|z|^2}$$

with

$$\omega = \frac{\ell}{2\pi i} \frac{dz \wedge d\bar{z}}{(1+|z|^2)^2}.$$

Using Weil's notation, for the construction of a holomorphic bundle $E = E_\ell$ we have transition functions

$$F_{jk}(z) = \exp\left(2\pi i\, f_{jk}(z)\right) = z^\ell$$

exactly if $\ell \in \mathbf{Z}$. We see immediately that for $\ell \in \mathbf{N}_0$ the holomorphic sections are given in the z-variable by

$$s(z) = z^\lambda, \quad \lambda = 0, 1, \ldots, \ell$$

(as exactly in these cases we have $(1/w)^\lambda w^\ell$ holomorphic in w), i.e. we find

$$\Gamma\left(E_\ell\right) \cong \mathbf{C}^{\ell+1}.$$

As described in a general procedure below, this space consists of the *polarized sections* of the prequantum bundle $B = B_\ell$, where the polarization is given by assigning to the point m with coordinate z the complex subspace spanned by ∂_z in the complexification

$$T_m \mathbf{P}^1(\mathbf{C}) \otimes \mathbf{C} \simeq \mathbf{C}\,\partial_z \oplus \mathbf{C}\,\partial_{\bar{z}}$$

of the tangent space $T_m \mathbf{P}^1(\mathbf{C})$ for each $m \in \mathbf{P}^1(\mathbf{C})$, i.e. $\Gamma(E_\ell)$ can be identified with the space of those sections of B_ℓ, which are constant "along the direction of the $\bar{z}$−coordinate".

We try to make this a bit more comprehensive. But to do it, again we need more general notions from complex and algebraic geometry. As seen above, these geometric notions can and shall be made later rather tangible by replacing them by algebraic ones in our concrete examples (and, hence, may be skipped at first try).

8.2.3   Quantization

The prequantization Hilbert space $\mathcal{H}$ built from C^∞–sections of the prequantum bundle B over M - as we just have seen - often turns out to be too large to be useful: This is to be seen again in the case of the coadjoint orbits, as the representations of the group G on $\mathcal{H}$ as a representation space are in general far from being irreducible. From representation theory procedures are known to construct appropriate representation spaces, for instance by parabolic or holomorphic induction. But in this approach here, it is "natural" to use a notion from the Kähler geometry to single out a subspace of $\mathcal{H}$, namely the notion of the *Kähler polarization*. We take over from [Wo] p.92 ff:

Definition 8.3: A *complex polarization* on a $2n$–dimensional symplectic manifold (M, ω) is a complex distribution P with

(CP1) For each $m \in M$, P_m is a complex Lagrangian subspace of $T_m^{\mathbb{C}} M$, i.e. with
$$P_m^\perp = P_m,$$

(CP2) $D_m := P_m \cap \overline{P_m} \cap T_m M$ has constant dimension for all $m \in M$,

(CP3) P is integrable.

The polarization P is *of type* (r, s) if the Hermitian form $<, >$ on P_m defined by

$$< Z, W > := -4i\omega(Z, \overline{W}) \text{ for } Z, W \in P_m$$

is of sign (r, s) for all $m \in M$.
P is a *Kähler polarization* if it is of type (r, s) with $r + s = n$. Such a polarization induces a complex structure J on M and (M, ω, J) is a *Kähler manifold* (in the broad sense that the real symmetric tensor g determined by

$$g(X, Y) = 2\omega(X, JY), \quad X, Y \in V(M)$$

may be indefinite, see the "note 1" to p.93 in [Wo]).

Perhaps the reader wants some more explanations. We follow [Wo] p.269:

– A *real distribution* on a real manifold M is a subbundle of the tangent bundle TM. The fiber P_m at $m \in M$ is a subspace of $T_m M$, which varies smoothly with m.
– A vector field X is *tangent* to P if $X(m) \in P_m$ for every m. The space of vector fields tangent to P is denoted by $V_P(M)$. A smooth function is *constant along P* if df vanishes on restriction to P. The space of smooth functions constant along P is denoted by $C_P^\infty(M)$.
– An immersed submanifold $N \subset M$ is an *integral manifold* of P if $P_m = T_m N$ for every $m \in N$.
– A distribution is *integrable* if $[X, Y]$ is tangent to P whenever X and Y are tangent to P.
– A *complex distribution* is similarly defined to be a subbundle of the complexified tangent bundle $T_{\mathbb{C}} M$. A complex distribution P is integrable if in some neighbourhood of each point, one has smooth complex functions $f_{k+1}, \ldots, f_l$ with gradients that are independent and annihilate all complex vector fields tangent to P, where $l - k$ is the fiber dimension of P.

– A *complex structure* on a real vector space V is a linear transformation $J : V \longrightarrow V$ such that $J^2 = -1$. Complex structures exist only in spaces of even dimension. On a real manifold a complex structure J is given if one has a complex structure J_m on each tangent space $T_m M$ and if these J_m vary smoothly with m, i.e. the complex distribution P spanned by the vector fields $X - iJX, X \in V(M)$, is integrable. A real manifold with a complex stucture comes out as a complex manifold.

– A *connection* $\triangledown$ on a vector bundle $\mathcal{B}$ is an operator that assigns a one-form $\triangledown s$ with values in $\mathcal{B}$ to each (smooth) section s of $\mathcal{B}$, i.e.,

$$\triangledown : \Gamma(\mathcal{B}) \longrightarrow \Gamma(T^*M \otimes \mathcal{B}),$$

such that for any function f and sections $s, s' \in \mathcal{C}^\infty_{\mathcal{B}}(M)$ one has

$$\triangledown(s + s') = \triangledown s + \triangledown s' \text{ and } \triangledown(fs) = (df)s + f \triangledown s.$$

Vice versa, a given *Kähler manifold* (M, ω, J), i.e. a manifold which is simultaneously in a compatible manner a symplectic and a complex manifold, carries two Kähler polarizations: the *holomorphic polarization* P spanned at each point by the vectors (∂_{z^a}) and its complex conjugate, the *anti–holomorphic polarization* $\overline{P}$ spanned by the $(\partial_{\overline{z}^a})$'s. Locally it is possible to find a real smooth function f such that

$$\omega = i\partial\overline{\partial}f$$

and

$$\Theta = -i\partial f \text{ resp. } \overline{\Theta} = i\overline{\partial}f$$

is a symplectic potential *adapted* to P resp. $\overline{P}$.

– A *Hermitian structure* on a complex vector bundle $\mathcal{B}$ is a Hermitian inner product $(\cdot, \cdot)$ on the fibers, which is smooth in the sense that $\mathcal{B} \longrightarrow \mathbf{C}, v \longmapsto (v, v)$ is a smooth function. It is *compatible with the connection* $\triangledown$ if for all sections s, s' and all real vector fields X

$$i(X)d(s, s') = (\triangledown_X s, s') + (s, \triangledown_X s'), \quad \triangledown_X s := i(X) \triangledown s.$$

Now, let be given a Kähler polarization P on the quantizable (M, ω) with prequantum line bundle B over M with connection $\triangledown$, Hermitian structure $<, >$ and Hilbert space $\mathcal{H}$ of square integrable sections of B. Then we define a space of *polarized* sections

$$\mathcal{C}^\infty_B(M, P) := \{s \in \mathcal{C}^\infty_B(M); \, \triangledown_{\overline{X}} s = 0 \text{ for all } X \in V_{\mathbf{C}}(M, P)\}.$$

Here, for $U \subset M$ open,

$$V_{\mathbf{C}}(U, P) := \{X \in V_{\mathbf{C}}(M), \, X_m \in P_m \text{ for all } m \in M\}$$

denotes the vector fields tangent to P on U. Then

$$\mathcal{H}_P := \mathcal{H} \cap \mathcal{C}^\infty_B(M, P)$$

comes out as our "new" Hilbert space. As remarked above, it consists of all integrable holomorphic sections of B.

Alternatively, it can be said that the prequantum bundle B (via P) is given a structure of a complex line bundle. Anyway, in important cases, this $\mathcal{H}_P$ is the representation space of a discrete series representation.

We leave aside the problem, which of the classical observables have a selfadjoint counterpart acting in $\mathcal{H}_P$, discussed to some length in [Wo] and the modification by the introduction of *half–densities* and go directly to the application in the construction of unitary representations via line bundles on the coadjoint orbits.

8.2.4   Coadjoint Orbits and Hamiltonian G-spaces

This is the central topic of this chapter. For background information we recommend the standard sources by Kirillov [Ki] §15, Kostant [Kos], (at least) the first pages of Kirillov's book [Ki1], or, for a comprehensive overview, the article by Vogan [Vo].

Let G be a (real) linear group with Lie algebra $\mathfrak{g}$ and its dual space $\mathfrak{g}^*$. As usual, we realize all this by matrices and then have the *adjoint representation* Ad of G on $\mathfrak{g}$ given by

$$\mathrm{Ad}(g)X := gXg^{-1}.$$

(This is the derived map of the inner automorphism $\kappa_g : G \longrightarrow G, g_0 \longmapsto gg_0g^{-1}$.) Following our preparation in section 1.4.3, we can construct the contragredient representation Ad^* defined on the dual $\mathfrak{g}^*$ to $\mathfrak{g}$ by

$$\mathrm{Ad}^*(g) := \mathrm{Ad}(g^{-1})^*.$$

The asterisk on the right-hand side indicates a *dual operator:* If $< \eta, X >$ denotes the value $\eta(X)$ of the linear functional η applied to the vector X and A is an operator on $\mathfrak{g}$ the dual A^* is defined by

$$< A^*\eta, X > := < \eta, AX > \quad \text{for all } X \in \mathfrak{g}, \eta \in \mathfrak{g}^*.$$

Hence we have as our central prescription

$$< \mathrm{Ad}^*(g)\eta, X > \,= \,< \eta, \mathrm{Ad}(g^{-1})X > \quad \text{for all } g \in G, X \in \mathfrak{g}, \eta \in \mathfrak{g}^*.$$

This representation Ad^* is called the *coadjoint representation*. Often we will abbreviate $g \cdot \eta := \mathrm{Ad}^*(g)\eta$.

In 6.6.1 we already worked with the trace form for $A, B \in M_n(\mathbf{C})$ given by

$$(A, B) \longmapsto \mathrm{Re}\,\mathrm{Tr}\,(AB) =: < A, B > .$$

(Here, unfortunately, one has a double meaning of $< .,. >$.) If $\mathfrak{g}$ is semisimple or reductive, one can use this bracket to identify $\mathfrak{g}^*$ with $\mathfrak{g}$ (G-equivariantly) via

$$\mathfrak{g}^* \ni \eta \longmapsto X_\eta \in \mathfrak{g}, \text{ defined by } \eta(Y) = < X_\eta, Y > \quad \text{for all } Y \in \mathfrak{g}.$$

Then the coadjoint representation is fixed by the simple prescription $X_\eta \longmapsto gX_\eta g^{-1}$.

Now we are prepared to define the *coadjoint orbit* $\mathcal{O}$ of $\eta \in \mathfrak{g}^*$ by

$$\mathcal{O}_\eta := G \cdot \eta.$$

If G_η denotes the stabilizing group $G_\eta := \{g \in G; g \cdot \eta = \eta\}$, one has $\mathcal{O}_\eta \simeq G/G_\eta$ and if we can identify $\mathfrak{g}^*$ with a matrix space as proposed above, we have

$$\mathcal{O}_\eta = \{gX_\eta g^{-1}, g \in G\}.$$

It is an outstanding and most wonderful fact that our coadjoint orbits are symplectic manifolds. We cite from [Vo] p.187:

Theorem 8.2: Suppose $M = \mathcal{O}_\eta$ is the coadjoint orbit of $\eta \in \mathfrak{g}^*$ and $\mathfrak{g}_\eta := \operatorname{Lie} G_\eta$. Then one has

1.) The tangent space to M at η is

$$T_\eta M \simeq \mathfrak{g}/\mathfrak{g}_\eta.$$

2.) The skew-symmetric bilinear form

$$\omega_\eta(X, Y) := \eta([X, Y]) \quad \text{for all } X, Y \in \mathfrak{g}$$

on $\mathfrak{g}$ has radical exactly $\mathfrak{g}_\eta$ and so defines a symplectic form on $T_\eta M$.
3.) The form ω_η makes M into a symplectic manifold.

The bilinear form ω_η is called *Kirillov-Kostant form*, sometimes also *Kirillov-Kostant-Souriau form* or any permutation of these names. It is also written as ([Ki] p.230 or [Wo] p.52)

(8.5) $\qquad \omega_\eta(\eta')(\xi_{\hat{X}}(\eta'), \xi_{\hat{Y}}(\eta')) = < \eta', [\hat{X}, \hat{Y}] > \quad$ for all $\hat{X}, \hat{Y} \in \mathfrak{g}, \eta' \in \mathcal{O}_\eta$

where $\xi_{\hat{X}}$ denotes the vector field associated to $\hat{X}$ in the form of a differential operator acting on smooth functions f on M by the definition

(8.6) $\qquad\qquad\qquad (\xi_{\hat{X}} f)(\eta) := \dfrac{d}{dt}(f(\exp t\hat{X} \cdot \eta))|_{t=0}.$

For a proof we refer to [Ki] 15.2 or [Kos] Theorem 5.4.1. It is obvious that ω_η is skew symmetric. Also it is, in principle, not deep that a form given for the tangent space of one point can be transported to the whole space if the space is homogeneous. But it may be an intricate task to give an explicit description of ω in local coordinates. We shall give an example soon how (8.5) can be used. The fact that ω comes out as a closed form can be proved by direct calculation or by more sophisticated means as in [Ki1] p.5-10.

The coadjoint orbits still have more power, they are *Hamiltonian G-spaces*:

Definition 8.4: Suppose (M, ω) is a symplectic manifold and $f \in C^\infty(M)$ is a smooth function with its Hamiltonian vector field X_f. Moreover suppose that G is a linear group endowed with a smooth action on M, which respects the form ω. We say that M is a *Hamiltonian G-space* if there is a linear map

$$\tilde{\mu} : \mathfrak{g} \longrightarrow C^\infty(M)$$

with the following properties:
– i) $\tilde{\mu}$ intertwines the adjoint action of G on $\mathfrak{g}$ with its action on $C^\infty(M)$.
– ii) For each $Y \in \mathfrak{g}$, $X_{\tilde{\mu}(Y)}$ is the vector field by which Y acts on M.
– iii) $\tilde{\mu}$ is a Lie algebra homomorphism.

$\tilde{\mu}$ can be reinterpreted as a *moment map*

$$\mu : G \longrightarrow \mathfrak{g}^*, \quad \mu(m)(Y) := \tilde{\mu}(Y)(m) \quad \text{for all } m \in M, Y \in \mathfrak{g}.$$

Then the highlight of this theory is the following result to be found also in the above cited sources. We shall not have occasion to use it here but state it to give a more complete picture.

Theorem 8.3: The homogeneous Hamiltonian G-spaces are the covering spaces of coadjoint orbits. More precisely, suppose M is such a space, with map $\tilde{\mu} : \mathfrak{g} \longrightarrow C^\infty(M)$ and moment map μ. Then μ is a G-equivariant local diffeomorphism onto a coadjoint orbit, and the Hamiltonian G-space structure on M is pulled back from that on the orbit by the map μ.

If G is reductive and we have identified $\mathfrak{g}$ with its dual $\mathfrak{g}^*$ as $n \times n$−matrices, one has the following classification of the coadjoint orbits, which later will be reflected in a mechanism to construct representations of G.

Definition 8.5: An element $X \in \mathfrak{g}$ is called *nilpotent* if it is nilpotent as a matrix, i.e. if $X^N = 0$ for N large enough.
X is called *semisimple* if X regarded as a complex matrix is diagonalizable.
X is called *hyperbolic* resp. *elliptic* if it is semisimple and all eigenvalues are real resp. purely imaginary.
An orbit $\mathcal{O} = G \cdot \eta$ inherits the name from $X = X_\eta$.

For instance, the real matrix

$$X = \begin{pmatrix} 0 & 1 \\ -1 & 0 \end{pmatrix}$$

is elliptic, since over $\mathbf{C}$ it is conjugate to $\begin{pmatrix} i & 0 \\ 0 & -i \end{pmatrix}$. Now, this finally is the moment to discuss an example.

Coadjoint Orbits of $\mathrm{SL}(2, \mathbf{R})$

We take

$$G = \mathrm{SL}(2, \mathbf{R}) \supset K = \mathrm{SO}(2)$$

and

$$\mathfrak{g}^* \simeq \mathfrak{g} = \mathrm{Lie}\, G = \left\{ U(x,y,z) := \begin{pmatrix} x & y+z \\ y-z & -x \end{pmatrix};\ x,y,z \in \mathbf{R} \right\} = <X,Y,Z>_{\mathbf{R}}$$

with

$$X = \begin{pmatrix} 1 & \\ & -1 \end{pmatrix},\ Y = \begin{pmatrix} & 1 \\ 1 & \end{pmatrix},\ Z = \begin{pmatrix} & 1 \\ -1 & \end{pmatrix}$$

$$[X,Y] = 2Z, \quad [X,Z] = 2Y, \quad [Y,Z] = -2X$$

and

$$X^2 + Y^2 - Z^2 = 3I_2.$$

Obviously, X and Y are hyperbolic, Z is elliptic, and $F = \begin{pmatrix} & 1 \\ & \end{pmatrix}$ as well as $G = \begin{pmatrix} & \\ 1 & \end{pmatrix}$ are nilpotent.

One can guess the form of the orbits generated by these matrices, but to be on the safe side we calculate:

$$gU(x, y, z)g^{-1} =: U(\tilde{x}, \tilde{y}, \tilde{z}) \text{ for } g = \begin{pmatrix} a & b \\ c & d \end{pmatrix}$$

is given by

$$
\begin{aligned}
\tilde{x} &= (ad + bc)x + (bd - ac)y - (bd + ac)z, \\
(8.7) \quad \tilde{y} &= (dc - ab)x + (1/2)(a^2 + d^2 - b^2 - c^2)y + (1/2)(a^2 + b^2 - c^2 - d^2)z, \\
\tilde{z} &= -(ab + cd)x + (1/2)(a^2 - b^2 + c^2 - d^2)y + (1/2)(a^2 + b^2 + c^2 + d^2)z.
\end{aligned}
$$

Hence, for $U = \alpha X, \alpha \neq 0$, as a hyperbolic orbit $\mathcal{O}_{\alpha X}$ we have the one-sheeted hyperboloid in $\mathbf{R}^3$ given by

$$x^2 + y^2 - z^2 = \alpha^2.$$

For $U = \alpha Z, \alpha \gtrless 0$, as elliptic orbits $\mathcal{O}_{\alpha Z}$ we have the upper resp. lower half of the two-sheeted hyperboloid given by

$$x^2 + y^2 - z^2 = -\alpha^2, z \gtrless 0.$$

For $U = F$ and $U = G$, as nilpotent orbits $\mathcal{O}_F$ and $\mathcal{O}_G$ we have the upper resp. lower half of the cone given by

$$x^2 + y^2 - z^2 = 0, z \gtrless 0.$$

Finally, for $U = 0$, there is the nilpotent orbit $\mathcal{O}_0$ consisting only of the origin in $\mathbf{R}^3$. Hence we have a disjoint decomposition of $\mathbf{R}^3$ into the orbits in their realization as subsets of $\mathbf{R}^3$

$$\mathbf{R}^3 = \coprod_{\alpha > 0} \mathcal{O}_{\alpha X} \coprod_{\alpha \gtrless 0} \mathcal{O}_{\alpha Z} \coprod (\mathcal{O}_F \cup \{0\} \cup \mathcal{O}_G).$$

It is an easy exercise to determine the stabilizing groups of the elements we chose to generate our orbits. We get (again with $\alpha \neq 0$)

$$G_{\alpha X} = \left\{ \begin{pmatrix} a & \\ & 1/a \end{pmatrix}; a \in \mathbf{R}^* \right\},$$

$$G_{\alpha Z} = \mathrm{SO}(2),$$

$$G_F = \left\{ \begin{pmatrix} \pm 1 & b \\ & \pm 1 \end{pmatrix}; b \in \mathbf{R} \right\},$$

$$G_G = \left\{ \begin{pmatrix} \pm 1 & 0 \\ c & \pm 1 \end{pmatrix}; c \in \mathbf{R} \right\}.$$

From the general theory we have the prediction that the coadjoint orbits are symplectic manifolds. We can verify this here directly:

Let us take the elliptic orbit

$$\mathcal{O}_{\alpha Z} = G \cdot (\alpha Z) \simeq G/\mathrm{SO}(2), \ \alpha > 0.$$

We want to give an explict expression for the Kirillov-Kostant form ω in this case. Here one has to be rather careful. We used the trace form for matrices $< A, B >= \mathrm{Tr}\, AB$ to identify $\mathfrak{g}^*$ with $\mathfrak{g}$ via $\eta \longmapsto X_\eta$ with $< X_\eta, U >= \eta(U)$ for all $U \in \mathfrak{g}$. Now, we write $\mathfrak{g}^* =< X_*, Y_*, Z_* >_{\mathbf{R}}$ with

$$X_* = \begin{pmatrix} 1 & \\ & -1 \end{pmatrix}, \ Y_* = \begin{pmatrix} & 1 \\ 1 & \end{pmatrix}, \ Z_* = \begin{pmatrix} & 1 \\ -1 & \end{pmatrix},$$

and then have as non-vanishing duality relations

$$< X_*, X >= 2, \ < Y_*, Y >= 2, \ < Z_*, Z >= -2.$$

We put

$$\mathfrak{g}^* \ni \eta = x_* X_* + y_* Y_* + z_* Z_*$$

and

$$\hat{X} = xX + yY + zZ, \ \hat{Y} = x'X + y'Y + z'Z \in \mathfrak{g}.$$

Hence, by the commutation relations for X, Y, Z, we have

$$[\hat{X}, \hat{Y}] = 2((xy' - x'y)Z + (xz' - x'z)Y - (yz' - y'z)X).$$

From the Kirillov-Kostant prescription we have

$$\omega_\eta(\hat{X}, \hat{Y}) = < \eta, [\hat{X}, \hat{Y}] >$$
$$= -2(yz' - yz')x_* + 2(xz' - x'z)y_* - 2(xy' - x'y)z_*,$$

i.e. in particular for the elliptic orbit through $\eta = kZ_*$

$$\omega_{kZ_*}(\hat{X}, \hat{Y}) = -(xy' - x'y)k.$$

There are several ways to go from here to an explicit differential two-form ω on

$$M := G \cdot kZ_*$$
$$= \{(x_*, y_*, z_*) \in \mathbf{R}^3; \ x_*^2 + y_*^2 - z_*^2 = -k^2\}.$$

It is perhaps not the shortest one but rather instructive to proceed like this: One knows that

$$\omega_1 = \frac{d\dot{x} \wedge d\dot{y}}{\dot{y}^2}$$

is a G-invariant symplectic form on $M' := \mathfrak{H} \simeq G/\mathrm{SO}(2)$ where

$$\dot{\tau} = \dot{x} + i\dot{y} := g(i)$$

with

$$\dot{x} = \frac{bd + ac}{c^2 + d^2}, \ \dot{y} = \frac{1}{c^2 + d^2} \ \text{ for } g = \begin{pmatrix} a & b \\ c & d \end{pmatrix}.$$

The coadjoint G-action on M and the usual action of G on $M' = \mathfrak{H}$ are equivariant (**Exercise 8.2:** Verify this). From (8.7) we have the description of the elements $(x_*, y_*, z_*) \in M$ as

$$x_* = -(ac + bd)k,$$
$$y_* = (1/2)(a^2 + b^2 - c^2 - d^2)k,$$
$$z_* = (1/2)(a^2 + b^2 + c^2 + d^2)k,$$

and, hence, a G-equivariant diffeomorphism

$$\psi : M \longrightarrow M' = \mathfrak{H}, \quad (x_*, y_*, z_*) \longmapsto (\dot{x}, \dot{y})$$

given by

$$\dot{x} = \frac{x_*}{y_* - z_*}, \quad \dot{y} = \frac{k}{z_* - y_*}.$$

We use ψ to pull back ω_1 to a symplectic form ω_0, which in local coordinates (x_*, y_*) with $z_* = \sqrt{x_*^2 + y_*^2 + k^2}$, i.e. $z_* dz_* = x_* dx_* + y_* dy_*$ comes out as

$$\omega_0 = \psi^* \omega_1 = \frac{dx_* \wedge dy_*}{kz_*}.$$

(Verify this as **Exercise 8.3.**)

Now we have to compare this with the Kirillov-Kostant prescription above and determine the right scalar factor to get the Kirillov-Kostant symplectic form ω. We shall see that one has

$$\omega = -\frac{dx_* \wedge dy_*}{z_*}.$$

Perhaps the more experienced reader can directly pick this off the Kirillov-Kostant prescription. As we feel this is a nice occasion to get some more practice in handling the notions we introduced here, we propose to evaluate the general formula (8.5)

$$\omega_\eta(\eta')(\xi_{\hat{X}}(\eta'), \xi_{\hat{Y}}(\eta')) := \; < \eta', [\hat{X}, \hat{Y}] > \quad \text{for all } \hat{X}, \hat{Y} \in \mathfrak{g}, \eta' \in \mathcal{O}_\eta$$

in this situation. For the vector fields $\xi_{\hat{X}}$ acting on functions f on M we have the general formula

$$(\xi_{\hat{X}} f)(\eta) = \frac{d}{dt} f((\exp t\hat{X}) \cdot \eta)|_{t=0}.$$

Putting $g_{\hat{X}}(t) := \exp t\hat{X}$ one has

$$g_X(t) = \begin{pmatrix} e^t & \\ & e^{-t} \end{pmatrix},$$
$$g_F(t) = \begin{pmatrix} 1 & t \\ & 1 \end{pmatrix},$$
$$g_G(t) = \begin{pmatrix} 1 & \\ t & 1 \end{pmatrix}.$$

And, putting $A(g)$ for the matrix of the coadjoint action $\eta \longmapsto g \cdot \eta$, one deduces from (8.7)

$$A(g_X(t)) = \begin{pmatrix} 1 & & \\ & (1/2)(e^{2t} + e^{-2t}) & (1/2)(e^{2t} - e^{-2t}) \\ & (1/2)(e^{2t} - e^{-2t}) & (1/2)(e^{2t} + e^{-2t}) \end{pmatrix},$$

$$A(g_F(t)) = \begin{pmatrix} 1 & t & -t \\ -t & (1/2)(2 - t^2) & (1/2)t^2 \\ -t & (1/2)t^2 & (1/2)(2 + t^2) \end{pmatrix},$$

$$A(g_G(t)) = \begin{pmatrix} 1 & -t & -t \\ t & (1/2)(2 - t^2) & -(1/2)t^2 \\ -t & (1/2)t^2 & (1/2)(2 + t^2) \end{pmatrix}.$$

From (8.6) we get the differential operators (with $Y = F + G, Z = F - G$)

$$\xi_X = 2(z_* \partial_{y_*} + y_* \partial_{z_*}),$$
$$\xi_Y = -2(z_* \partial_{x_*} + x_* \partial_{z_*}),$$
$$\xi_Z = 2(y_* \partial_{x_*} - x_* \partial_{y_*}).$$

When we treat the functions f as functions in the two local coordinates (x_*, y_*) with $z_*^2 = x_*^2 + y_*^2 + k^2$, these operators reduce to

$$\xi_X = 2z_* \partial_{y_*},$$
$$\xi_Y = -2z_* \partial_{x_*},$$
$$\xi_Z = 2(y_* \partial_{x_*} - x_* \partial_{y_*}).$$

And a vector field in these local coordinates comes as

$$\hat{X} = \alpha \partial_{x_*} + \beta \partial_{y_*} = x\xi_X + y\xi_Y + z\xi_Z$$

with

$$\alpha = 2(zy_* - yz_*), \quad \beta = 2(xz_* - zx_*)$$

and analogously for

$$\hat{Y} = \alpha' \partial_{x_*} + \beta' \partial_{y_*} = x'\xi_X + y'\xi_Y + z'\xi_Z.$$

One has

$$\alpha\beta' - \alpha'\beta = 4((zx' - xz')y_* z_* + (xy' - yx')z_*^2 + (yz' - y'z)x_* z_*)$$

and hence in (8.5) for $\omega = (1/z_*)dx_* \wedge dy_*$ the left hand side is

$$(1/z_*)dx_* \wedge dy_*(\xi_{\hat{X}}, \xi_{\hat{Y}}) = (1/z_*)dx_* \wedge dy_*(\alpha \partial_{x_*} + \beta \partial_{y_*}, \alpha' \partial_{x_*} + \beta' \partial_{y_*})$$
$$= 4((zx' - xz')y_* + (xy' - yx')z_* + (yz' - y'z)x_*).$$

And this is exactly the negative of the outcome of the right hand side evaluated for $\eta' = x_* X_* + y_* Y_* + z_* Z_*$

$$< \eta', [\hat{X}, \hat{Y}] >= -4(xy' - x'y)z_* + 4(xz' - x'z)y_* - 4(xy' - yx')x_*.$$

Coadjoint Orbits for SU(2)

As already done several times, we take

$$G := \mathrm{SU}(2) = \{g = \begin{pmatrix} a & b \\ -\bar{b} & \bar{a} \end{pmatrix}; \ a, b \in \mathbf{C}, \ |a|^2 + |b|^2 = 1\}.$$

Thus, as a (real) manifold, G is the unit 3–sphere $S^3 \subset \mathbf{C}^2$. One has

$$\begin{aligned}
\mathfrak{g} = \mathfrak{su}(2) &= \{X \in M_2(\mathbf{C}); \ \bar{X} = -{}^t X\} \\
&= \{X = (1/2) \sum_{j=1}^3 a_j H_j; \ \mathbf{a} = (a_1, a_2, a_3) \in \mathbf{R}^3\}
\end{aligned}$$

with (see (6.5) in Section 6.5)

$$H_1 = \begin{pmatrix} -i & \\ & i \end{pmatrix}, \ H_2 = \begin{pmatrix} & -i \\ -i & \end{pmatrix}, \ H_3 = \begin{pmatrix} & -1 \\ 1 & \end{pmatrix},$$

and

$$[H_1, H_2] = 2H_3, \quad [H_2, H_3] = 2H_1, \quad [H_3, H_1] = 2H_2.$$

$\mathfrak{su}(2)$ is isomorphic to $\mathfrak{so}(3)$ (see the exercises 6.4, 6.5 and 6.6 in Section 6.1) and, as usual, we identify $X \in \mathfrak{su}(2)$ and $\mathbf{a} = (a_1, a_2, a_3) \in \mathbf{R}^3$ with

$$A = \begin{pmatrix} 0 & -a_3 & a_2 \\ a_3 & 0 & -a_1 \\ -a_2 & a_1 & 0 \end{pmatrix} =: a_1 A_1 + a_2 A_2 + a_3 A_3 \in \mathfrak{so}(3)$$

Using this identification the Lie bracket corresponds to the vector product in $\mathbf{R}^3$

$$[A, B] = \mathbf{a} \wedge \mathbf{b}$$

and the trace form $<,>$ in $M_3(\mathbf{R})$ to the scalar product $(\cdot, \cdot)$ in $\mathbf{R}^3$

$$< A, B > = -2(\mathbf{a}, \mathbf{b}) = -2\,{}^t\mathbf{ab}.$$

Since $<,>$ is proportional to the euclidean scalar product on $\mathbf{R}^3$, we can use the trace form to identify $\mathfrak{so}(3)^*$ with $\mathfrak{so}(3)$ via the prescription

(8.8) $\quad \mathfrak{so}(3)^* \ni \eta \longmapsto A_\eta$ such that $\eta(B) =< A_\eta, B >$ for all $B \in \mathfrak{so}(3)$:

We choose a basis (A_1^*, A_2^*, A_3^*) with $< A_i^*, A_j > = \delta_{ij}$ and for an element $\eta \in \mathfrak{so}(3)^*$ we write $\eta \equiv A_\eta = a_1^* A_1^* + a_2^* A_2^* + a_3^* A_3^*$.

As is well known (see the proof of the fact that SU(2) is a twofold covering of SO(3) in section 4.3), using these identifications the coadjoint operations of SU(2) on $\mathfrak{su}(2)^*$ resp. $\mathfrak{so}(3)^*$ are simply the rotations of $\mathbf{R}^3$. The orbits are the spheres of radius s, $M_s := G \cdot \eta_s$ for $\eta_s := s A_3^*$ with $s > 0$, centered at the origin and the origin itself. One has a symplectic structure on the sphere M_s of radius s given by the volume form, which is given by restriction to M_s of the 2–form

$$\omega_0 := (1/s) \sum_{\substack{\{i, j, k\} \\ \text{cyclic}}} a_i^* da_j^* \wedge da_k^*.$$

As local coordinates we use (a_1^*, a_2^*) such that M_s is given by the parametrization

$$\psi : J := \{(a_1^*, a_2^*); (a_1^*)^2 + (a_2^*)^2 \leq s^2\} \longrightarrow M_s,$$
$$(a_1^*, a_2^*) \longmapsto (a_1^*, a_2^*, a_3^*), (a_3^*)^2 = s^2 - (a_1^*)^2 - (a_2^*)^2.$$

Exercise 8.4: Verify that one has

$$\omega_0 = (s/a_3^*)da_1^* \wedge da_2^*$$

and

$$vol_2(M_s) = 2 \int_J \psi^* \omega_0 = 4\pi s^2.$$

We want to compare this with the Kirillov-Kostant recipe

$$\omega_\eta(A, B) = \eta([A, B]),$$

resp. (8.5)

$$\omega(\xi_A, \xi_B) = < \eta', [A, B] > .$$

We have

$$A = a_1 A_1 + a_2 A_2 + a_3 A_3, \quad B = b_1 A_1 + b_2 A_2 + b_3 A_3,$$

i.e.

$$[A, B] = (a_1 b_2 - a_2 b_1)A_3 + (a_2 b_3 - a_3 b_2)A_1 + (a_3 b_1 - a_1 b_3)A_2,$$

and with $\eta = a_1^* A_1^* + a_2^* A_2^* + a_3^* A_3^*$

(8.9) $$\omega_\eta(A, B) = (a_1 b_2 - a_2 b_1)a_3^* + (a_2 b_3 - a_3 b_2)a_1^* + (a_3 b_1 - a_1 b_3)a_2^*.$$

As in the preceding example we determine the associated vector fields ξ_{A_j}. Namely using here $g_j(t) = \exp t A_j, j = 1, 2, 3,$

$$\xi_{A_j} f(\eta) = \frac{d}{dt} f(g_j(t) \cdot \eta)|_{t=0}$$

leads to

$$\xi_{A_1} = -a_3^* \partial_{a_2^*} + a_2^* \partial_{a_3^*},$$
$$\xi_{A_2} = a_3^* \partial_{a_1^*} - a_1^* \partial_{a_3^*},$$
$$\xi_{A_3} = -a_2^* \partial_{a_1^*} + a_1^* \partial_{a_2^*}.$$

and hence in the local coordinates fixed above

$$\xi_A = a_1 \xi_{A_1} + a_2 \xi_{A_2} + a_3 \xi_{A_3}$$

$$= \alpha \partial_{a_1^*} + \beta \partial_{a_2^*},$$

with

$$\alpha = (a_2 a_3^* - a_3 a_2^*), \quad \beta = (a_3 a_1^* - a_1 a_3^*)$$

and the corresponding expressions α', β' for ξ_B replacing the a_j by b_j. We get

$$\alpha\beta' - \alpha'\beta = (a_2 b_3 - a_3 b_2)a_1^* a_3^* + (a_1 b_2 - a_2 b_1)a_3^* a_3^* + (a_3 b_1 - a_1 b_3)a_2^* a_3^*.$$

By evaluation at $\eta = sA_3^*$ one obtains

$$\omega_0(\xi_A, \xi_B) = (s/a_3^*)(\alpha\beta' - \alpha'\beta)$$
$$= s^2(a_1 b_2 - a_2 b_1).$$

We compare this with (8.9) and see that the Kirillov-Kostant form in this case is

$$\omega = s\omega_0 = (s^2/a_3^*)da_1^* \wedge da_2^*.$$

The isotropy group of η is

$$G_\eta = \left\{ h = \begin{pmatrix} e^{it} & 0 \\ 0 & e^{-it} \end{pmatrix}, \ t \in \mathbf{R} \right\} \simeq U(1).$$

G/G_η is identified with the sphere $M_s \simeq S^2$ via the important *Hopf map*

$$\psi : G = \mathrm{SU}(2) \longrightarrow M_s$$

given by

$$(a, b) \longmapsto s(b\bar{a} + a\bar{b}, ib\bar{a} - ia\bar{b}, a\bar{a} - b\bar{b})$$

By some more calculation using the relation

$$ad\bar{a} + \bar{a}da + bd\bar{b} + \bar{b}db = 0,$$

which is a consequence of $a\bar{a} + b\bar{b} = 1$, we get for the drawback of the Kirillov-Kostant form by the Hopf map

$$\psi^*((s/a_3^*)da_1^* \wedge da_2^*) = 2is(da \wedge d\bar{a} + db \wedge d\bar{b}).$$

As we already constructed bundles on $\mathbf{P}^1(\mathbf{C})$ (in 7.6 Example 7.19 and in 8.2.2) it is quite fruitful to compare this description of the coadjoint $\mathrm{SU}(2)$-orbits with the following alternative: We discussed (in 7.1 Example 7.3) the identification

$$\mathrm{SU}(2)/U(1) \simeq \mathbf{P}^1(\mathbf{C}) = \{(u : v); \ u, v \in \mathbf{C}^2 \setminus \{(0,0)\}\}$$

essentially given by

$$g \longmapsto z = u/v = b/\bar{a},$$

which is a consequence of g acting by multiplication from the left on the column ${}^t(u, v)$. The same way, one has an identification $\tilde{\psi}$ via

(8.10) $$g \longmapsto z = v/u = b/a$$

coming from multiplication from the right to the row (u, v). If we take the symplectic form ω studied in 8.2.2

$$\omega = 2ni\frac{dz \wedge d\bar{z}}{(1 + |z|^2)^2}, \quad n \in \mathbf{Z},$$

and via $\tilde{\psi}$ draw it back to $\mathrm{SU}(2)$, we get again the form obtained above

$$2ni(da \wedge d\bar{a} + db \wedge d\bar{b})$$

and see that we have quantizable orbits for half integral s resp. n.

Now that we have examples for coadjoint orbits and bundles living on them, we continue in the program to use the sections of these bundles as representation spaces and we have to discuss how the general polarization business works in practical cases.

8.2.5 Construction of an Irreducible Unitary Representation by an Orbit

If we want to construct representations from a coadjoint orbit of a group, the orbit has to be *admissible*, i.e. "integral" and been provided with a "nice" polarization. Then one can construct a complex line bundle, whose polarized sections lead to the space for a unitary (and perhaps even irreducible) representation of the group. Or, in another formulation, this representation can be got by an induction procedure from a subgroup arising from the polarized orbit. Up to now, all this is not in a final form if one strives for maximal generalilty (which we do not do in this text). Though there are now several more refined procedures using and discussing Duflo's version of admissibility as for instance in Vogan [Vo] p.193 ff or in Torasso [To], we will follow the procedure designed by Kirillov [Ki] p.235 ff and [Ki1] (but in some places adapting our notation to the one used by Vogan). Similar presentations can be found in [GS1] p.235 and [Wo] p.103.

Step 1. For each orbit $\mathcal{O}_\eta \subset \mathfrak{g}^*$ we look for real subalgebras $\mathfrak{n} \subset \mathfrak{g}$ which are *subordinate* to η. This means that the condition

$$< \eta, [\check{X}, \check{Y}] > \, = 0 \quad \text{for all } \check{X}, \check{Y} \in \mathfrak{n}$$

holds or, equivalently (this looks quite different but is not too hard to prove), the map

$$\check{Y} \longmapsto 2\pi i < \eta, \check{Y} >$$

defines a one-dimensional representation $\varrho = \varrho_\eta$ of $\mathfrak{n}$.

Step 2. We say that $\mathfrak{n}$ is a *real algebraic polarization* of η if in addition the condition

$$2 \dim(\mathfrak{g}/\mathfrak{n}) = \dim \mathfrak{g} + \dim \mathfrak{g}_\eta$$

is satisfied. The notion of a *complex algebraic polarization* is defined in the same way: we extend η to $\mathfrak{g}_c = \mathfrak{g} \otimes \mathbf{C}$ by complex linearity and consider complex subalgebras $\mathfrak{n} \subset \mathfrak{g}_c$ that satisfy the corresponding conditions as in the real case.
An algebraic polarization is called *admissible* if it is invariant under $\operatorname{Ad} G_\eta$. In the sequel, we only look for admissible algebraic polarizations.

The relation of these "algebraic" polarizations to the "geometric" ones defined earlier is explained in [Ki1] p.28/9 and will be made explicit in a simple example later.

Step 3. Moreover, we suppose (as in [Ki] p.238) that in the complex case $\mathfrak{m}_c = \mathfrak{n} + \bar{\mathfrak{n}}$ is a subalgebra of $\mathfrak{g}_c$ such that we have closed subgroups Q and M of G with Lie algebras $\mathfrak{q}$ and $\mathfrak{m}$ fulfilling the relations

$$L := G_\eta \subset Q = G_\eta Q^\circ \subset M, \ \mathfrak{n} \cap \bar{\mathfrak{n}} = \mathfrak{q}_c, \ \mathfrak{n} + \bar{\mathfrak{n}} = \mathfrak{m}_c.$$

In the real case, we have $\mathfrak{q} = \mathfrak{n}$.

Step 4. Finally, and this is our practical criterium to guarantee the integrality or quantizability of the orbit, we suppose that ϱ_η can be integrated to a unitary character χ_Q of Q with $d\chi_Q = \varrho_\eta$. Then we get a unitary representation π of our group G on (the completion of) the space $\mathcal{F}_+ := C^\infty(G, \mathfrak{n}, Q, \varrho, \chi_Q)$ of smooth **C**–valued functions ϕ on G with

$$\phi(gl) = \triangle_Q(l)^{-1/2}\chi_Q(l^{-1})\phi(g) \quad \text{for all } l \in L, \; g \in G$$

and

$$\mathcal{L}_{\check{Y}}\phi = 0 \quad \text{for all } \check{Y} \in \mathfrak{u}, \; \mathfrak{n} =: \mathfrak{g}_\eta + \mathfrak{u},$$

where $\triangle_Q$ is the usual modular function and $\mathcal{L}_{\check{Y}}\phi(g) = \frac{d}{dt}\phi(g\exp t\check{Y})|_{t=0}$.
(Here we follow [Vo] p.194 and p.199. For an essentially equivalent approach see [Ki] p.199ff).

Though this looks (and is) rather intricate, we shall see that the scheme can easily be realized in our simple examples, in particular for the Heisenberg group in the next but one section. As already said, there are a lot of subtle questions hidden here. For instance (see [Ki] p.236), as another condition *Pukanszky's condition*, i.e. the additional property $\eta + \mathfrak{n}^\perp \subset \mathcal{O}_\eta$ comes in where $\mathfrak{n}^\perp$ denotes the annihilator of $\mathfrak{n}$ in $\mathfrak{g}^*$. And, as discussed in [Ki] p.239, the fundamental group $\pi_1(\mathcal{O})$ of the orbit plays an important role in the integration of the representation ϱ_η of the Lie algebra $\mathfrak{q}$ to a representation of the group Q. This leads to the notion of *rigged orbits*, i.e. orbits in $\mathfrak{g}^*_{rigg}$, the set of pairs (η, χ) where $\eta \in \mathfrak{g}^*$ and χ is here a one-dimensional representation of G_η such that $d\chi = 2\pi i\eta|_{\mathfrak{g}_\eta}$ (see [Ki1] p.123 (and Vogan's criticism of Kirillov's book on his homepage)). Here we only want to give an impression, and (from [Ki] p.241) we cite a theorem by Kostant and Auslander:

Theorem 8.4: Assume G to be a connected and simply connected solvable Lie group. Then representations that correspond to different orbits or different characters of the fundamental group of the orbit are necessarily inequivalent.

And (a special version of) the famous *Borel-Weil-Bott theorem* says:

Theorem 8.5: All irreducible representations of a compact connected, and simply connected Lie group G correspond to integral G-orbits of maximal dimension in $\mathfrak{g}^*$.

In more detail, in our special examples the result of this procedure is as follows.

8.3   The Examples SU(2) and SL(2, **R**)

1. $G = \mathrm{SU}(2)$

We begin by taking up again the here already often treated example $G = \mathrm{SU}(2)$. Using (6.5) in Section 6.5, we take

$$\mathfrak{g} = \{X = \Sigma a_j H_j; \; a_1, a_2, a_3 \in \mathbf{R}\}$$

and, similar to (8.8) in 8.2.4, we choose (H_j^*) such that $< H_j^*, H_k > = \delta_{jk}$ and

$$\mathfrak{g}^* = \{\eta = \Sigma a_j^* H_j^*; \; a_1^*, a_2^*, a_3^* \in \mathbf{R}\}.$$

From (8.9) for the elliptic orbit $\mathcal{O}_\eta \simeq SU(2)/U(1)$ passing through to $\eta = sH_1^*$ we have the Kirillov-Kostant prescription

$$\omega_\eta(X, Y) = s(a_2 b_3 - a_3 b_2).$$

One has $\mathfrak{g}_\eta = \operatorname{Lie} U(1) \simeq \langle H_1 \rangle$. As $\mathfrak{g}$ has no two-dimensional subalgebras $\mathfrak{n}$, there is no real polarization. But one has two complex polarizations given by the subalgebras $\mathfrak{n}_+ = \langle H_0, H_+ \rangle$ and $\mathfrak{n}_- = \langle H_0, H_- \rangle$ of the complexification $\mathfrak{g}_c = \langle H_0, H_\pm \rangle$ with

$$H_0 = iH_1, \ H_\pm = H_2 \pm iH_3.$$

As one easily verifies, both algebras are subordinate to η in the sense defined above in Subsection 8.2.5 and, for $s = k \in \mathbf{N}_0$, the character of $\mathfrak{g}_\eta$

$$\mathfrak{g}_\eta \ni Y \longmapsto 2\pi i < \eta, Y >$$

integrates to a unitary character χ_k of $H = U(1)$ identified as a subgroup of $G = SU(2)$. In this case the space $\mathcal{F}_+$ is the space belonging to the induced representation $\operatorname{ind}_H^G \chi_k$ restricted by the polarization condition

$$\mathcal{L}_{H_\pm} \phi = 0$$

since one has $\mathfrak{n}_+ = \mathfrak{g}_\eta + < H_+ >$ or $\mathfrak{n}_- = \mathfrak{g}_\eta + < H_- >$. And this space can be understood exactly as the subspace of holomorphic resp. antiholomorphic smooth sections of the real bundle B over $\mathbf{P}^1(\mathbf{C})$ from 8.2 and the end of 7.6: To show this, we determine the appropriate differential operators acting on the smooth sections f of B resp. the functions ϕ on G with $\phi(gh) = \chi_{2s}(h^{-1})\phi(g)$ where we denote $\phi(g) =: \phi(a, b) =: \phi(\alpha, \alpha', \beta, \beta')$ decomposing our standard coordinates for g into real coordinates

$$a =: \alpha + i\alpha', \ b =: \beta + i\beta'.$$

We put $g_j(t) := \exp tH_j$ and get

$$g_1(t) = \begin{pmatrix} e^{-it} & \\ & e^{it} \end{pmatrix}, \ g_2(t) = \begin{pmatrix} \cos t & -i\sin t \\ -i\sin t & \cos t \end{pmatrix}, \ g_3(t) = \begin{pmatrix} \cos t & -\sin t \\ \sin t & \cos t \end{pmatrix}.$$

Application of the prescription $\mathcal{L}_{H_j}\phi(g) := \frac{d}{dt}\phi(gg_j(t))|_{t=0}$ to ϕ as a function of the four real variables $\alpha, \alpha', \beta, \beta'$ leads to the operators

$$\mathcal{L}_{H_1} = \alpha' \partial_\alpha - \alpha \partial_{\alpha'} - \beta' \partial_\beta + \beta \partial_{\beta'},$$
$$\mathcal{L}_{H_2} = \beta' \partial_\alpha - \beta \partial_{\alpha'} + \alpha' \partial_\beta - \alpha \partial_{\beta'},$$
$$\mathcal{L}_{H_3} = \beta \partial_\alpha + \beta' \partial_{\alpha'} - \alpha \partial_\beta - \alpha' \partial_{\beta'},$$

and hence, using the Wirtinger relations $\partial_a = (1/2)(\partial_\alpha - i\partial_{\alpha'})$, $\partial_{\bar{a}} = (1/2)(\partial_\alpha + i\partial_{\alpha'})$, we get the operators

$$\mathcal{L}_{H_2+iH_3} = 2i(\bar{b}\partial_{\bar{a}} - a\partial_b), \ \mathcal{L}_{H_2-iH_3} = -2i(b\partial_a - \bar{a}\partial_{\bar{b}})$$

acting on the functions ϕ here written in the coordinates $a, \bar{a}, b, \bar{b}$. If ϕ comes from a section f of the bundle B over $\mathbf{P}^1(\mathbf{C})$, for $b \neq 0$ it is of the form (see (8.10) in 8.2)

$$\phi(a, \bar{a}, b, \bar{b}) = f(z, \bar{z}), \ z = a/\bar{b}.$$

Hence we get

$$\mathcal{L}_{H_2+iH_3}\phi = (2i/b^2)f_{\bar{z}}, \quad \mathcal{L}_{H_2-iH_3}\phi = (2i/\bar{b}^2)f_z,$$

and we see that the polarization by $\mathfrak{n}_\pm$ provides that the section f has to be holomorphic resp. antiholomorphic.

Summary: For $G = \mathrm{SU}(2)$ (and $\mathrm{SO}(3)$) all irreducible unitary representations can be constructed by the orbit method.
Each integral orbit $\mathcal{O}_\eta$ carries a line bundle and has two complex polarizations.
The correponding polarized sections provide equivalent representations.

2. $G = \mathrm{SL}(2, \mathbf{R})$

We continue by completing the results obtained for the orbits from the SL(2)-theory in the last section. The orbits are all realized as subsets of

$$M_\beta := \{(x_*, y_*, z_*) \in \mathbf{R}^3; \; x_*^2 + y_*^2 - z_*^2 = \beta\}$$

with convenient $\beta \in \mathbf{R}$.

2.1. Hyperbolic Orbits

For the *hyperbolic* orbit $\mathcal{O}_\eta = G \cdot \alpha X_* = M_{\alpha^2} \simeq G/L$, $L = G_\eta = MA$, $\alpha \neq 0$ we have Ad L-invariant real polarizations given by

$$\mathfrak{q}_1 := \langle X, Y + Z \rangle \quad \text{and} \quad \mathfrak{q}_2 := \langle X, Y - Z \rangle.$$

The associated characters ρ_1 and ρ_2 of $\mathfrak{q}_1$ resp. $\mathfrak{q}_2$ given by

$$\rho_1(\check{Y}) = 2\pi i \langle \alpha X_*, xX + \check{y}(Y + Z) \rangle = 4\pi i \alpha x \quad \text{for} \quad \check{Y} = xX + \check{y}(Y + Z)$$

and

$$\rho_2(\check{Y}) = 2\pi i \langle \alpha X_*, xX + \check{z}(Y - Z) \rangle = 4\pi i \alpha x \quad \text{for} \quad \check{Y} = xX + \check{z}(Y - Z)$$

integrate uniquely to unitary characters $\chi_{j,s}$ $(j = 1, 2)$ of the respective groups

$$Q_1 = MNA \quad \text{and} \quad Q_2 = M\bar{N}A,$$

which are trivial on $Q_j \cap K$. For

$$b = \varepsilon\, n(x)\, t(y) \quad \text{resp.} \quad = \varepsilon\, \bar{n}(x)\, t(y)$$

these characters are given by

$$\chi_{j,s}(b) := (y^{1/2})^{is} \quad \text{with} \quad s = 4\pi i \alpha.$$

Then by inducing these characters from Q_j to G one gets irreducible unitary representations. This coincides quantitatively with the fact that one has the principal series representations $\mathcal{P}^{\pm,is} := \pi_{is,\pm}$, which we constructed in 7.2 and which have models consisting of (smooth) functions $\phi = \phi(g)$ on G with

$$\phi\left(mn(x)t(y)g\right) = \gamma(m)\,(y^{1/2})^{(is+1)}\,\phi(g)\,,\ \gamma\left(\begin{array}{cc}\varepsilon & \\ & \varepsilon\end{array}\right) = \begin{array}{ll}\varepsilon & \text{for}\ - \\ 1 & \text{for}\ +\end{array}\,,$$

i.e. in a space spanned by functions

$$\phi_{s,2k}\,,\quad k \in \mathbf{Z}\ \text{ for }\ +\,.$$

and

$$\phi_{s,2k+1}\,,\quad k \in \mathbf{Z}\ \text{ for }\ -$$

with

(8.11) $$\phi_{s,j}(g) = y^{(is+1)/2}\,e^{ij\theta},\ j \in \mathbf{Z}.$$

The unitary equivalence $\mathcal{P}^{\pm,is} \simeq \mathcal{P}^{\pm,-is}$ can be translated into the fact that both polarizations produce equivalent representations (see the discussion in [GS1] p. 299 f). Thus, by the orbit method, the construction above yields only one half of the representations. This arises from the fact that the orbit M_{α^2} is not simply connected but has here a fundamental group $\pi_1\,(M_{\alpha^2}) \simeq \mathbf{Z}$, so that we have one representation (the even one) belonging to $id \in \pi_1\,(M_{\alpha^2})$ and the other one to a generator of $\pi_1\,(M_{\alpha^2})$ (see the discussion in [Ki3] p.463). This may serve as motivation to introduce the notion of rigged orbits already mentioned at the end of our Subsection 8.2.5.

2.2. Elliptic Orbits

For the *elliptic* orbits $M^{\pm}_{-\alpha^2} := \mathcal{O}_\eta = G\cdot\alpha Z_* \simeq G/SO(2)$, $\alpha \neq 0$, consisting of the upper half $M^{+}_{-\alpha^2}$ (i.e. with $z_* > 0$) of the two-sheeted hyperboloid M_{α^2} for $\alpha > 0$ and the lower half $M^{-}_{-\alpha^2}$ (with $z_* < 0$) for $\alpha < 0$, we have two purely complex polarizations : We take

$$Z_1 = -iZ = \left(\begin{array}{cc} & -i \\ i & \end{array}\right)\,,\ X_\pm = (1/2)\,(X \pm iY) = (1/2)\left(\begin{array}{cc} 1 & \pm i \\ \pm i & 1 \end{array}\right)$$

with

$$[Z_1, X_\pm] = \pm 2X_\pm\,,\quad [X_+, X_-] = Z_1\,.$$

and have

$$\mathfrak{g}_c = \langle Z_1, X_+, X_- \rangle.$$

This establishes the complex bilinear form

$$B_\eta([\check{X}, \check{Y}]) = i\alpha(a_+ b_- - a_- b_+)$$

for

$$\check{X} := a_0 Z_1 + a_+ X_+ + a_- X_-\,,\ \check{Y} := b_0 Z_1 + b_+ X_+ + b_- X_-\,.$$

We have $B_\eta \equiv 0$ on the $\operatorname{Ad} G_\eta$-invariant subalgebras

$$\mathfrak{n} := \mathfrak{n}_\pm = \langle Z_1, X_\pm \rangle_{\mathbf{C}}$$

of $\mathfrak{g}_c$ and $\mathfrak{n} + \bar{\mathfrak{n}} = \mathfrak{g}_c$, $\mathfrak{n} \cap \bar{\mathfrak{n}} = \mathfrak{k}_c = \langle Z_1 \rangle$

$$\mathfrak{n}_\pm \cap \mathfrak{g}_1 = \langle Y - Z \rangle_{\mathbf{R}} = \mathfrak{k} .$$

At first we see that a one-dimensional representation of $\mathfrak{k}$ is defined by

$$\rho(a_0 Z_1) := i \langle \begin{pmatrix} & \alpha \\ -\alpha & \end{pmatrix}, a_0 Z_1 \rangle = i\, 2\alpha\, a_0$$

and this integrates to a representation χ_k of SO(2) (with $\chi_k(r(\theta)) = e^{ik\theta}$) exactly for $2\alpha =: k \in \mathbf{Z} \setminus 0$. Thus the elliptic orbits are quantizable for $2\alpha = k \in \mathbf{Z} \setminus \{0\}$. Then we use $\mathfrak{n}_+$ and $\mathfrak{n}_-$ to define a complex structure on $M = M_{\alpha^2}^\pm \simeq G_1/\mathrm{SO}(2) \simeq \mathfrak{H}^\pm$. And by the usual procedure we get the discrete series representations $\pi_k^\pm$, $k \in \mathbf{N}$, realized on the L^2-spaces of holomorphic sections of the holomorphic line bundles $\mathcal{L}$ on $M_{-\alpha^2}^\pm$. With similar calculations as we did for SU(2), these can also be interpretated as spaces of polarized sections ϕ of the bundle belonging to χ_k on $M_{-\alpha^2}^\pm$, i.e. with

$$\mathcal{L}_{X_+}\phi = 0 \quad \text{resp.} \quad \mathcal{L}_{X_-}\phi = 0 .$$

These representations are also representations of $K = \mathrm{SO}(2)$ and we have

$$\operatorname{mult}(\chi_{\tilde{k}}, \pi_k^+) = 1 \text{ for } \tilde{k} = k + 2l, l \in \mathbf{N}_0 \text{ and } = 0 \text{ else.}$$

2.3. Nilpotent Orbits

In the *nilpotent* case $\mathcal{O}_\eta = G \cdot (Y_* + Z_*) = M_0^+ = G/\bar{N}$ we have one real polarization. The form $\omega_{Y_* + Z_*}$ is zero on the $\operatorname{Ad} \bar{N}$- invariant subalgebra of $\mathfrak{g}$ given by

$$\mathfrak{n} = < X, Y - Z > .$$

The standard representation ρ of $\mathfrak{n}$ is trivial

$$\rho(xX + y(Y - Z)) = < Z_*, xX + y(Y - Z) > = 0.$$

The group $B = MNA = \{\pm(\begin{smallmatrix} a & b \\ 0 & a^{-1} \end{smallmatrix}); a > 0, b \in \mathbf{R}\}$ has Lie $B = \mathfrak{n}$. We can (at least formally) proceed as follows. If $\chi = id$ denotes the trivial representation of B, one has here the representation $\operatorname{ind}_B^G id$ given by right translation on the space of functions $\phi : G \longrightarrow \mathbf{C}$ with

$$\phi(bg) = y^{1/2}\phi(g) \qquad \text{for} \quad b \in B, g \in G,$$

(spanned by $\phi(g) = y^{1/2}e^{i\ell\theta}, \ell \in 2\mathbf{Z}$), i.e. the representation $\mathcal{P}^{+,0}$. As in the previous case, here we have a fundamental group $\pi_1 \simeq \mathbf{Z}$, so there is a second representation $\mathcal{P}^{-,0}$ consisting of odd functions and spanned by $\phi(g) = y^{1/2}e^{i\ell\theta}, \ell \in 1 + 2\mathbf{Z}$), which decomposes into irreducible halves

$$\mathcal{P}^{-,0} = \pi_1^+ \oplus \pi_1^-$$

(see for instance [Kn] p.36). The lower half of the cone $M_0^- = G_1 \cdot Y_*$ generates an equivalent situation.

All this is a small item in the big and important theme of how to attach representations to nilpotent orbits (see in particular the discussion in Example 12.3 of Vogan [Vo1] p.386).

Summary

For the representations of $G = \mathrm{SL}(2, \mathbf{R})$, we get a correspondence between
– elliptic orbits and discrete series representations.
– (rigged) hyperbolic orbits and principal series representations,
We do not get the complementary series.
For the nilpotent orbits in the $\mathrm{SL}(2, \mathbf{R})$-theory, it seems still not clear how the correspondence is to be fixed (see the discussion in [Vo1] p.386). To me it seems most probable (see in particular [Re] p.I.110f) to associate to the nilpotent cone the principal series $\mathcal{P}^{+,0}$ and the (decomposing) $\mathcal{P}^{-,0} = D_1 \oplus D_{-1}$.

Let us finally point to a remark by Kirillov in [Ki3] 8.4 and [Ki1] 6.4 that he sees the possibility to extend the correspondence framework such that even the complementary series fits in. This should be further elucidated. Up to now one has here an example of an imperfect matching between the unitary dual and the set of coadjoint orbits. But there is a conjecture that by the orbit method one gets all representations appearing in the decomposition of the regular representation of a group ([Ki1] p 204): "Indeed, according to the ideology of the orbit method, the partition of $\mathfrak{g}^*$ into coadjoint orbits corresponds to the decomposition of the regular decomposition into irreducible components".

8.4 The Example Heis($\mathbf{R}$)

The treatment of the Heisenberg group is even easier than that of the two previous examples. But one has one additional complication as one can not any longer $\mathrm{Ad}\,G$-equivariantly identify the Lie algebra with its dual via the trace or Killing form. So we use the original definition of the coadjoint representation to determine the set $\mathcal{O}(G)$ of coadjoint orbits and compare it with the unitary dual $\hat{G}$ of the Heisenberg group $G = \mathrm{Heis}(\mathbf{R})$, which we determined in Example 7.11 in 7.4 as the disjoint union of the plane $\mathbf{R}^2$ with a real line without its origin $\mathbf{R} \setminus \{0\}$. We shall get a perfect matching between $\hat{G}$ and $\mathcal{O}(G)$.

As the formulae are a bit more smooth, we work with the realization of the Heisenberg group by three-by-three matrices, i.e. we take

$$G := \mathrm{Heis}'(\mathbf{R}) = \{g = \begin{pmatrix} 1 & a & c \\ & 1 & b \\ & & 1 \end{pmatrix} ; \ a, b, c \in \mathbf{R}\}$$

and

$$\mathfrak{g} := \{\check{X} = \begin{pmatrix} 0 & x & z \\ & 0 & y \\ & & 0 \end{pmatrix} = xX + yY + zZ; \ x, y, z \in \mathbf{R}\}.$$

The Coadjoint Action

As in 8.2, we have the adjoint representation Ad of G given on $\mathfrak{g}$ by

$$(\operatorname{Ad} g)\check{X} = g\check{X}g^{-1} = \begin{pmatrix} 1 & a & c \\ & 1 & b \\ & & 1 \end{pmatrix} \begin{pmatrix} 0 & x & z \\ & 0 & y \\ & & 0 \end{pmatrix} \begin{pmatrix} 1 & -a & -c+ab \\ & 1 & -b \\ & & 1 \end{pmatrix}$$

$$= \begin{pmatrix} 0 & x & z+ay-bx \\ & 0 & y \\ & & 0 \end{pmatrix}.$$

The coadjoint representation Ad^* is the contragredient of Ad (see 1.4.3) and given on $\mathfrak{g}^*$ by $\operatorname{Ad}^* g = (\operatorname{Ad} g^{-1})^*$, i.e. if we write

$$\langle \eta, \check{X} \rangle := \eta(\check{X})$$

for all $\eta \in \mathfrak{g}^*$ and $\check{X} \in \mathfrak{g}$, then $\operatorname{Ad}^* g$ is fixed by the relation

$$\langle \operatorname{Ad}^* g\, \eta, \check{X} \rangle = \langle \eta, (\operatorname{Ad} g^{-1})\check{X} \rangle \quad \text{for all } \check{X} \in \mathfrak{g}.$$

We have

$$(\operatorname{Ad} g^{-1})\check{X} = \begin{pmatrix} 0 & x & z-ay+bx \\ & 0 & y \\ & & 0 \end{pmatrix}.$$

And if we take X_*, Y_*, Z_* as a basis for $\mathfrak{g}^*$, dual to X, Y, Z with respect to $<,>$, and write

$$\mathfrak{g}^* \ni \eta = x_* X_* + y_* Y_* + z_* Z_*,$$

we have

$$\operatorname{Ad}^* g\, \eta = \tilde{x}_* X_* + \tilde{y}_* Y_* + \tilde{z}_* Z_*$$

with

(8.12) $$\tilde{x}_* = x_* + bz_*, \quad \tilde{y}_* = y_* - az_*, \quad \tilde{z}_* = z_*.$$

There is a more elegant approach to this result: In [Ki1] p.61 Kirillov uses the trace form on $M_3(\mathbf{R})$ to identify $\mathfrak{g}^*$ with the space of lower-triangular matrices of the form

$$\eta = \begin{pmatrix} * & * & * \\ x & * & * \\ z & y & * \end{pmatrix}.$$

Here the stars remind us that one actually considers the quotient space of $M_3(\mathbf{R})$ by the subspace $\mathfrak{g}^\perp$ of upper-triangular matrices (including the diagonal). This way, one gets

$$\operatorname{Ad}^* g\, \eta = g \begin{pmatrix} * & * & * \\ x & * & * \\ z & y & * \end{pmatrix} g^{-1} = \begin{pmatrix} * & * & * \\ x+bz & * & * \\ z & y-az & * \end{pmatrix}.$$

Exercise 8.5: Do the same for the realization of the Heisenberg group and its Lie algebra by four-by-four matrices as in 0.1 and 6.3. Verify that Ad^*g is given by

$$(p_*, q_*, r_*) \longmapsto (p_* + 2\mu r_*, q_* - 2\lambda r_*, r_*)$$

for $g = (\lambda, \mu, \kappa)$ and $\eta = p_* P_* + q_* Q_* + r_* R_*$, where P_*, Q_*, R_* is a basis of $\mathfrak{g}^*$ dual to P, Q, R.

The Coadjoint Orbits

From (8.12) it is easy to see that one has just two types of coadjoint orbits $\mathcal{O}_\eta = G \cdot \eta \simeq G/G_\eta$.

i) For $\eta = mZ_*, m \neq 0$ we have a two-dimensional orbit

$$\mathcal{O}_m := \mathcal{O}_{mZ_*} = \{(x_*, y_*, m); \ x_*, y_* \in \mathbf{R}\} \simeq \mathbf{R}^2.$$

ii) For $\eta = rX_* + sY_*, r, s \in \mathbf{R}$, as orbit we have the point

$$\mathcal{O}_{rs} := \mathcal{O}_{rX_*+sY_*} = \{(r, s, 0)\}.$$

Obviously $\mathfrak{g}^* \simeq \mathbf{R}^3$ is the disjoint union of all these orbits and the set of orbits is in one-by-one correspondence with the unitary dual $\hat{G}$ described above. Now, we construct representations following the general procedure outlined in 8.2:

Orbits as Symplectic Manifolds

For $M := \mathcal{O}_m \simeq \mathbf{R}^2, m \neq 0$, we have as symplectic forms all multiples of $\omega_0 = dx_* \wedge dy_*$. As it is a nice exercise in the handling of our notions, we determine the factor for the Kirillov-Kostant form ω: Again we use (8.5)

$$\omega_\eta(\eta')(\xi_{\hat{X}}(\eta'), \xi_{\hat{Y}}(\eta')) = \ <\eta', [\hat{X}, \hat{Y}]> \quad \text{for all } \hat{X}, \hat{Y} \in \mathfrak{g}, \ \eta' \in \mathfrak{g}_\eta{}^*$$

We put

$$\hat{X} := xX + yY + zZ, \ \hat{Y} := x'X + y'Y + z'Z,$$

and have

$$[\hat{X}, \hat{Y}] = (xy' - x'y) Z$$

and

$$<\eta', [\hat{X}, \hat{Y}]> \ = (xy' - x'y)z_* \ \text{for } \eta' = x_* X_* + y_* Y_* + z_* Z_*.$$

We determine the vector fields ξ using again (8.6)

$$(\xi_{\hat{X}} f)(\eta) = \frac{d}{dt}(f(\exp t\hat{X} \cdot \eta))|_{t=0}.$$

One has

$$g_X(t) := \exp tX = \begin{pmatrix} 1 & t & \\ & 1 & \\ & & 1 \end{pmatrix},$$

(8.13)
$$g_Y(t) := \exp tY = \begin{pmatrix} 1 & & \\ & 1 & t \\ & & 1 \end{pmatrix},$$

$$g_Z(t) := \exp tZ = \begin{pmatrix} 1 & & t \\ & 1 & \\ & & 1 \end{pmatrix},$$

and, by (8.12),

$$g_X(t) \cdot \eta = x_* X_* + (y_* - t z_*) Y_* + z_* Z_*,$$
$$g_Y(t) \cdot \eta = (x_* + t z_*) X_* + y_* Y_* + z_* Z_*,$$
$$g_Z(t) \cdot \eta = x_* X_* + y_* Y_* + z_* Z_*.$$

and, hence,

$$\xi_X = -z_* \partial_{y_*}, \ \xi_Y = z_* \partial_{x_*}, \ \xi_Z = 0,$$

that is

$$\xi_{\hat{X}} = x \xi_X + y \xi_Y + z \xi_Z,$$
$$= \alpha \partial_{x_*} + \beta \partial_{y_*}$$

with

$$\alpha = y z_*, \ \beta = -x z_*.$$

Thus we have

$$\omega_0(\xi_{\hat{X}}, \xi_{\hat{Y}}) = (\alpha \beta' - \alpha' \beta) = z_*^2 (x y' - x' y).$$

If we compare this with the Kirillov-Kostant prescription above, we see that one has the symplectic form ω on $M = \mathcal{O}_m$ given by

$$\omega = (1/m) dx_* \wedge dy_*.$$

Polarizations of the Orbits $\mathcal{O}_m$

The algebraic approach to real polarizations asks for subalgebras $\mathfrak{n}$ of $\mathfrak{g} = \langle X, Y, Z \rangle$ subordinate to $\eta = m Z_*$. Obviously one has the (abelian) subalgebras $\mathfrak{n}_1 = \langle X, Z \rangle$ and $\mathfrak{n}_2 = \langle Y, Z \rangle$ and more generally, for $\alpha, \beta \in \mathbf{R}$ with $\alpha\beta \neq 0$, $\mathfrak{n} = \langle \alpha X + \beta Y, Z \rangle$. In the geometric picture these correspond to a polarization given by a splitting of $\mathcal{O}_m \simeq \mathbf{R}^2$ into the union of parallel lines $\alpha x_* + \beta y_* = $ const., which are the orbits of $\mathrm{Ad}^* Q$ where Q is the subgroup of G with Lie $Q = \mathfrak{n}$, i.e. $Q = \{(\alpha t, \beta t, c); t, c \in \mathbf{R}\}$.

A complex polarization of $\mathcal{O}_m$, which is translation invariant, is generated by a constant vector field $\xi = \partial_{x_*} + \tau \partial_{y_*}, \tau \in \mathbf{C} \setminus \mathbf{R}$, in particular $\tau = i$. The functions F satisfying $\xi F = 0$ are simply holomorphic functions in the variable $w := x_* + \tau y_*$. In the algebraic version the complex polarizations are given by the subalgebra $\mathfrak{n} = \langle (X + \tau Y), Z \rangle$.

Construction of Representations

We want to follow the scheme outlined at the end of 8.2. As up to now our representations of the Heisenberg group were realized by right translations and the recipe in 8.2.5 is given for left translations, we have to make the appropriate changes. For the zero-dimensional orbits $\mathcal{O}_{r,s}, r, s \in \mathbf{R}$ there is not much to be done: One has $\eta = r X_* + s Y_*$ and $G_\eta = G$ and hence the one-dimensional representation π given by $\pi(a, b, c) = \exp(2\pi i(ra + sb))$. For the two-dimensional orbits $\mathcal{O}_m, m \neq 0$ we have $\eta = m Z_*$, $G_\eta = \{(0,0,c), c \in \mathbf{R}\} = L$ and for each polarization given by $\mathfrak{n}$ the representation $\check{Y} \longmapsto 2\pi i < \eta, \check{Y} >= 2\pi i m z$ integrates to a character χ_m of the group Q belonging to $\mathfrak{n}$.

Hence each orbit $\mathcal{O}_m$ is integral and we get a representation space consisting of sections of the real line bundle over $M = \mathcal{O}_m$ (constructed using the character χ_m of G_η with $\chi_m((0,0,c)) = \exp(2\pi i m c) = e(mc))$ which are constant along the directions fixed by the polarization $\mathfrak{n}$. This leads to consider the following functions: We look at $\phi : G \longrightarrow \mathbf{C}$ with

$$\phi(lg) = \chi_m(l)\phi(g),$$

i.e. ϕ is of the form $\phi(a,b,c) = e(mc)F(a,b)$ with a smooth function $F : \mathbf{R}^2 \longrightarrow \mathbf{C}$ restricted by the polarization condition, which in our case for $\mathfrak{n} = <U, Z>$ is given by

$$\mathcal{R}_U \phi = 0$$

where $\mathcal{R}_U$ is the right-invariant differential operator

$$\mathcal{R}_U \phi(g) = \frac{d}{dt}\phi(g_U(t)^{-1}g)|_{t=0}, \quad g_U(t) := \exp tU.$$

We have

$$\begin{array}{llll}
g_X(t) &=& (t,0,0), & \text{and} \quad g_X(t)^{-1}g = (a-t, b, c-tb), \\
g_Y(t) &=& (0,t,0), & \qquad\quad g_Y(t)^{-1}g = (a, b-t, c), \\
g_Z(t) &=& (0,0,t), & \qquad\quad g_Z(t)^{-1}g = (a, b, c-t).
\end{array}$$

and hence

(8.14) $$\mathcal{R}_X = -\partial_a - b\partial_c, \quad \mathcal{R}_Y = -\partial_b, \quad \mathcal{R}_Z = -\partial_c.$$

– i) The polarization given by $\mathfrak{n} = <Y, Z>$ leads to

$$0 = \mathcal{R}_Y \phi = -\partial_b \phi = -F_b e(mc),$$

i.e. ϕ is of the form $\phi(a,b,c) = f(a)e(mc)$. As to be expected, via

$$\begin{aligned}
\pi(g_0)\phi(g) = \phi(gg_0) &= f(a+a_0)e(m(c+c_0+ab_0)) \\
&= e(mc)e(m(c_0+ab_0))f(a+a_0),
\end{aligned}$$

we recover the Schrödinger representation $\pi_m =: \pi_S$

$$f(t) \longmapsto \pi_m(g_0)f(t) = e(m(c_0 + b_0 t))f(t + a_0).$$

– ii) $\mathfrak{n} = <X, Z>$ leads to

$$0 = \mathcal{R}_X \phi = -(\partial_a + b\partial_c)\phi = -(F_a + 2\pi i m b F)e(mc),$$

i.e. ϕ is of the form $\phi(a,b,c) = e(-mab)f(b)e(mc)$. We get the representation

$$f(t) \longmapsto e(mc_0 - a_0(b_0 + t))f(t + b_0) =: \pi'_m(g_0)f(t).$$

From the Stone-von Neumann Theorem one knows that this representation (and any other constructed this way via another polarization) is equivalent to the Schrödinger representation. In our context we can see this directly using the Fourier transform as an intertwining operator:

We abbreviate again $e(u) := \exp(2\pi i u)$ and write as *Fourier transform*

$$\hat{f}(b) := \int_{\mathbf{R}} f(t)e(mbt)dt.$$

We get

$$\pi'_m(g_0)\hat{f}(b) = e(m(c_0 - a_0(b + b_0)))\int_{\mathbf{R}} f(t)e(mt(b + b_0))dt$$

and see that this coincides with the Fourier transform of the application of the Schrödinger representation $\pi_m(g_0)f$, namely with $a + a_0 =: t$ we get

$$(\widehat{\pi_m(g_0)}f)(b) = \int e(m(c_0 + ab_0)f(a + a_0)e(mab)da$$

$$= e(mc_0)e(-m(a_0(b + b_0)))\int e(mtb_0)f(t)e(mbt)dt.$$

– iii) As another example of this procedure we use a complex polarization to construct the *Fock representation*, which (among other things) is fundamental for the construction of the *theta functions*. The Fock representation has as its representation space the space $\mathcal{H}_m$ of holomorphic functions f on $\mathbf{C}$ subjected to the condition

$$\| f \|^2 = \int_{\mathbf{C}} | f(z) |^2 d\mu(z) < \infty, \ d\mu(z) := (1/(2i))dz \wedge d\bar{z}.$$

To adopt our presentation to the more general one given in Igusa [Ig] p.31ff, we introduce complex coordinates for our Heisenberg group

$$z := -ia + b, \ \bar{z} = ia + b,$$

i.e.

$$a = (\bar{z} - z)/(2i), \ b = (z + \bar{z})/2$$

and

$$\partial_z = (1/2)(i\partial_a + \partial_b), \ \partial_{\bar{z}} = (1/2)(-i\partial_a + \partial_b).$$

We use the complexification $\mathfrak{g}_c =< Y_\pm, Z_0 >$ with $Y_\pm := \pm iX + Y, Z_0 = 2iZ$ and the complex polarization given by $\mathfrak{n} =< Y_-, Z_0 >$. Then, in order to construct a representation space following our general procedure, we come to look at smooth complex functions ϕ on G, which are of the form

$$\phi(a, b, c) = F(z, \bar{z})e(mc)$$

and subjected to the polarization condition

$$\mathcal{R}_{Y_-}\phi = 0.$$

From (8.14) we have $\mathcal{R}_X = -(\partial_a + b\partial_c), \mathcal{R}_Y = -\partial_b$. We deduce

$$\mathcal{R}_{Y_-} = -2\partial_{\bar{z}} + ib\partial_c$$

and, hence, that one has $0 = \mathcal{R}_{Y_-}\phi = (-2F_{\bar{z}} - 2\pi mbF)e(mc)$, i.e.

$$\partial_{\bar{z}} \log F = -\pi mb = -\pi m(z + \bar{z})/2.$$

This shows that F is of the form

$$F(z, \bar{z}) = e^{-(\pi m/2)(z\bar{z} + \bar{z}^2/2)} \tilde{f}(z)$$

with a holomorphic function $\tilde{f}$. As we soon will see, to get a unitary representation on the space $\mathcal{H}_m$ introduced above, we have to choose here $\tilde{f}(z) = \exp(\pi m z^2/4) f(z)$ with $f \in \mathcal{H}_m$, i.e. our ϕ is of the form

$$\phi(a, b, c) = e(mc) e^{-(\pi m/2)(z\bar{z} + \bar{z}^2/2 - z^2/2)} f(z).$$

By a small computation we see that the representation by right translation

$$\pi(g_0)\phi(g) = \phi(g g_0)$$

leads to the formula of the *Fock representation* π_F

(8.15) $\qquad f(z) \longmapsto (\pi_F(g_0)f)(z) = e(mc_0) e^{-\pi m((z + z_0/2)\bar{z}_0 + (\bar{z}_0^2 - z_0^2)/4)} f(z + z_0).$

with $g_0 = (a_0, b_0, c_0)$, $z_0 = b_0 - ia_0$. Now it is easy to verify that π_F acts unitarily in $\mathcal{H}_m$ (**Exercise 8.6**).

Perhaps it is not a complete waste of time and energy to show here how this representation comes up also (as it must) by the induction procedure we outlined in 7.1: We have the standard projection

$$G = \mathrm{Heis}'(\mathbf{R}) \longrightarrow H \backslash G \simeq \mathcal{X}, \ g = (a, b, c) \longmapsto Hg = x = (a, b) \leftrightarrow (z, \bar{z})$$

with the Mackey section s given by $s(x) = (a, b, 0)$, i.e. $g = (0, 0, c)(a, b, 0)$, and the master equation (7.7)

$$s(x)g_0 = (a, b, 0)(a_0, b_0, c_0) = h^* s(x g_0)$$

with

$$s(x g_0) = (a + a_0, b + b_0, 0), \ h^* = (0, 0, c_0 + a b_0) = (0, 0, c_0 + (z z_0 - z \bar{z}_0 + \bar{z}_0 z_0 - \bar{z}\bar{z}_0)/(4i)).$$

$d\mu(x) = d\mu(z)$ is a quasi-invariant measure on $\mathcal{X} = \mathbf{C}$ with Radon-Nikodym derivative

$$\frac{d\mu(x g_0)}{d\mu(x)} = e^{-m\pi(z\bar{z}_0 + \bar{z}z_0 + |z_0|^2)} = \delta(h^*).$$

Hence, injecting these data into the formula from Theorem 7.4 in 7.1.2 for the representation in the second realization

$$\tilde{\pi}(g_0)f(x) = \delta(h^*)^{1/2} \pi_0(h^*) f(x g_0)$$

provides the formula (8.15) for the Fock representation obtained above.

We know that the Fock representation π_F is equivalent to the Schrödinger representation π_S but it is perhaps not so evident how the intertwining operator looks like.

Proposition 8.1: The prescription

$$f(t) \longmapsto F(z) = If(z) := \int_{\mathbf{R}} k(t, z)f(t)dt$$

with the *kernel*

$$k(t, z) = 2^{1/4}e^{-\pi t^2}e(tz)e^{(\pi/2)z^2}$$

provides an isometry I between the spaces $L^2(\mathbf{R})$ and $\mathcal{H}_m$ for $m = 1$ and intertwines the corresponding Schrödinger and Fock representations.

The fact that I is an intertwining operator is verified by a straightforward computation (**Exercise 8.7**). For the verification of the isometry we refer to [Ig] p.31-35 where one goes back to explicit Hilbert bases for the spaces.

Summary

For the Heisenberg group (as for each solvable group), one has a perfect matching between coadjoint orbits and irreducible unitary representations.
Everything goes through as well for the higher-dimensional Heisenberg groups if one changes the notation appropriately.

8.5   Some Hints Concerning the Jacobi Group

In Kirillov's book [Ki1] one can find a lot of examples and a thorough discussion of the merits and demerits of the orbit method. As orbits already turned up naturally in our presentation of Mackey's method in 7.1 where we applied it to Euclidean groups and the Poincaré group, it is a natural topic to inspect the coadjoint orbits for semidirect products, in particular for the Euclidean groups.

Exercise 8.8: Determine the coadjoint orbits of the Euclidean group

$$G^E(3) = \mathrm{SO}(3) \ltimes \mathbf{R}^3.$$

The reader will find material for this in [GS] p.124ff. As the author of this text is particularly fond of the *Jacobi group* G^J, i.e. – in its simplest version – a semidirect product of $\mathrm{SL}(2, \mathbf{R})$ with $\mathrm{Heis}(\mathbf{R})$, as another exercise we propose here to treat some questions in this direction concerning the Jacobi group. Most anwers to these can be found in [BeS] and [Be1] where we collected some elements of the representation theory of G^J and the application of the orbit method to G^J. But at a closer look there are enough rather easily accessible open problems to do some original new work:

The Jacobi group is in general the semidirect product of the symplectic group with an appropriate Heisenberg group. Here we look at

$$G^J(\mathbf{R}) := \mathrm{SL}(2, \mathbf{R}) \ltimes \mathrm{Heis}(\mathbf{R}).$$

In this case we fix the multiplication law by the embedding into the symplectic group $\mathrm{Sp}(2, \mathbf{R})$ given by

$$\mathrm{Heis}(\mathbf{R}) \ni (\lambda, \mu, \kappa) \longmapsto \begin{pmatrix} 1 & & & \mu \\ \lambda & 1 & \mu & \kappa \\ & & 1 & -\lambda \\ & & & 1 \end{pmatrix},$$

$$\mathrm{SL}(2, \mathbf{R}) \ni M = \begin{pmatrix} a & b \\ c & d \end{pmatrix} \longmapsto \begin{pmatrix} a & & b & \\ & 1 & & \\ c & & d & \\ & & & 1 \end{pmatrix}.$$

We write

$$g = (p, q, \kappa)M \quad \text{or} \quad g = M(\lambda, \mu, \kappa) \in G^J(\mathbf{R}).$$

As in [BeS] or [Ya], we can describe the Lie algebra $\mathfrak{g}^J$ as a subalgebra of $\mathfrak{g}_2 = \mathfrak{sp}(2, \mathbf{R})$ by

$$G(x, y, z, p, q, r) := \begin{pmatrix} x & 0 & y & q \\ p & 0 & q & r \\ z & 0 & -x & -p \\ 0 & 0 & 0 & 0 \end{pmatrix}$$

and denote

$$X := G(1, 0, \ldots, 0), \ldots, R := G(0, \ldots, 0, 1).$$

We get the commutators

$$\begin{array}{lll} [X, Y] = 2Y, & [X, Z] = -2Z, & [Y, Z] = X, \\ [X, P] = -P, & [X, Q] = Q, & [P, Q] = 2R, \\ [Y, P] = -Q, & [Z, Q] = -P, & \end{array}$$

all others are zero. Hence, we have the complexified Lie algebra given by

$$\mathfrak{g}_c^J = < Z_1, X_{\pm}, Y_{\pm}, Z_0 >$$

where as in [BeS] p.12

$$Z_1 := -i(Y - Z), \quad Z_0 := -iR,$$

$$X_{\pm} := (1/2)(X \pm i(Y + Z)), \quad Y_{\pm} := (1/2)(P \pm iQ)$$

with the commutation relations

$$[Z_1, X_{\pm}] = \pm 2X_{\pm}, [Z_0, Y_{\pm}] = \pm Y_{\pm}, \text{ etc.}$$

Exercise 8.9 : Verify all this and compute left invariant differential operators

$$\mathcal{L}_{Z_0} = i\partial_\kappa$$

$$\mathcal{L}_{Y_{\pm}} = (1/2)y^{-1/2}e^{\pm i\theta}(\partial_p - (x \pm iy)\partial_q - (p(x + iy) + q)\partial_\kappa)$$

$$\mathcal{L}_{X_{\pm}} = \pm(i/2)e^{\pm 2i\theta}(2y(\partial_x \mp i\partial_y) - \partial_\theta)$$

$$\mathcal{L}_{Z_1} = -i\partial_\theta$$

acting on differentiable functions $\phi = \phi(g)$ with the coordinates coming from

$$g = (p, q, \kappa) n(x) t(y) r(\theta)$$

where

$$n(x) = \begin{pmatrix} 1 & x \\ & 1 \end{pmatrix}, \ t(y) = \begin{pmatrix} y^{1/2} & \\ & y^{-1/2} \end{pmatrix}, \ r(\theta) = \begin{pmatrix} \alpha & \beta \\ -\beta & \alpha \end{pmatrix}$$

with

$$\alpha = \cos\theta, \ \beta = \sin\theta,$$

and

$$g(i, 0) = (\tau, z) = (x + iy, p\tau + q).$$

As usual, we put

$$N := \{n(x); x \in \mathbf{R}\}, \ A := \{t(a); a \in \mathbf{R}_{>0}\},$$

$$K := \mathrm{SO}(2) = \{r(\theta); \theta \in \mathbf{R}\}, \ M := \{\pm E\}.$$

Elements of the representation theory of $G^J(\mathbf{R})$ can be found in the work of Pyatetskii–Shapiro, Satake, Kirillov, Howe, Guillemin, Sternberg, Igusa, Mumford and many others. To a large part, they are summed up in [BeS], the essential outcome being that the representations π of $G^J(\mathbf{R})$ with nontrivial central character are essentially products of projective representations of $\mathrm{SL}(2, \mathbf{R})$ with a fundamental (projective) representation π_{SW}^m of G^J called the *Schrödinger–Weil representation* in [BeS], as it is in turn the product of the Schrödinger representation π_S of the Heisenberg group $\mathrm{Heis}(\mathbf{R})$ and the *(Segal–Shale–) Weil* or *oscillator representation* π_W of $\mathrm{SL}(2, \mathbf{R})$. This representation π_{SW}^m is a representation of lowest weight $k = 1/2$ and index m. It is characterized by the fact that all its K^J–types are one-dimensional ($K^J := \mathrm{SO}(2) \times C(\mathrm{Heis}(\mathbf{R})) \simeq \mathrm{SO}(2) \times \mathbf{R}$), which in turn is characterized by the fact that the lowest weight vector for π_{SW}^m satisfies the heat equation. To be a bit more explicit, the irreducible unitary representations π of $G^J(\mathbf{R})$ with central character ψ^m (with $\psi^m(x) := e^m(x)$) for $m \neq 0$ are infinitesimally equivalent to

$$\hat{\pi}_{m,s,\nu}, \ s \in i\mathbf{R}, \ \nu = \pm 1 \quad (\hat{\pi}_{m,s,\nu} \simeq \hat{\pi}_{m,-s,\nu}),$$
$$\hat{\pi}_{m,s,\nu}, \ s \in \mathbf{R}, \ s^2 < 1/4,$$
$$\hat{\pi}_{m,k}^+, \ k \in \mathbf{N},$$
$$\hat{\pi}_{m,k}^-, \ k \in \mathbf{N}.$$

Here and in the following, we restrict our treatment to $m > 0$, but there is a "mirror image" for $m < 0$ which, to save space, we will not discuss. These infinitesimal representations are of the form

$$\hat{\pi}_{m,s,\nu} = \hat{\pi}_{SW}^m \otimes \hat{\pi}_{s,\nu},$$

$$\hat{\pi}_{m,k}^\pm = \hat{\pi}_{SW}^m \otimes \hat{\pi}_{k_0}^\pm, \ k = k_0 + 1/2.$$

The Schrödinger–Weil representation $\hat{\pi}_{SW}^m$ is given on the space

$$V_m^{1/2} := \langle v_j \rangle_{j \in N_0}$$

by (see [BeS] p.33)

$$Z v_j := (1/2 + j) v_j, \ Z_0 v_j := \mu v_j, \ Y_+ v_j := v_{j+1}, \ X_+ v_j := -1/(2\mu) v_{j+2}, \ldots.$$

And, for instance, the discrete series representation $\hat{\pi}^+_{k-1/2}$ of $\mathfrak{sl}(2)$ is given on

$$W_{k-1/2} := \langle w_l \rangle_{l \in 2\mathbf{N}_0}$$

by

$$Z w_l := (k - 1/2 + l) w_l, \quad X_+ w_l := w_{l+2}, \quad X_- w_l := (l/2)(k - 3/2 + l) w_{l-2}.$$

That is, we have as a space for $\hat{\pi}^+_{m,k}$

$$V^+_{m,k} := V^{1/2}_m \otimes W_{k-1/2} = \langle v_j \otimes w_l \rangle_{j \in \mathbf{N}, l \in 2\mathbf{N}_0}$$

with

$$
\begin{aligned}
Z_0(v_j \otimes w_l) &= \mu(v_j \otimes w_l), \\
Z(v_j \otimes w_l) &= (k + j + l)(v_j \otimes w_l), \\
Y_+(v_j \otimes w_l) &= v_{j+1} \otimes w_l, \\
X_+(v_j \otimes w_l) &= -(1/(2\mu))v_{j+2} \otimes w_l + v_j \otimes w_{l+2}.
\end{aligned}
$$

In particular, for the spaces of Z-weight $k + \lambda, \lambda \geq 0$,

$$V^{(\lambda)} := \{ v \in V^+_{m,k}; \; Z v = (k + \lambda) v \}$$

we have

$$\dim V^{(0)} = 1, \; \dim V^{(1)} = 1, \; \dim V^{(2)} = 2, \; \dim V^{(3)} = 2, \; \dim V^{(4)} = 3, \ldots$$

If we denote by $\rho_{m,\tilde{k}}$ the (one-dimensional) representation of K^J given by

$$\rho_{m,\tilde{k}}(r(\theta), (0, 0, \kappa)) = e^{(i\tilde{k}\theta) + 2\pi i m \kappa}, \; \theta \in [0, 2\pi], \kappa \in \mathbf{R},$$

we have

$$
\begin{aligned}
\text{mult}\,(\rho_{m,\tilde{k}}, \pi^+_{m,k}) &= 1 && \text{for } \tilde{k} = k, \\
&= 1 + l && \text{for } \tilde{k} = k + 2l \text{ or } = k + 2l + 1, l \in \mathbf{N}_0 \\
&= 0 && \text{for } \tilde{k} < k,
\end{aligned}
$$

and

$$\text{mult}\,(\rho_{m,\tilde{k}}, \pi) = \infty$$

in the other cases listed above. Now, as to be seen for instance in [BeS] p.28ff, a *lowest weight* or *vacuum vector* ϕ_0 for a discrete series representation $\pi^+_{m,k}$ of $G^J(\mathbf{R})$ realized by right translation in a space $\mathcal{H}^+_{m,k}$ of smooth functions $\phi = \phi(g)$ on $G^J(\mathbf{R})$ is characterized by the equations

$$\mathcal{L}_{X_-}\phi_0 = \mathcal{L}_{Y_-}\phi_0 = 0, \; \mathcal{L}_{Z_0}\phi_0 = 2\pi m \phi_0, \; \mathcal{L}_Z \phi_0 = k \phi_0 \, .$$

The Jacobi group is not semisimple (and indeed a prominent example for a non-reductive group). We describe its coadjoint orbits using our embedding of G^J into $\text{Sp}(2, \mathbf{R})$:

We identify the dual $\mathfrak{sp}^*$ of $\mathfrak{sp}$ by the $\text{Sp}(2, \mathbf{R})$ invariant isomorphism $\eta \mapsto X_\eta$ given by

$$\eta(\check{Y}) = \text{Tr}(X_\eta \check{Y}) =:< X_\eta, \check{Y} > \text{ for all } \check{Y} \in \mathfrak{sp}.$$

$\eta_1, \ldots, \eta_6$ denotes a basis of $(\mathfrak{g}^J)^*$ dual to

$$(X_1, \ldots, X_6) := (X, \ldots, R).$$

We realize $(\mathfrak{g}^J)^*$ as a subspace of $\mathfrak{sp}$ by the matrices

$$M(x, y, z, p, q, r) := \begin{pmatrix} x & p & z & 0 \\ 0 & 0 & 0 & 0 \\ y & q & -x & 0 \\ q & r & -p & 0 \end{pmatrix}, x, \ldots, r \in \mathbf{R},$$

and put

$$X_* := M(1, 0, \ldots, 0), \ldots, R_* := M(0, \ldots, 0, 1).$$

Then

$$X_{\eta_1} = X_*, \ldots, X_{\eta_6} = R_*$$

is a basis of $(\mathfrak{g}^J)^*$ with

$$< X_*, X >= 2, \; < Y_*, Y >= 1, \; < Z_*, Z >= 1$$
$$< P_*, P >= 2, \; < Q_*, Q >= 2, \; < R_*, R >= 1.$$

By a straightforward computation one obtains the following result:

Lemma: If g^{-1} is denoted as

$$g^{-1} = \begin{pmatrix} a & 0 & b & a\mu - b\lambda \\ \lambda & 1 & \mu & \kappa \\ c & 0 & d & c\mu - d\lambda \\ 0 & 0 & 0 & 1 \end{pmatrix},$$

the coadjoint action of g on $M(x, \ldots, r)$ is given by

$$\mathrm{Ad}^*(g)M(x, \ldots, r) = M(\tilde{x}, \ldots, \tilde{r})$$

with

$$\tilde{x} = (ad + bc)x + bdy - acz + (2ac\mu - (ad + bc)\lambda)p$$
$$+ ((ad + bc)\mu - 2bd\lambda)q + r(a\mu - b\lambda)(c\mu - d\lambda),$$

$$\tilde{y} = 2dcx + d^2 y - c^2 z + 2(c\mu - d\lambda)(cp + dq) + r(c\mu - d\lambda)^2,$$

(8.16) $\quad \tilde{z} = -2abx - b^2 y + a^2 z - 2(a\mu - b\lambda)(ap + bq) - r(a\mu - b\lambda)^2,$

$$\tilde{p} = ap + bq + r(a\mu - b\lambda),$$

$$\tilde{q} = cp + dq + r(c\mu - d\lambda),$$

$$\tilde{r} = r.$$

Exercise 8.10: Verify this and determine the coadjoint orbits. For instance, show that for

$$U_* = mR_* + \alpha X_*, m \neq 0, \alpha \neq 0,$$

one has the 4-dimensional *hyperbolic* orbit $\mathcal{O}_{U_*}$ contained in

$$M_{m,-\alpha^2} := \{(x, y, z.p, q, r) \in \mathbf{R}^6; m(x^2 + yz - \alpha^2) = 2pqx - p^2y + q^2z, r = m\}.$$

Do you get $\mathcal{O}_{U_*} = M_{m,-\alpha^2}$?

It is natural to try to apply the general procedure outlined in this section to realize representations of G^J by bundles carried by these orbits and to ask, which ones you get and, if not, why? As already said above, there are some answers in [Be1] and [Ya], but far from all.

Chapter 9

Epilogue: Outlook to Number Theory

Representations of groups show up in many places in Number Theory, for instance as representations of Galois groups. There seems to be no doubt that the relationship of the two topics culminates in the Langlands Program, which (roughly said) seeks to establish a correspondence between Galois and automorphic representations. Many eminent mathematicians have worked and work on this program, and it is now even interesting for physicists; see Frenkel's *Lectures on the Langlands Program and Conformal Field Theory* [Fr]. We cannot dare to go into this here, but we will try to introduce at least some initial elements by presenting some representation spaces for the special groups we treated in our examples, consisting of, or at least containing, theta functions, and modular and automorphic forms. And we will also introduce the notions of zeta and *L*-functions, which ultimately are the foundations of the bridge between the Galois and automorphic representations. There are many useful books available. We can only cite some of them, which by now are classic: *Representation Theory and Automorphic Functions* by Gelfand, Graev and Pyatetskii-Shapiro [GGP], *Automorphic Forms on* GL(2) by Jacquet and Langlands [JL], *Automorphic Forms on Adele Groups* by Gelbart [Ge], *Analytic Properties of Automorphic L-Functions* by Gelbart and Shahidi [GS], *Automorphic Forms and Representations* by Bump [Bu], *Theta Functions* by Igusa [Ig], *The Weil representation, Maslov index and Theta Series* by Lion and Vergne [LV], *Fourier Analysis on Number Fields* by Ramakrishnan and Valenza [RV], and Mumford's *Tata Lectures on Theta* I, II, *and* III [Mu].

The goal of these last sections will be to lead a way to *automorphic representations*: While we already encountered the decomposition of the regular representation on the space $L^2(G)$, we shall here take a discrete subgroup Γ of G and ask for the representations appearing in the decomposition of $L^2(\Gamma \setminus G)$, i.e., we consider representation spaces consisting of functions with a certain periodicity, invariance or covariance property. All this is most easily understood for the case of the Heisenberg group and leads to the notion of the theta functions. So we will start by treating this case, and afterwards will look at the group SL(2, **R**) and introduce modular forms as examples for more general automorphic forms. Finally, we shall consider several kinds of *L*-functions appearing in connection with representation theory.

Since this is an epilogue and an outlook and the material has such a great extension and depth, the reader will excuse (we hope) that we shall not be able to give more than some indications, and even less proofs, than in the preceding chapters.

9.1   Theta Functions and the Heisenberg Group

Whatever we do in this section can easily be extended to higher dimensional cases by conveniently adopting the notation. As already done before, we denote again $z \in \mathbf{C}$ and $\tau \in \mathfrak{H}$ and moreover, as usual in this context,

$$q := e(\tau) = e^{2\pi i \tau}, \ \epsilon := e(z) = e^{2\pi i z}.$$

Then one has the *classic Jacobi theta function* ϑ given by

(9.1) $$\vartheta(z, \tau) := \sum_{n \in \mathbf{Z}} e^{\pi i(n^2 \tau + 2nz)} = \sum_{n \in \mathbf{Z}} q^{n^2/2} \epsilon^n$$

going back to Jacobi in 1828. One has to verify convergence:

Lemma: ϑ converges absolutely and uniformly on compact parts of $\mathbf{C} \times \mathfrak{H}$.

Proof: Exercise 9.1.

Moreover, today, one has variants, which unfortunately are differently normalized by different authors. For instance in [Ig] p.V, one finds the *theta function with characteristics* $m = (m', m'') \in \mathbf{R}^2$

$$\theta_m(\tau, z) := \sum_{n \in \mathbf{Z}} e((1/2)(n + m')^2 \tau + (n + m')(z + m'')).$$

In our text, we shall follow the notation of [EZ] p.58 where, for $2m \in \mathbf{N}$, $\mu = 0, 1, ., 2m-1$, one has

(9.2) $$\theta_{m,\mu}(\tau, z) \ := \sum_{r \in \mathbf{Z},\, r \equiv \mu \ \text{mod}\ 2m} q^{r^2/(4m)} \epsilon^r$$

$$= \sum_{n \in \mathbf{Z}} e^m((n + \mu/(2m))^2 \tau + 2(n + \mu/(2m))z).$$

Obviously, we have $\vartheta(z, \tau) = \theta_{1/2,0}(\tau, z)$. Theta functions have a rather evident *quasiperiodicity* property concerning the complex variable z and a deeper *modular* property concerning the variable $\tau \in \mathfrak{H}$, which makes them most interesting for applications in number theory and algebraic geometry, and – concerning both variables – ϑ satisfies the "one-dimensional" *heat equation*

$$4\pi i \vartheta_\tau = \vartheta_{zz}.$$

At first, we treat the dependance on the complex variable where, as we shall see, the Heisenberg group comes in. But to give already here at least a tiny hint, for instance, theta functions are useful if one asks for the number of ways a given natural number can be written as, say, four squares of integers (see [Ma] p.354 or our Example 9.3 in 9.2).

As we keep in mind that the group really behind theta functions is the Jacobi group introduced in 8.5, we take the version of the Heisenberg group, which comes from its realization by four-by-four matrices though in other sources different coordinizations are used (like we did in the last chapter). Thus, we take

$$G := \mathrm{Heis}(\mathbf{R}) = \{g = (\lambda, \mu, \kappa) \in \mathbf{R}^3\} \text{ with } gg' := (\lambda + \lambda', \mu + \mu', \kappa + \kappa' + \lambda\mu' - \lambda'\mu).$$

For fixed $\tau \in \mathfrak{H}$, $G = \mathrm{Heis}(\mathbf{R})$ acts on $\mathbf{C}$ by

$$\mathbf{C} \ni z \longmapsto g(z) := z + \lambda\tau + \mu, \ g = (\lambda, \mu, \kappa) \in G.$$

Obviously, the stabilizer of $z \in \mathbf{C}$ is the center $C(\mathbf{R}) = \{(0, 0, \kappa); \kappa \in \mathbf{R}\} \simeq \mathbf{R}$ of the Heisenberg group and one has

$$\mathcal{X} := G(\mathbf{R})/C(\mathbf{R}) \simeq \mathbf{C}.$$

This action induces an action on functions F on $\mathbf{C}$ in the usual way

$$F \longmapsto F^g \text{ with } F^g(z) := F(g(z)) = F(z + \lambda\tau + \mu).$$

Recalling the cocycle condition (7.8) in 7.1 and the first example of an automorphic factor in 7.2, we refine this action by introducing for $m \in \mathbf{R}$ as an *automorphic factor for the Heisenberg group*

$$j_m(g, z) := e^m(\lambda^2\tau + 2\lambda z + \lambda\mu + \kappa).$$

Exercise 9.2: Verify the relation

$$j_m(gg', z) = j_m(g, g'(z))j_m(g', z).$$

Then we define $F|[g]$ by

$$F|[g](z) := j_m(g, z)F(g(z))$$

and introduce as our first example of an *automorphic form* a holomorphic function F on $\mathbf{C}$ with

(9.3) $F|[\gamma] = F$ for all $\gamma = (r, s, t) \in \Gamma = \mathrm{Heis}(\mathbf{Z}) = \{\gamma = (r, s, t) \in \mathbf{Z}^3\}.$

We denote by $\Theta(m)$ the vector space of all these functions F.

Thus, an entire function F (i.e. holomorphic on $\mathbf{C}$) is an element of $\Theta(m)$ iff

$$F(z + s) = F(z) \text{ and } F(z + r\tau) = e^{-m}(r^2\tau + 2rz)F(z) \text{ for all } r, s \in \mathbf{Z}.$$

If we relate this to the theta functions defined above, we are led to the following central result, which is not too difficult to prove (using the periodicity and the functional equation (9.3)).

Theorem 9.1: For $2m \in \mathbf{N}$, $\Theta(m)$ is spanned by the theta functions

$$\theta_{m,\alpha}, \ \alpha = 0, 1, \ldots, 2m - 1$$

and hence has dimension $2m$.

To any function F on $\mathbf{C}$ we associate a *lifted* function $\phi = \phi_F$ defined on $G = \mathrm{Heis}(\mathbf{R})$ by

$$\phi_F(g) := (F|[g])(0) = F(\lambda\tau + \mu)e^m(\lambda^2\tau + \lambda\mu + \kappa).$$

This prescription is to be seen on the background of the relation between the First and the Second Realization in the induction procedure (see 7.1). Then we come to understand immediately the first two items in the

Proposition 9.1: Under the map $F \longmapsto \phi_F$, the space $\Theta(m)$ is isomorphic to the space $\mathcal{H}(m)$ of functions $\phi : G \longrightarrow \mathbf{C}$ satisfying

i) $\phi(\gamma g) = \phi(g)$ for all $\gamma \in \Gamma$,
ii) $\phi(g\kappa) = \phi(g)e^m(\kappa)$ for all $\kappa := (0,0,\kappa) \in C(\mathbf{R})$,
iii) ϕ is smooth and satisfies $\mathcal{L}_{Y_-}\phi = 0$.

The third statement is result of a discussion parallel to the one of the Fock representation in 8.4: here we use $Y_- := P - \tau Q$ to express the holomorphicity of F by a differential equation for ϕ: For

$$g_P(t) = (t,0,0),$$
$$g_Q(t) = (0,t,0),$$
$$g_R(t) = (0,0,t)$$

one has

$$gg_P(t) = (\lambda + t, \mu, \kappa - \mu t),$$
$$gg_Q(t) = (\lambda, \mu + t, \kappa + \lambda t),$$
$$gg_R(t) = (\lambda, \mu, \kappa + t)$$

and for the left invariant operators $\mathcal{L}_U$ acting on functions ϕ living on G defined by

$$\mathcal{L}_U\phi(g) := \frac{d}{dt}\phi(gg_U(t))|_{t=0},$$

we get

$$\mathcal{L}_P = \partial_\lambda - \mu\partial_\kappa,$$
$$\mathcal{L}_Q = \partial_\mu + \lambda\partial_\kappa,$$
$$\mathcal{L}_R = \partial_\kappa,$$

and

$$\mathcal{L}_{Y_-} = \mathcal{L}_{P-\tau Q} = \partial_\lambda - \tau\partial_\mu - (\lambda\tau + \mu)\partial_\kappa.$$

In the prescription for the lifting from $\mathbf{C}$ to G we assume $z = \lambda\tau + \mu, \bar{z} = \lambda\bar{\tau} + \mu$ and (via Wirtinger calculus) deduce

$$\partial_{\bar{z}} = (\tau\partial_\mu - \partial_\lambda)/(\tau - \bar{\tau}).$$

If we have

$$\phi(\lambda, \mu, \kappa) = F(z, \bar{z})e^m(\kappa + \lambda^2\tau + \lambda\mu),$$

a tiny computation shows that the condition $\mathcal{L}_{Y_-}\phi = 0$ translates into the holomorphicity condition $\partial_{\bar{z}}F = 0$.

This leads us directly to the following representation theoretic interpretation. The conditions i) and ii) of the space $\mathcal{H}(m)$ show up if one looks at still another standard representation of the Heisenberg group, namely the *lattice representation*:

We take the subgroup
$$H := \{h = (r, s, t);\ r, s \in \mathbf{Z}, t \in \mathbf{R}\}$$
and, for $m \in \mathbf{Z} \setminus \{0\}$, its character given by
$$\pi_0(h) := e^m(t).$$

Then we have $\mathcal{X} = H \setminus G \simeq \mathbf{Z}^2 \setminus \mathbf{R}^2$. The action of G on $\mathbf{C}$ given by
$$z \longmapsto z + \lambda \tau + \mu$$

restricts to the action of H on $0 \in \mathbf{C}$ producing the points $r\tau + s$ forming a lattice L in $\mathbf{C}$. By our general procedure from Mackey's approach in 7.1, we get a representation by right translation on the space of functions ϕ on G, which satisfy the functional equation
$$\phi(hg) = e^m(t)\phi(g) \quad \text{for all } h = (r, s, t),\ r, s \in \mathbf{Z}, t \in \mathbf{R}, g \in G.$$

(As $\mathcal{X}$ is compact, we have no problem with the condition of the finite norm integral.) This induced representation is not irreducible and we proceed as follows. For $\Gamma = \text{Heis}(\mathbf{Z})$ we denote by $L^2(\Gamma \setminus G)$ the Hilbert space of Γ-invariant measurable functions $G \longrightarrow \mathbf{C}$ with the scalar product
$$\langle \phi, \psi \rangle := \int_{\Gamma \setminus G} \overline{\phi(g)} \psi(g) dg.$$

This is a Hilbert space direct sum
$$L^2(\Gamma \setminus G) = \oplus_{m \in \mathbf{Z}} L^2(\Gamma \setminus G)_m$$

where $L^2(\Gamma \setminus G)_m$ denotes the subspace of functions ϕ that satisfy
$$\phi(g\kappa) = e^m(\kappa)\phi(g) \quad \text{for all } \kappa = (0, 0, \kappa) \in Z(\mathbf{R}), g \in G.$$

Now for $m > 1/2$, this induced representation on $L^2(\Gamma \setminus G)_m$ is not irreducible. We find here the first example of an important general statement.

Theorem 9.2 (*Duality Theorem for the Heisenberg group*): For $m \in \mathbf{N}$, the multiplicity of the Schrödinger representation π_m in $L^2(\Gamma \setminus G)$ is equal to the dimension of the space $\Theta(m)$.

Proof: There are several different approaches. In the our context, we can argue like this:
The Schrödinger representation π_m transforms a function $f \in L^2(\mathbf{R})$ into
$$\pi_m(g_0)f(x) = e^m(\kappa_0 + (2x + \lambda_0)\mu_0)f(x + \lambda_0)$$

and is intertwined with the Heisenberg representation (see Example 7.1 at the beginning of 7.1) given by right translation on functions ϕ on G via
$$f(x) \longmapsto e^m(\kappa + \lambda\mu)f(\lambda) = \phi_f(g).$$

Obviously, one has $\phi(\kappa g) = e^m(\kappa)\phi(g)$ but as the Heisenberg representation is induced by the two-dimensional subgroup $\{(0, \mu, \kappa); \mu, \kappa \in \mathbf{R}\}$ of G, one also has invariance concerning the μ-variable. Now, to produce a subspace of the lattice representation, one has to create in- or covariance concerning the λ-variable: We define for $n \in \mathbf{Z}$ and $\alpha = 0, 1, \ldots, 2m - 1$

$$\lambda_{n,\alpha} := (n + \alpha/(2m), 0, 0)$$

and get

$$\lambda_{n,\alpha}g = (\lambda + n + \alpha/(2m), \mu, \kappa + \mu(n + \alpha/(2m))).$$

Hence for every α, we have a map $\vartheta_\alpha : f \longmapsto \phi_{f,\alpha}$, called *theta transform*, with

$$\phi_{f,\alpha}(g) := \sum_{n \in \mathbf{Z}} \phi_f(\lambda_{n,\alpha}g)$$

$$= \sum_{n \in \mathbf{Z}} e^m(\kappa + \mu(n + \alpha/(2m)) + (\lambda + n + \alpha/(2m))\mu)f(\lambda + n + \alpha/(2m)).$$

It is not difficult to verify that one has for all $\gamma \in \Gamma$ and $\kappa \in \mathbf{R}$

$$\phi_{f,\alpha}(\gamma g) = \phi_{f,\alpha}(g) \text{ and } \phi_{f,\alpha}(g\kappa) = \phi_{f,\alpha}(g).$$

Thus we see that for $\alpha = 0, 1, \ldots, 2m - 1$ every ϑ_α intertwines the Schrödinger representation with the lattice representation. We still have to show that these are all different and that there are no others. To do so, we realize the Schrödinger representation π_m by choosing as vacuum vector f_0 not $f_0(x) = \exp(-2\pi m x^2)$, as in Remark 6.13 in 6.5, but

$$f_0(x) = e^m(\tau x^2).$$

Then this function is annihilated by the differential operator $\hat{Y}_- = \partial_x - (4\pi i m)\tau x$ from the derived representation $d\pi_m$ (as in Example 6.9 in 6.3) belonging to $Y_- = P - \tau Q$ introduced above. If one applies the theta transform ϑ_α to f_0 one gets

$$\vartheta_\alpha f_0(g) = e^m(\kappa + \tau\lambda^2 + \lambda\mu)\theta_{m,\alpha}(\tau, \lambda\tau + \mu),$$

i.e. exactly the theta basis of the space $\mathcal{H}(m)$ appearing in Proposition 9.1 above, which characterizes the lifting of $\Theta(m)$ to functions on G.

The reader could try to find another version of the proof using the bundle approach to induced representations. One can prove (Theorem of Appell-Humbert) that every line bundle on the torus $\mathcal{X} := L \backslash \mathbf{C}$, $L = \mathbf{Z}\tau + \mathbf{Z}$ is of the form $\mathcal{L}(H, \beta)$, i.e. the quotient of the trivial bundle $\mathbf{C} \times \mathbf{C} \longrightarrow \mathbf{C}$ by the action of L given by

$$(\ell, (v, z)) \longmapsto (\beta(\ell)e^{iH(z,\ell)+(1/2)H(\ell,\ell)}v, z + \ell) \quad \text{for all } \ell \in L, v, z \in \mathbf{C}.$$

Here $H : \mathbf{C} \times \mathbf{C} \longrightarrow \mathbf{C}$ is a hermitian form such that $E = \text{Im}(H)$ is integer valued on L and $\beta : L \longrightarrow S^1$ a map with

$$\beta(\ell_1 + \ell_2) = e^{i\pi E(\ell_1, \ell_2)}\beta(\ell_1)\beta(\ell_2).$$

A computation shows that these hermitian forms are exactly the forms $H_m, m \in \mathbf{Z}$, given by

$$H_m(z, w) = m\bar{z}w/\text{Im}(\tau).$$

Up to now, we only treated the theta functions as quasiperiodic functions in the complex variable z. There is also a remarkable *modular* behaviour concerning the variable $\tau \in \mathfrak{H}$. Before we discuss this, we return to some more general considerations.

9.2 Modular Forms and SL(2,**R**)

Besides theta functions and strongly intertwined with it, modular forms is another topic relating classical function theory to number theory and algebraic and analytic geometry. There are many competent introductions. A very brief one is given in Serre [Se1]. Moreover we recommend the books by Schöneberg [Sch] and Shimura [Sh] containing more relevant details.

Classical Theory (Rudiments)

We already met several times with the action of $G = \mathrm{SL}(2, \mathbf{R})$ on the upper half plane $\mathcal{X} = \mathfrak{H} := \{\tau = x + iy \in \mathbf{C}; y > 0\}$ given by

$$\left(g = \begin{pmatrix} a & b \\ c & d \end{pmatrix}, \tau\right) \longmapsto g(\tau) := \frac{a\tau + b}{c\tau + d}.$$

As g and $-g$ produce the same action, it is more convenient to treat the group

$$\bar{G} = \mathrm{PSL}(2, \mathbf{R}) := \mathrm{SL}(2, \mathbf{R})/\{\pm E_2\}.$$

But since we worked so far with G and its representation theory can be rather easily adopted to $\bar{G}$, we stay with this group.

For $k \in \mathbf{R}$ (but mainly $k \in \mathbf{Z}$) we have (see 7.2.2) the *automorphic factor*

$$j(g, \tau) := (c\tau + d)^{-1} \text{ resp. } j_k(g, \tau) := (c\tau + d)^{-k}$$

fulfilling the cocycle relation (7.8) $j(gg', \tau) = j(g, g'(\tau))j(g', \tau)$. Let f be a smooth function on $\mathfrak{H}$. Then one writes

$$f|_k[g](\tau) := f(g(\tau))j_k(g, \tau)$$

(so that $\pi_k(g)f(\tau) := f(g^{-1}(\tau))j_k(g^{-1}, \tau)$ at least formally is the prescription of a representation).

There are several important types of discrete subgroups Γ of G and the theory becomes more and more beautiful if one treats more and more of these. For our purpose to give an introduction, it is sufficient to restrict ourselves to the (full) *modular group*

$$\Gamma = \mathrm{SL}(2, \mathbf{Z})$$

and eventually to the *main congruence group of level N*

$$\Gamma = \Gamma(N) := \{\gamma \in \mathrm{SL}(2, \mathbf{Z}); \gamma \equiv E_2 \mod N\}, \ N \in \mathbf{N},$$

Hecke's *special congruence subgroup*

$$\Gamma = \Gamma_0(N) := \{\gamma = \begin{pmatrix} a & b \\ c & d \end{pmatrix} \in \mathrm{SL}(2, \mathbf{Z}); c \equiv 0 \mod N\}, \ N \in \mathbf{N},$$

and the *theta group* Γ_ϑ generated by T^2 and S, i.e.

$$\Gamma = \Gamma_\vartheta := <T^2, S> \text{ with } S = \begin{pmatrix} 0 & -1 \\ 1 & 0 \end{pmatrix} \text{ and } T = \begin{pmatrix} 1 & 1 \\ 0 & 1 \end{pmatrix}.$$

There is a lot to be said about these groups. We mention only two items:

Remark 9.1: $\Gamma = \mathrm{SL}(2, \mathbf{Z})$ is generated by just two elements, namely S and T.

Exercise 9.3: Prove this and realize that T acts on $\mathfrak{H}$ as a translation $\tau \longmapsto T(\tau) = \tau + 1$ and S as a reflection at the unit circle $\tau \longmapsto S(\tau) = -1/\tau$.
Determine the fixed points for the action of Γ and their stabilizing groups.

Remark 9.2: $\Gamma = \mathrm{SL}(2, \mathbf{R})$ has the following standard *fundamental domain*

$$\mathcal{F} := \{\tau \in \mathfrak{H}; \ |\tau| > 1, \ -1/2 < \mathrm{Re}\ \tau < 1/2\}.$$

Here we use the definion from [Sh] p.15: For any discrete subgroup Γ of $G = \mathrm{SL}(2, \mathbf{R})$, we call $\mathcal{F}$ a *fundamental domain* for $\Gamma \backslash \mathfrak{H}$ (or simply for Γ) if
– i) $\mathcal{F}$ is a connected open subset of $\mathfrak{H}$,
– ii) no two points of $\mathcal{F}$ are equivalent under Γ,
– iii) every point of $\mathfrak{H}$ is equivalent to some point of the closure of $\mathcal{F}$ under Γ.
(Other authors sometimes use different definitions.)

Exercise 9.4: Verify this and show that the set of Γ–orbits on $\mathfrak{H}$ is in bijection to

$$\mathcal{F}' := \mathcal{F} \cup \{\tau \in \mathbf{C}; |\tau| \geq 1, \ \mathrm{Re}\ \tau = -1/2\} \cup \{\tau \in \mathbf{C}; |\tau| = 1, -1/2 \leq \ \mathrm{Re}\ \tau \leq 0\}.$$

If you have trouble, search for help in [Sh] p.16 or any other book on modular forms.

We look at functions $f : \mathfrak{H} \longrightarrow \mathbf{C}$ with a certain invariance property with respect to the discrete subgroup. Here we take $\Gamma = \mathrm{SL}(2, \mathbf{Z})$ and $k \in \mathbf{Z}$.

Definition 9.1: f is called (*elliptic*) *modular form of weight* k iff one has

– i) $f|_k[\gamma] = f$ for all $\gamma \in \Gamma$,
– ii) f is holomorphic,
– iii) For $\mathrm{Im}\ \tau \gg 0$, f has a Fourier expansion or (*q–development*)

$$f(\tau) = \sum_{n \geq 0} c(n)q^n, q = e(\tau) = e^{2\pi i \tau}, c(n) \in \mathbf{C}.$$

f is called *cusp form* (*forme parabolique* in French) if one has moreover $c(0) = 0$.
We denote by $A_k(\Gamma)$ the space of all elliptic modular forms of weight k and by $S_k(\Gamma)$ the space of cusp forms. Both are $\mathbf{C}$ vector spaces and their dimensions are important characteristic numbers attached to Γ. From i) taken for $\gamma = -E_2$, one sees immediately that for odd k there are no non-trivial forms. For even $k = 2\ell$ one has

$$\begin{aligned} \dim A_{2\ell}(\Gamma) &= [\ell/6] & &\text{if } \ell \equiv 1 \mod 6, \ \ell \geq 0 \\ &= [\ell/6] + 1 & &\text{if } \ell \not\equiv 1 \mod 6, \ \ell \geq 0 \end{aligned}$$

where $[x]$ denotes the *integral part* of x, i.e. the largest integer n such that $n \leq x$. For a proof see for instance [Se] p.88 or (with even more general information) [Sh] p.46.

Remark 9.3: One also studies *meromorphic modular forms* where in ii) holomorphic is replaced by meromorphic and in iii) one has a q–development $f(\tau) = \sum_{n > -\infty} c(n)q^n$.

As Γ is generated by S and T, it is sufficient to demand the two conditions

$$f(\tau + 1) = f(\tau), \ f(-1/\tau) = \tau^k f(\tau).$$

Hence a modular form of weight k is given by a series $f(\tau) = \sum_{n \geq 0} c(n) q^n$ converging for $|q| < 1$ and fulfilling the condition

$$f(-1/\tau) = \tau^k f(\tau).$$

Remark 9.4: The condition iii) concerning the Fourier expansion can also be understood in several different ways, for instance as a certain boundedness condition for $y \longmapsto \infty$ (in particular for $f \in S_k(\mathrm{SL}(2, \mathbf{Z})), |f(\tau)| y^{k/2}$ is bounded for all $\tau \in \mathfrak{H}$), or, if k is even, as holomorphicity of the differential form $\alpha = f(\tau) d\tau^{k/2}$ at ∞: one can compactify the upper half plane $\mathfrak{H}$ by adjoining as *cusps* the rational numbers and the "point" ∞, i.e. one takes $\mathfrak{H}^* := \mathfrak{H} \cup \mathbf{Q} \cup \{\infty\}$. Then one extends the action of $\mathrm{SL}(2, \mathbf{R})$ on $\mathfrak{H}$ to an action on $\mathfrak{H}^*$ and identifies $\Gamma \setminus \mathfrak{H}^*$ with the Riemann sphere or $\mathbf{P}^1(\mathbf{C})$. In this sense modular forms are understood as holomorphic $k/2$-forms on $\mathbf{P}^1(\mathbf{C})$.

All this is made precise in books on the theory of modular forms. Here, for our purpose, the essential point is that, similar to the theta functions in the previous section, modular forms are realizations of dominant weight vectors of discrete series representations of $G = \mathrm{SL}(2, \mathbf{R})$. Before we go into this, let us at least give some examples of modular forms and the way they can be constructed.

Example 9.1: The most well-known elliptic modular form is the $\Delta-$*function* given by its $q-$expansion

$$(2\pi)^{-12} \Delta(q) := q \prod_{n=1}^{\infty} (1 - q^n)^{24}$$

$$= \sum_{n=1}^{\infty} \tau(n) q^n = q - 24q^2 + 252q^3 - 1472q^4 + \dots.$$

This is a cusp form of weight 12. The function $n \longmapsto \tau(n)$ is called the *Ramanujan function*. It has many interesting properties, for instance there is the *Ramanujan conjecture*

$$\tau(n) = O(n^{(11/2)+\varepsilon}), \ \varepsilon > 0,$$

which has been proved by Deligne in 1974 (as consequence of his proof of the much more general *Weil conjecture* concerning Zeta functions of algebraic varieties (see Section 9.6)). But it is still an open question whether all $\tau(n)$ are non-zero.

A modular form in a slightly more refined sense is the $\eta-$*function*, also appearing in the physics literature in several contexts,

$$\eta(\tau) := q^{1/24} \prod_{n=1}^{\infty} (1 - q^n).$$

Example 9.2: A more comprehensive construction principle leads to the *Eisenstein series* E_k resp. G_k. For $k = 4, 6, \dots, G_k$ is defined by

$$G_k(\tau) := {\sum}'_{m,n} (m\tau + n)^{-k}$$

where the sum is taken over all $m, n \in \mathbf{Z}$ with exception of $(m, n) = (0, 0)$.

It is not too much trouble to verify that this is really a modular form of weight k and has a Fourier expansion (see [Sh] p.32)

$$G_k(\tau) = 2\zeta(k) + 2\frac{(2\pi i)^k}{(k-1)!}\sum_{n=1}^{\infty}\sigma_{k-1}(n)q^n$$

where ζ is *Riemann's Zeta function* defined for complex s with $\mathrm{Re}\,s > 1$ by

$$\zeta(s) := \sum_{n=1}^{\infty}n^{-s}$$

and $\sigma_k(n)$ denotes the sum of d^k for all positive divisors d of n. These functions appear naturally in the discussion of lattices $L = \tau\mathbf{Z} + \mathbf{Z}$ in $\mathbf{C}$. They can also be introduced by a far reaching averaging concept similar to the theta transform we used in the previous section to understand the construction of theta functions:

Every function φ on $\mathfrak{H}$ formally gives rise to an object satisfying the covariance property i) of a modular form by the averaging

$$\sum_{\gamma\in\Gamma}\varphi|_k[\gamma].$$

One has the problem whether this series converges (and has the properties ii) and iii)). In general this is not to be expected. But one can use the following fact: For $g_0 = T^\ell$ our automorphic factor has the property $j(g_0,\tau) = 1$, and, hence, one has

$$j(g_0 g,\tau) = j(g,\tau) = (c\tau + d)^{-1}.$$

We see that for a constant function φ_0, say for $\varphi_0(\tau) = 1$, we have

$$\varphi_0|_k[T^\ell] = \varphi_0.$$

Thus to get a covariant expression, one has only to sum up

$$E_k(\tau) := \sum_{\gamma\in\Gamma_\infty\backslash\Gamma}(\varphi_0|_k[\gamma])(\tau)$$

with $\Gamma_\infty := \{T^\ell; \ell \in \mathbf{Z}\}$. The map

$$\Gamma \longrightarrow \mathbf{Z}^2,\ g = \begin{pmatrix} a & b \\ c & d \end{pmatrix} \longmapsto (c,d)$$

induces a bijection

$$\Gamma_\infty \backslash \Gamma \simeq \{(c,d) \in \mathbf{Z}^2;\ (c,d) = 1\}$$

where here (c,d) denotes the greatest common divisor. Hence, one can write

$$E_k(\tau) = \sum_{c,d\in\mathbf{Z},\ (c,d)=1}(c\tau + d)^{-k}.$$

This E_k has the properties we wanted and is related to G_k by

$$\zeta(k)E_k(\tau) = G_k(\tau).$$

The same procedure goes through for every Γ_∞-invariant φ. For instance, $\varphi(\tau) = e(m\tau)$ leads to the so called *Poincaré series*, which are used to construct bases for spaces of cusp forms.

Example 9.3: As already mentioned in 9.1, Jacobi's theta function ϑ has a modular behaviour concerning the variable τ. Here we restrict the function to its value at $z = 0$

$$\vartheta(\tau) := \vartheta(0, \tau) = \sum_{n \in \mathbf{Z}} e^{\pi i n^2 \tau}.$$

One has the relations

$$\vartheta(\tau + 2) = \vartheta(\tau), \ \ \vartheta(-1/\tau) = (\tau/i)^{1/2}\vartheta(\tau).$$

The first one is easy to see and the second one requires some work (see for instance [FB] p.344/5). These relations entail that this ϑ is a modular form with respect to the theta group Γ_ϑ if one generalizes the definition of modular forms appropriately. There are more general theta series associated to quadratic forms resp. lattices leading directly to modular forms:

Let $S \in M_n(\mathbf{Z})$ be a symmetric *positive* matrix, i.e. with

$$S[x] := {}^t x S x > 0 \ \ \text{for all } 0 \neq x \in \mathbf{R}^n$$

then one defines an associated theta series by

$$\vartheta(S; \tau) := \sum_{x \in \mathbf{Z}^n} e^{\pi i S[x] \tau}.$$

We get a function with period 2 whose Fourier development

$$\vartheta(S; \tau) = \sum_{m=0}^{\infty} a(S, m) e^{\pi i m \tau}$$

encodes the numbers

$$a(S, m) := \#\{x \in \mathbf{Z}^n; \ S[x] = m\}$$

of representations of the natural number m by the quadratic form, which belongs to S. If we assume that S is moreover *unimodular*, i.e. S is invertible with $S^{-1} \in M_n(\mathbf{Z})$, *even*, i.e. $S[x] \in 2\mathbf{Z}$ for all $x \in \mathbf{Z}^n$, and if n is divisible by 8, then $\vartheta(S; \tau)$ is an elliptic modular form of weight $n/2$. For a proof we refer to [FB] p.352 or [Sel] p.53.

In particular, for

$$a_n(m) := a(E_n, m) = \#\{x \in \mathbf{Z}^n; \sum_{j=1}^{n} x_j^2 = m\}$$

one has results obtained by Jacobi in 1829 resp. 1828, namely

$$a_8(m) = 16 \sum_{d|m} (-1)^{m-d} d^3 \ \text{ and } \ a_4(m) = 8 \sum_{d|m, 4 \nmid d} d.$$

Proofs can be found for instance in [FB] or [Mu] I. The second relation can be verified using the nice identity

$$\vartheta^4(\tau) = \pi^{-2}(4G_2(2\tau) - G_2(\tau/2)).$$

Example 9.4: As a final example we present the j−invariant or *modular invariant*, which is a meromorphic modular form: We put

$$g_2(\tau) := 60G_4(\tau), \ g_3(\tau) := 140G_6(\tau).$$

Then, we have $\Delta(\tau) = g_2(\tau)^3 - 27g_3^2$ and the j−function is defined by

$$j(\tau) := 12^3 g_2(\tau)^3/\Delta(\tau).$$

j is a modular function with Fourier expansion (see for instance [Sh] p.33)

$$j(\tau) = q^{-1}(1 + \sum_{n=1}^{\infty} c_n q^n), \ n \in \mathbf{Z}.$$

This function is of special importance because it induces a bijection $\Gamma\backslash\mathfrak{H}^* \simeq \mathbf{P}^1(\mathbf{C})$. And the coefficients c_n hide some deep mysteries (essentially assembled under the heading of *monstrous moonshine* (see [CN], [FLM])).

Modular Forms as Functions on SL(2,R)

Parallel to the relation between the Second and First Realization in the induction procedure one defines a *lifting* of functions f from the homogeneous space $\mathcal{X} = G/K \simeq Gx_0$ with $K = G_{x_0}$ to the group using the automorphic factor $j(g,\tau)$ satisfying the cocycle condition (7.8) by $f \longmapsto \phi_f$ where we put

$$\phi_f(g) := f(g(x_0))j(g, x_0) = f|[g](x_0).$$

Our aim is to translate the conditions i) to iii) from the definition of modular forms f to appropriate conditions characterizing the lifts of these forms among the functions ϕ on G. Classic sources giving all details for this are [GGP], [Ge] and [Bu].

Remark 9.5: If for a subgroup Γ of G and a function f on $\mathcal{X}$ one has

(9.4) $(f|[\gamma])(x) = f(\gamma(x))j(\gamma, x) = f(x)$ for all $\gamma \in \Gamma$,

the lifted function $\phi = \phi_f$ satisfies

ia) $\phi(\gamma g) = \phi(g)$ for all $\gamma \in \Gamma$

and

ib) $\phi(g\kappa) = \phi(g)j(\kappa, x_0)$ for all $\kappa \in K$.

Proof: By definition, (9.4), and (7.8), we have

$$\phi(\gamma g) = f(\gamma g(x_0))j(\gamma g, x_0),$$
$$= f(g(x_0))j(\gamma, g(x_0))^{-1}j(\gamma g, x_0),$$
$$= f(g(x_0))j(g, x_0).$$

And by definition, (7.8), and $K = G_{x_0}$, we have as well

$$
\begin{aligned}
\phi(g\kappa) &= f(g\kappa(x_0))j(g\kappa, x_0) \\
&= f(g(x_0))j(g, x_0)j(\kappa, x_0) \\
&= \phi(g)j(\kappa, x_0).
\end{aligned}
$$

Remark 9.6: In our case $G = \mathrm{SL}(2, \mathbf{R}), x_0 = i \in \mathcal{X} = \mathfrak{H} = G/K, K = \mathrm{SO}(2)$ and $j(g, \tau) = (c\tau + d)^{-k}$, for the lifted function we have

$$
\phi(g) = f(g(i))(ci + d)^{-k} = f(\tau)y^{k/2}e^{ik\vartheta}
$$

if g is given by the Iwasawa decomposition (7.12)

$$
g = n(x)t(y)r(\vartheta).
$$

Proof: In this case one has

$$
ci + d = y^{-1/2}(-i \sin \vartheta + \cos \vartheta) = y^{-1/2}e^{-i\vartheta}.
$$

The translation of the holomorphicity condition ii) to a condition restricting a function ϕ on G follows the scheme we already used several times: For $\mathfrak{sl}(2, \mathbf{C})_c = <Z, X_\pm>$ one has the left invariant differential operators (7.16)

$$
\begin{aligned}
\mathcal{L}_{X_\pm} &= \pm(i/2)e^{\pm 2i\vartheta}(2y(\partial_x \mp i\partial_y) - \partial_\vartheta), \\
\mathcal{L}_Z &= -i\partial_\vartheta,
\end{aligned}
$$

and, hence, for a function $\phi = \phi_f$ with $\phi_f(g) = c_k(g)f(x, y), c_k(g) := y^{k/2}e^{ik\vartheta}$

$$
\begin{aligned}
\mathcal{L}_Z\phi &= k\phi, \\
\mathcal{L}_{X_\pm}\phi &= c_{k\pm 2}\tilde{D}_\pm f
\end{aligned}
$$

where

$$
\tilde{D}_+ = i(\partial_x - i\partial_y - ik/y), \quad \tilde{D}_- = -iy^2(\partial_x + i\partial_y).
$$

Furthermore, using Wirtinger calculus with

$$
\partial_\tau = (1/2)(\partial_x - i\partial_y), \quad \partial_{\bar{\tau}} = (1/2)(\partial_x + i\partial_y)
$$

we get

$$
\begin{aligned}
\mathcal{L}_Z\phi &= k\phi, \\
\mathcal{L}_{X_\pm}\phi &= c_{k\pm 2}D_\pm f
\end{aligned}
$$

where

$$
D_+ = 2i(\partial_\tau + k/(\tau - \bar{\tau})), \quad D_- = (i/2)(\tau - \bar{\tau})^2\partial_{\bar{\tau}}.
$$

Hence in particular, we have found:

Remark 9.7: Let $\phi = \phi_f$ be the lift of f. The condition ii) that f is holomorphic on $\mathfrak{H}$ is equivalent to

ii') $\quad \mathcal{L}_{X_-}\phi = 0.$

The last condition iii) is easily seen to be equivalent to the boundedness condition

$$|f(\tau)| < M \text{ for } \operatorname{Im} \tau \gg 0.$$

With a bit more work, one translates the cuspidality of a modular form, i.e. $f \in S_k(\Gamma)$, to the following

Lemma: $f \in A_k(\Gamma)$ is in $S_k(\Gamma)$ iff for every $y_0 > 0$ there is an $M > 0$ such that

$$y^{k/2}|f(\tau)| < M \quad \text{for all } y = \operatorname{Im} \tau > y_0.$$

Hence one has for the lifted functions $\phi = \phi_f$ the following property.

Remark 9.8: The condition iii) translates to

$$|\phi(x, y, \vartheta)y^{-k/2}| < M \text{ for } y \gg 0$$

and the cusp condition $c(0) = 0$ to

$$|\phi(g)| < M \quad \text{for all } g \in G.$$

The remarks 9.5 to 9.8 together give proof to a fundamental statement.

Theorem 9.3: Via the lifting φ_k, the space $A_k(\Gamma)$ of modular forms on $\mathfrak{H}$ is isomorphic to the space $\mathcal{A}_k$ of smooth functions ϕ on G with

ia)　　$\phi(\gamma g) = \phi(g)$　for all $\gamma \in \Gamma = \mathrm{SL}(2, \mathbf{Z})$,

ib)　　$\phi(gr(\vartheta)) = \phi(g)e^{ik\vartheta}$　for all $r(\vartheta) \in K = \mathrm{SO}(2)$,

ii')　　$\mathcal{L}_{X_-}\phi = 0$,

iii')　　$|\phi(g)y^{-k/2}| < M$　for all $y \gg 0$.

Moreover, the space of cusp forms $S_k(\Gamma)$ is isomorphic to

$$\mathcal{A}_k^0 = \{\phi \in \mathcal{A}_k;\ |\phi| < M\}.$$

Remark 9.9: The cusp condition in Remark 9.8 and in the Theorem above is equivalent to

(cusp)　　$\int_{\Gamma \cap N \backslash N} \phi(n(x)g)dx = 0$　for all $g \in G$.

This follows by identifying $N = \{n(x); x \in \mathbf{R}\}$ with $\mathbf{R}$ and computing

$$\int_{\Gamma \cap N \backslash N} \phi(n(x)g)dx = \int_0^1 \phi(n(x)g)dx$$
$$= \int_0^1 f(\tau + x)\, e^{ik\vartheta} y^{k/2} dx$$
$$= e^{ik\vartheta} y^{k/2} c(0).$$

Now we can refine Theorem 9.3 above:

Theorem 9.4: The space of cusp forms $\mathcal{A}_k^0$ is characterized by the following conditions

ia) $\phi(\gamma g) = \phi(g)$ for all $\gamma \in \Gamma = SL(2, \mathbf{Z})$,

ib) $\phi(gr(\vartheta)) = \phi(g)e^{ik\vartheta}$ for all $r(\vartheta) \in K = SO(2)$,

ii) $\Delta\phi = \mathcal{L}_\Omega\phi = k((k/2) - 1)\phi$,

iii) ϕ is bounded,

iv) $\int_{\Gamma \cap N \backslash N} \phi(n(x)g)dx = 0$ for all $g \in G$.

Here
$$\Delta = \mathcal{L}_\Omega = 2y^2(\partial_x^2 + \partial_y^2) - 2y\partial_x\partial_\vartheta$$

is the Laplacian (7.17) from 7.2 and, by a small calculation, it is easy to see that the lift ϕ_f of a cusp form $f \in S_k(\Gamma)$ satisfies the condition ii) and hence all the conditions from the theorem. The proof of the fact that the lifting map φ_k is a surjection onto the space given by the conditions ia) to iv) reduces to the verification that for $\phi \in \mathcal{A}_k^0$ the function f with

(9.5) $f(\tau) := \phi(g)y^{-k/2}e^{-ik\vartheta}$

is well defined for each g with $g(i) = \tau$ (which is quite obvious) and that f is a cusp form. This assertion is based on a considerably deeper theorem, the *Theorem of the discrete spectrum*, which we shall state below.

Decomposition of $L^2(\Gamma \backslash G)$

For $G = SL(2, \mathbf{R}) \supset \Gamma = SL(2, \mathbf{Z})$, we put $\mathcal{H} = L^2(\Gamma \backslash G)$ with respect to the biinvariant measure $d\dot{g}$ on $\mathcal{X} = \Gamma \backslash G$ fixed by

$$dg = \frac{1}{2\pi}\frac{dxdy}{y^2}d\vartheta.$$

We want to discuss the decomposition of the representation ρ of G given by right translation on $\mathcal{H}$ and show how the cusp forms come in here. A first indication is given by the

Remark 9.10: For $f \in S_k(\Gamma)$, the lift $\phi_f = \varphi_k(f)$ is in $\mathcal{H}$.

This follows from

$$\int_{\mathcal{X}} |\phi_f(\dot{g})|^2 d\dot{g} = \int_{\mathcal{F}} |f(\tau)|^2 y^{k-2}dxdy < \infty$$

by the boundedness condition iii') for the cusp form f.

We need another tool from functional analysis (for background see for instance [La1] p.389ff):

Remark 9.11: The minimal closed extension of the Laplacian Δ is a selfadjoint operator $\hat{\Delta}$. As Δ does by constuction, the extension $\hat{\Delta}$ also commutes with the right translation and hence, by Schur's Lemma ("the unitary Schur" mentioned in 3.2) $\Delta|_{\mathcal{H}_0}$ is scalar for each G-invariant subspace $\mathcal{H}_0$ of $\mathcal{H}$. From functional analysis we take over that $\hat{\Delta}$ and ρ have a discrete spectrum only for the subspace of *cuspidal functions*

$$\mathcal{H}^0 := \{\phi \in L^2(\Gamma \backslash G)\,;\, \int_{\Gamma \cap N \backslash N} \phi(n(x)g)dx = 0 \text{ for almost all } g \in G\}.$$

The central fact is as follows.

Theorem 9.5 (*Theorem of the discrete spectrum*): The right regular representation ρ of $SL(2, \mathbf{R})$ on $\mathcal{H}^0$ is completely reducible

$$\mathcal{H}^0 = \oplus_{j\in\mathbf{Z}}\mathcal{H}_j$$

and each irreducible component has finite multiplicity.

As this leads too much into (most fascinating) functional analysis, we can not go into the proof of this theorem and refer to [La1] and [GGP] p.94ff. A main point is a multiplicity-one statement going back to Godement, which we already mentioned (without proof) in 6.4. But we try to give an indication of the power of this theorem: For the proof of characterization of cusp forms by the Laplacian in the theorem above, we have to show that the function f from (9.5)

$$f(\tau) = \phi(g)y^{-k/2}e^{-ik\vartheta}$$

is holomorphic. If one decomposes ϕ as in Theorem 9.5

$$\phi = \sum_j \phi_j,$$

the components ϕ_j have to fulfill the same relation ib) as ϕ does. And one has

$$\Delta\phi_j = \lambda_k\phi_j, \quad \lambda_k = k((k/2) - 1),$$

because by the last remark above, Δ is scalar on $\mathcal{H}_j$, i.e. one has a λ_j with

$$\Delta\phi_j = \lambda_j\phi_j.$$

And then using the inner product in $\mathcal{H}$ one has

$$< \phi, \Delta\phi > \, = \lambda_j < \phi, \phi_j >$$

and, as $\hat{\Delta}$ is self adjoint,

$$< \phi, \Delta\phi_j > \, = \, < \Delta\phi, \phi_j > \, = \lambda_k < \phi, \phi_j >$$

and therefore $\lambda_j = \lambda_k$. Hence, for the restriction of ρ to its irreducible subrepresentations the components ϕ_j are vectors of lowest weight k for the discrete series representation π_k^+. Such vectors are annihilated by $\mathcal{L}_{X_-}$ and, hence, f is holomorphic.

As another consequence we come to the statement parallel to the one we proved in 9.1 for theta functions.

Theorem 9.6 (*Duality Theorem for the discrete series of* $SL(2, \mathbf{R})$): The multiplicity of π_k^- in the right regular representation ρ on the space $\mathcal{H}^0$ of cuspidal functions is equal to the number of linear independant cusp forms of weight k

$$\text{mult}(\pi_k^-, \rho) = \dim S_k(\Gamma).$$

Proof: There is a nice and comprehensive classic proof in [GGP] p.53-57. As we already have at hand the necessary tools, we reproduce its highlights changing only the parametrization of $G = \mathrm{SL}(2, \mathbf{R})$ used by these authors to the one used here all the time.

a)At first, we show that $m_k := \mathrm{mult}(\pi_k^-, \rho)$ is equal to the number $\tilde{m}_k$ of linear independent functions Ψ on $X = \Gamma \backslash G$ in $\mathcal{H}^0$ with

(9.6) $\qquad \Delta\Psi = \lambda_k \Psi, \lambda_k := k((k/2) - 1), \; \rho(r(\varphi))\Psi = e^{-ik\varphi}\Psi \quad \text{for all } \varphi \in \mathbf{R}.$

ai) For each irreducible subspace $\mathcal{H}_j$ of $\mathcal{H}^0$, which is a model for π_k^-, there is - up to a nonzero factor - exactly one highest weight vector where ρ restricted to $\mathcal{H}_j$ has highest weight $-k$ and, hence Δ has eigenvalue λ_k.
aii) Every $\Psi \in \mathcal{H}^0$ satisfying (9.6) is a linear combination of vectors of highest weight from irreducible subspaces of $\mathcal{H}^0$ equivalent to a representation space of π_k^-.

b) In the second step, we show that $\tilde{m}_k$ is equal to the number of cusp forms: The space of functions Ψ as above is in bijection to the space of functions Φ on G with

(9.7) $\qquad \Delta\Phi = \lambda_k \Phi, \; \Phi(gr(\varphi)) = e^{-ik\varphi}\Phi(g), \; \Phi(\gamma^{-1}g) = \Phi(g) \quad \text{for all } \varphi \in \mathbf{R}, \gamma \in \Gamma.$

We describe G by our parameters τ, ϑ with $\tau = g(i), e^{-i\vartheta} = (ci + d)/ \mid ci + d \mid$.

Hence, for g_0^{-1} with components a_0, b_0, c_0, d_0, the matrix $g_0^{-1}g$ has parameters τ_1, ϑ_1 with $\tau_1 = g_0^{-1}(\tau), \vartheta_1 = \vartheta - \arg(c_0\tau + d_0)$. And $gr(\varphi)$ has the parameters $\tau, \vartheta + \varphi$. Thus the second equation in (9.7) shows that $\Phi(g)$ is a function of type $f_1(\tau)e^{-ik\vartheta}$. Guided by the proof of our Remark 9.7, in order to also fulfill the other conditions in (9.7) we try the ansatz

$$\Phi(g) = e^{-ik\vartheta}y^{k/2}\overline{f(\tau)}.$$

Using our previous notation, this can also be written as $\Phi(g) = \overline{j(g, i)}^{-k}\overline{f(\tau)}$. We see that using this ansatz, the third condition in (9.7) is equivalent to the functional equation for modular forms

$$f(\gamma^{-1}(\tau))j(\gamma^{-1}, \tau)^{-k} = f(\tau) \quad \text{for all } \gamma \in \Gamma.$$

For the real coordinates x, y of τ the first equation in (9.7) comes down to

$$y^{2+(k/2)}e^{-ik\vartheta}(\bar{f}_{xx} + \bar{f}_{yy}) + iky^{1+(k/2)}e^{-ik\vartheta}(\bar{f}_x - i\bar{f}_y) = 0$$

resp.

$$y(f_{xx} + f_{yy}) - ik(f_x + if_y) = 0$$

If f is a modular form, ergo holomorphic, one has $f_{\bar{\tau}} = 0$ and this equation is fulfilled. On the other hand, if Φ fulfills (9.7), by the Decomposition Theorem, it is a linear combination of highest weight vectors with higest weight $-k$, which are annihilated by $\mathcal{L}_{X_+} = e^{2i\vartheta}(2y(\partial_x - i\partial_y) - \partial_\vartheta)$. And this again leads just to the condition $f_{\bar{\tau}} = 0$ showing that Φ corresponds to a linear combination of cusp forms.

The reader again is invited to watch very closely whether all signs are correct. It has to be emphasized that the notions *highest* and *lowest* weight depend on our parametrization of $K = \mathrm{SO}(2)$ and interchange if one uses $r(-\vartheta)$ in place of our $r(\vartheta)$ (as Gelbart does in [Ge]). [GGP] wisely use the term vector of *dominant* weight.

Remark 9.12: It is a natural question to ask under which conditions the lift $\phi = \phi_f$ of a cusp form f is a cyclic vector, i.e. generates an irreducible representation. There is a nice answer to this: if f is an eigenfunction to the *Hecke operators* $T(p)$. We will introduce these operators later in the context of associating L−functions to modular forms and here only mention that the space of cusp forms has a basis consisting of such eigenfunctions.

The big difference between this duality statement and the one for the Heisenberg group in 9.1 is that in this case we have no way to an explicit construction of the modular forms starting from their property of being a dominant weight vector. Apparently, this reflects the fact that the structure of the group at hand is considerably more complicated than that of the Heisenberg group. Some progress here should give a deep insight into the modular group.

The characterization theorem is the starting point for several generalizations leading to the notion of *automorphic forms*. It can be extended to other groups, even *adelic* ones as to be found for instance in [Ge]. There it is also pointed out that the right habitat for the modern theory of modular forms is not SL(2) but GL(2). Here we can not be too ambitious and we stay with $G = \mathrm{SL}(2,\mathbf{R})$, $K = \mathrm{SO}(2)$ and $\Gamma = \mathrm{SL}(2,\mathbf{Z})$ and only treat the following generalization.

Definition 9.2: The smooth function ϕ on G is called an *automorphic form* for Γ iff ϕ satisfies the following conditions

ia) $\phi(\gamma g) = \phi(g)$ for all $\gamma \in \Gamma$,

ib) ϕ is *right K−finite*, i.e. $\dim \langle \phi(gr(\vartheta)); \ r(\vartheta) \in K \rangle < \infty$,

ii) ϕ is an eigenfunction of Δ,

iii) ϕ fulfills the growth condition that there are C, M with $|\phi(g)| \leq Cy^M$ for $y \longmapsto \infty$.

ϕ is called *cusp form* if moreover one has

(cusp) $\int_{\Gamma \cap N \backslash N} \phi(n(x)g)dx = 0$.

Besides the cusp forms realizing highest weight vectors of π_k^- we just treated, as their counterparts there come in forms realizing the lowest weight vectors of π_k^+, corresponding to antiholomorphic modular forms on $\mathfrak{H}$ and the famous *Maass wave forms*: these are bounded right K-invariant eigenfunctions of the Laplacian $\Delta\phi = \lambda_s\phi$ with

$$\lambda_s = -(1 - s^2)/2, \ s \in i\mathbf{R}.$$

In analogy to the duality theorem stated above, one has here the fact that the dimension of the space $W_s(\Gamma)$ of these wave forms is equal to the multiplicity of the principal series representation π_{is} in ρ (for this and in general more precise information see [GGP] in particular p.50). In [Ge] p.29 we find the statement: one knows that $W_s(\Gamma)$ is non-trivial for infinitely many values of s, but almost nothing is known concerning the specific values of s for which $W_s(\Gamma)$ is non-trivial. Still now, Maass wave forms are an interesting subject. For information on more recent research we recommend the paper by Lewis and Zagier [LZ].

We finish this section by some remarks concerning the continuous spectrum of the Laplacian, resp. the orthocomplement $\mathcal{H}^c$ of the subspace $\mathcal{H}^d$ of $\mathcal{H} = L^2(\Gamma \backslash G)$ consisting of the cuspidal and the constant functions. One has rather good control about this space as it can be spanned by the so called *non-holomorphic Eisenstein series*. As sources for this we go back to [Ku] and [La1] p.239ff.

The classical object is fixed as follows. One takes

$$f_\mu(\tau) := y^\mu, \ \mu \in \mathbf{C}$$

and for $k \in \mathbf{N}_0$

$$E_k(\tau, \mu) := \sum_{\gamma \in \Gamma_\infty \backslash \Gamma} (f_{(\mu-k)/2}|_k[\gamma])(\tau),$$

$$= \sum_{c,d \in \mathbf{Z}, \ (c,d)=1} \frac{y^{(\mu-k)/2}}{|c\tau + d|^{\mu-k}} \frac{1}{(c\tau + d)^k}$$

with $\Gamma_\infty = \{\gamma = \begin{pmatrix} 1 & b \\ & 1 \end{pmatrix}; b \in \mathbf{Z}\}$.

This prescription formally has the transformation property of a modular form of weight k and it can be verified that for μ with Re $\mu > 2$ it defines a (real analytic) function on $\mathfrak{H}$, which is non-cuspidal. Using the standard lift φ_k, this function can be lifted to a function on G

$$E_k(g, \mu) := \varphi_k E_k(\tau, \mu),$$

$$= y^{\mu/2} e^{ik\vartheta} \sum_{c,d \in \mathbf{Z}, \ (c,d)=1} (\frac{c\bar\tau + d}{c\tau + d})^{k/2} \frac{1}{|c\tau + d|^\mu},$$

$$= \sum_{\gamma \in \Gamma_\infty \backslash \Gamma} \phi_{s,k}(\gamma g),$$

where for $\mu = is + 1$ the functions (8.11) from 8.3 $\phi_{s,k}, k \in \mathbf{Z}$, with

$$\phi_{s,k}(g) = y^{(is+1)/2} e^{ik\vartheta}$$

come in, which span the space of the principal series representations. We have

$$\Delta E_k(., \mu) = -((s^2 + 1)/2) E_k(., \mu)$$

and these functions are further examples for the definition of automorphic forms above.

Already in our (brief) chapter on abelian groups, we met with a continuous decomposition of the regular representation for $G = \mathbf{R}$ by throwing in elements of Fourier analysis without really explaining the background from functional analysis. Now, this gets worse here: There is the central statement ([Ge] p.33):

Theorem 9.7: The restriction ρ^c of ρ to $\mathcal{H}^c$ is the continuous sum of the principal series representations $\pi_{is,+}$, i.e.

$$\rho^c(g) = \int_0^\infty \pi_{is,+}(g) ds.$$

Using the inner product for functions $f, \tilde{f}$ on $\mathfrak{H}$ given by

$$(f, \tilde{f}) := \int_{\mathcal{F}} \overline{f(\tau)} \tilde{f}(\tau) y^{-2} dx dy$$

one can say that any slowly increasing function f on $\mathfrak{H}$, which is orthogonal to all cusp forms, may be expressed by a generalized Fourier integral

$$f(\tau) = \int_0^\infty \hat{f}(s) E(\tau, s) ds, \quad \hat{f}(s) := (f, E(., s)), \quad E(\tau, s) := \sum_{c, d \in \mathbf{Z}; (c,d)=1} \frac{y^{(is)/2}}{|c\tau + d|^{2s}}.$$

Though we will not try to get to the bottom of the proofs, we indicate the following also otherwise useful tools:

The Theta Transform θ

We already used the averaging procedure while introducing Eisenstein series: If φ is a function on $N \setminus G$, one gets, at least formally, a function $\theta\varphi$ on $\Gamma \setminus G$ by the prescription

$$\theta\varphi(g) := \sum_{\gamma \in \Gamma_\infty \setminus \Gamma} \varphi(\gamma g).$$

It is part of nice analysis to discuss under which conditions this leads to convergent objects and it is no wonder that it works at least if φ is a Schwartz function.

The Constant Term θ^*

Let ϕ be an element of $\mathcal{H} := L^2(\Gamma \setminus G)$. To this ϕ we associate $\theta^*\phi$ (again at least formally) by

$$\theta^*\phi(g) := \int_{\Gamma_\infty \setminus N} \phi(ng) dn.$$

As ϕ can be assigned a periodic function, which one denotes again by ϕ, $\theta^*\phi$ may be understood as the constant term in a Fourier expansion

$$\phi(n(x)g) = \sum_j \phi_j(g) e(jx).$$

θ and θ^* are adjoint operators if we introduce an inner product

$$< \varphi, \tilde{\varphi} >_{N \setminus G} := \int_{N \setminus G} \overline{\varphi(\dot{g})} \tilde{\varphi}(\dot{g}) d\dot{g},$$

i.e. we have for functions assuring the convergence of our prescriptions the relation

$$< \theta\varphi, \phi >_{\Gamma \setminus G} = < \varphi, \theta^*\phi >_{N \setminus G}.$$

Exercise 9.5: Verify this.

The Zeta Transform Z

This is a prescription to produce elements in the representation space $\mathcal{H}_{is,\pm}$ of a principal series representation $\pi_{is,\pm}$: For $\mu \in \mathbf{C}$ and φ a function on $N \setminus G$, we put (again at least formally)

$$Z(\varphi, g, \mu) := \int_0^\infty \varphi(t(y)g)y^{-\mu}\frac{dy}{y}.$$

For $\mu = (is + 1)/2$, we get

$$Z(\varphi, n(x_0)t(y_0)g, \mu) = \int_0^\infty \varphi(t(y_0y)g)y^{-\mu}\frac{dy}{y} = y_0^{(is+1)/2}Z(\varphi, g, \mu),$$

i.e. an element fulfilling the equation (7.11) in 7.2 characterizing the elements of $\mathcal{H}_{is,+}$.

The following statements are easy to prove.

Remark 9.13: For φ in the Schwartz space $\mathcal{S}(N \setminus G)$ and $\operatorname{Re}\mu > 0$, the integral defining $Z(\varphi, g, \mu)$ converges absolutely and $Z(\varphi, g, \mu)$ is an entire function in μ if moreover φ has finite support.

With some more work composing the Zeta and the Theta transform, one obtains a Γ−invariant function:

Remark 9.14: For $\varphi \in \mathcal{S}(N \setminus G)$ and $\operatorname{Re}\mu > 1$, the *Eisenstein series*

$$E(\varphi, g, \mu) := \theta Z(\varphi, g, \mu) = \sum_{\gamma \in \Gamma_\infty \setminus \Gamma} Z(\varphi, \gamma g, \mu)$$

is absolutely convergent.

By a small computation, we get the relation between this Eisenstein series and the one we defined above.

Remark 9.15: For

$$\varphi(g) := \varphi_{\ell,k}(g) = y^{-\ell}e^{-\pi/y}e^{ik\vartheta}, \ell \in \mathbf{N}_0, k \in \mathbf{Z},$$

one has

$$E(\varphi, g, \mu) = \pi^{-(\ell+\mu)}\Gamma(\ell + \mu)E_k(g, \mu).$$

The functions $\varphi_{\ell,k}, \ell \in \mathbf{N}_0, k \in \mathbf{Z}$ are a Hilbert basis of the space $L^2(N \setminus G)$ with respect to the measure $d\nu = y^{-2}dyd\vartheta$. For every fixed ℓ and $\mu = is + 1$, the images of these $\varphi_{\ell,k}$ by the Zeta transform are a basis for the representation spaces $\mathcal{H}_{is,\pm}$ if one takes all even resp. odd $k \in \mathbf{Z}$.

There is still much to be said here. We shall come back to the Zeta transform in the context of a discussion of $\zeta-$ and $L-$functions. But before doing this, we shall finish our outlook to automorphic forms by returning to theta functions.

9.3   Theta Functions and the Jacobi Group

In 9.1 we introduced theta functions and studied their behaviour as functions of the complex variable z. But as we already indicated and partially explored, there also is an important dependance on the modular variable $\tau \in \mathfrak{H}$. This leads to the ultimate interpretation of (lifts of) theta functions as automorphic forms for the Jacobi group G^J, which we introduced in 8.5. We defined

$$q := e(\tau) = e^{2\pi i \tau}, \ \epsilon := e(z) = e^{2\pi i z}.$$

and the *classic Jacobi theta function* ϑ (9.1) given by

$$\vartheta(z,\tau) := \sum_{n \in \mathbf{Z}} e^{\pi i (n^2 \tau + 2nz)} = \sum_{n \in \mathbf{Z}} q^{n^2/2} \epsilon^n,$$

the *theta function with characteristics* $m = (m', m'') \in \mathbf{R}^2$

$$\theta_m(\tau, z) := \sum_{n \in \mathbf{Z}} e((1/2)(n + m')^2 \tau + (n + m')(z + m'')),$$

and its variant (9.2) (following the notation of [EZ] p.58) with $2m \in \mathbf{N}$, $\mu = 0, 1, \ldots, 2m - 1$

$$\theta_{m,\mu}(\tau, z) := \sum_{r \in \mathbf{Z}, \, r \equiv \mu \mod 2m} q^{r^2/(4m)} \epsilon^r$$

$$= \sum_{n \in \mathbf{Z}} e^m((n + \mu/(2m))^2 \tau + 2(n + \mu/(2m))z).$$

One has $\vartheta(z,\tau) = \theta_{1/2,0}(\tau, z)$. We already explored the quasiperiodicity property concerning the complex variable z

$$\vartheta(z + r\tau + s, \tau) = e^{-\pi i (r^2 \tau + 2rz)} \vartheta(z, \tau)$$

and (in Example 9.3 in 9.2) the modular property of the value $\vartheta(\tau) := \theta(0, \tau)$ for $z = 0$ (also called by the German word *Thetanullwert*)

$$\vartheta(\tau + 2) = \vartheta(\tau), \ \vartheta(-1/\tau) = (\tau/i)^{1/2} \vartheta(\tau).$$

To be consistent with [EZ] and the usual Jacobi Theory, we have to change (from Jacobi's and Mumford's) notation and put

$$\theta(\tau, z) := \vartheta(z, \tau).$$

Then one has the fundamental modular transformation relation.

Theorem 9.8: For $a, b, c, d \in \mathbf{Z}$ with $ad - cd = 1$ and $ab, cd \in 2\mathbf{Z}$, there is an eighth root of unity ζ such that one has

$$(9.8) \qquad \theta(\frac{a\tau + b}{c\tau + d}, \frac{z}{c\tau + d}) = (c\tau + d)^{1/2} \, \zeta \, e((1/2)cz^2/(c\tau + d))\theta(\tau, z).$$

The unit root ζ can be determined as follows:

We may assume $c > 0$ or $c = 0$ and $d > 0$ (otherwise one would multiply the matrix $\begin{pmatrix} a & b \\ c & d \end{pmatrix}$ by -1) and hence have $(c\tau + d)^{1/2}$ in the first quadrant, i.e. with real- and imaginary part > 0.

We take the *Jacobi symbol* $(\frac{a}{b})$, i.e. the multiplicative extension of the *Legendre symbol*, which for $a, p \in \mathbf{Z}$, p prime is given by

$$
\begin{aligned}
(\tfrac{a}{p}) \quad &= \quad 0 \quad \text{if} \quad p \,|\, a, \\
&= \quad 1 \qquad p \nmid a \quad \text{and there is an } \; x \in \mathbf{Z} \text{ with } x^2 \equiv a \quad \bmod p, \\
&= \quad -1 \qquad p \nmid a \quad \text{and there is no } \; x \in \mathbf{Z} \text{ with } x^2 \equiv a \quad \bmod p.
\end{aligned}
$$

Then one has

$$
\begin{aligned}
\zeta \quad &= \quad i^{(d-1)/2}(\tfrac{c}{|d|}) \qquad \text{for } c \text{ even and } d \text{ odd}, \\
&= \quad e^{-\pi i c/4}(\tfrac{d}{c}) \qquad \text{for } c \text{ odd and } d \text{ even}.
\end{aligned}
$$

The proof of this theorem is a very nice piece of analysis but too lengthy for this text. So we refer to the books of Eichler [Ei], in particular p.59-62, Lion-Vergne [LV] p.145ff, or Mumford [Mu] I.

Another most remarkable fact is the following characterization of θ, which comes about as an exercise in complex function theory.

Theorem 9.9: Let

$$
f : \mathfrak{H} \times \mathbf{C} \longrightarrow \mathbf{C}
$$

be a holomorphic function satisfying

a) $f(\tau, z + 1) = f(\tau, z)$,
b) $f(\tau, z + \tau) = f(\tau, z)e(-(1/2)\tau - z)$,
c) $f(\tau + 1, z + 1/2) = f(\tau, z)$,
d) $f(-1/\tau, z/\tau) = f(\tau, z)(-i\tau)^{1/2}e((1/2)z^2/\tau)$ for all $(z, \tau) \in \mathbf{C} \times \mathfrak{H}$,
e) $\lim_{\mathrm{Im}\tau \longmapsto \infty} f(\tau, z) = 1$.

Then one has $f = \theta$.

By a closer look one sees that the quasi-periodicity and the modular transformation property of θ can be combined to a transformation property under an action of the Jacobi group G^J, which is a semi-direct product of $\mathrm{SL}(2, \mathbf{R})$ and $\mathrm{Heis}(\mathbf{R})$ and in 8.5 was realized as a subgroup of $\mathrm{Sp}(2, \mathbf{R})$ given by four-by-four matrices. As in 8.5, we use two notations

$$
G^J \ni g = \begin{pmatrix} a & b \\ c & d \end{pmatrix} (\lambda, \mu, \kappa) = (p, q, r) \begin{pmatrix} a & b \\ c & d \end{pmatrix}.
$$

By a short calculation one can verify that for $(\tau, z) \in \mathfrak{H} \times \mathbf{C}$

$$
(\tau, z) \longmapsto g(\tau, z) := \left(\frac{a\tau + b}{c\tau + d}, \frac{z + \lambda\tau + \mu}{c\tau + d} \right)
$$

defines an action of G^J on $\mathfrak{H} \times \mathbf{C}$. The stabilizing group of $(i, 0)$ is $K^J := \mathrm{SO}(2) \times C(\mathbf{R})$, $C(\mathbf{R})$ the center of the Heisenberg group. One has

$$
G^J/K^J \simeq \mathfrak{H} \times \mathbf{C}, \; g(i, 0) = (\tau = x + iy, z = p\tau + q) \text{ where } g = (p, q, r)n(x)t(y)r(\vartheta).
$$

And for $k, m \in \mathbf{Q}$ one has (see the first pages of [EZ]) an automorphic factor

(9.9) $j_{k,m}(g; (\tau, z)) := (c\tau + d)^{-k} e^m \left(-\dfrac{c(z + \lambda\tau + \mu)^2}{c\tau + d} + \lambda^2\tau + 2\lambda z + \lambda\mu + \kappa\right).$

This leads to an action of G^J on functions f living on $\mathfrak{H} \times \mathbf{C}$ via

$$f \longmapsto f|_{k,m}[g] \text{ with } f|_{k,m}[g](\tau, z) := f(g(\tau, z)) j_{k,m}(g; (\tau, z)).$$

Similar to modular forms and theta functions as automorphic forms for $\mathrm{SL}(2, \mathbf{R})$ resp. $\mathrm{Heis}(\mathbf{R})$, for the Jacobi group G^J, its discrete subgroup $\Gamma^J := \mathrm{SL}(2, \mathbf{Z}) \ltimes \mathbf{Z}^3$, and $k, m \in \mathbf{N}_0$ one defines (as in [EZ]):

Definition 9.3: A function $f : \mathfrak{H} \times \mathbf{C} \longrightarrow \mathbf{C}$ is called a *Jacobi form of weight k and index m* if one has

– i) $f|_{k,m}[\gamma](\tau, z) = f(\tau, z)$ for all $\gamma \in \Gamma^J$,

– ii) f is holomorphic,

– iii) for $\mathrm{Im}\,\tau \gg 0$, f has a Fourier development

$$f(\tau, z) = \sum_{n, r \in \mathbf{Z},\, 4mn - r^2 \geq 0} c(n, r) q^n \varepsilon^r, \ q = e(\tau), \ \varepsilon = e(z), \ c(n, r) \in \mathbf{C}.$$

f is called a *cusp form* if it satisfies moreover

– iii') $c(n, r) = 0$ unless $4mn > r^2$.

The vector spaces of all such functions f are denoted by $J_{k,m}$ resp. $J_{k,m}^{\mathrm{cusp}}$. They are finite dimensional by Theorem 1.1 of [EZ]. As usual, one lifts a function f on $\mathfrak{H} \times \mathbf{C}$ to a function $\phi = \phi_f$ on G^J by

$$\phi_f(g) := f(g(i, 0)) j_{k,m}(g; (i, 0)) = f(x, y, p, q) e^m(\kappa + pz) e^{ik\vartheta} y^{k/2}, \ g = (p, q, \kappa) n(x) t(y) r(\vartheta).$$

Similarly to the analysis in the former cases, one can characterize the images of Jacobi forms.

Proposition 9.2: $J_{k,m}$ is isomorphic to the space $\mathcal{A}_{k,m}$ of complex functions $\mathcal{C}^\infty(G^J)$ with

– i) $\phi(\gamma g) = \phi(g)$ for all $\gamma \in \Gamma^J$,

– ii) $\phi(gr(\vartheta)(0, 0, \kappa)) = \phi(g) e^m(\kappa) e^{ik\vartheta}$,

– iii) $\mathcal{L}_{X_-}\phi = \mathcal{L}_{Y_-}\phi = 0$,

– iv) $\phi(g) y^{-k/2}$ is bounded in domains of type $y > y_0$.

$J_{k,m}^{\mathrm{cusp}}$ is isomorphic to the subspace $\mathcal{A}_{k,m}^0$ of $\mathcal{A}_{k,m}$ with

– iv') $\phi(g)$ is bounded.

For a proof of this and the following statement see [BeB] or [BeS] p. 76ff.

Theorem 9.10 (*Duality Theorem for the Jacobi Group*): For $m, k \in \mathbf{N}$, dim $J_{k,m}^{\text{cusp}}$ is equal to the multiplicity of the appropriate discrete series representation of G^J in the right regular representation ϱ^0 on the space $\mathcal{H}_m^0$ of *cuspidal functions of type* m, which consists of the elements $\phi \in L^2(\Gamma^J \setminus G^J)$ satisfying the functional equation

$$\phi(g(0, 0, \kappa)) = e(m\kappa)\phi(g), \quad \text{for all } \kappa \in \mathbf{R}$$

and fulfilling the *cusp condition*

$$\int_{N^J \cap \Gamma^J \setminus N^J} \phi(g_{\lambda/(2m)}g)dn = 0 \text{ for almost all } g \in G^J \text{ and } \lambda = 0, \dots, 2m - 1$$

where $N^J := \{(0, q, 0)n(x); q, x \in \mathbf{R}\}$ and $g_\lambda := (\lambda, 0, 0)$.

Further information concerning the continuous part of the decomposition of $L^2(\Gamma^J \setminus G^J)$ is collected in [BeS].

At a closer look, one can see that the classic ϑ has the transformation property of a Jacobi form of weight and index $1/2$ for a certain subgroup of Γ^J. As already the infinitesimal considerations indicated, one has to multiply with modular forms of half-integral weight to get Jacobi forms in the sense of the definition above. This is very nicely exploited in [EZ]. The appearance of the half-integral weights can be explained by the fact that, somewhat similar to the relation of the representation theory of SO(3) and SU(2) resp. SO(3, 1) and SL(2, $\mathbf{C}$), here one has coming in a double cover of SL(2, $\mathbf{R}$) called the *metaplectic group* Mp(2, $\mathbf{R}$). But this group is no more a linear group and hence, unfortunately, outside the scope of this book.

9.4 Hecke's Theory of L−Functions of Modular Forms

It is a classical topic of analysis to study the relation of a power series $f := \sum_{n=1}^{\infty} c(n)q^n$ to a Dirichlet series

$$L_f(s) := \sum_{n=1}^{\infty} \frac{c(n)}{n^s}.$$

In our context, we take a cusp form $f \in S_k(\Gamma)$ and get an associated Dirichlet series convergent for s with Re $s > k$ as the coefficients $c(n)$ of the Fourier expansion of f fulfill an appropriate growth condition. The most important Dirichlet series have an *Euler product* similar to the prototype of a Dirichlet series, namely the *Riemann Zeta series*

$$\zeta(s) := \sum_{n=1}^{\infty} n^{-s},$$

which is convergent for s with Re $s > 1$ and has the Euler product (taken over all prime numbers p)

$$\zeta(s) = \Pi \frac{1}{1 - p^{-s}}.$$

Moreover, it is known that ζ fulfills a functional equation under $s \longmapsto 1 - s$ and can be continued to a meromorphic function on the whole plane $\mathbf{C}$ with a simple pole at $s = 1$.

Roughly said, Hecke's theory analyses how these and more general properties are related to properties of modular forms. The main tools are the following operators.

Hecke Operators

Definition 9.4: For a natural number n one introduces the *Hecke operator* $T(n)$ as an operator acting on functions f on the upper half plane $\mathfrak{H}$ by

$$(9.10) \qquad\qquad T(n)f(\tau) := n^{k-1} \sum_{a\geq 1, ad=n, 0\leq b<d} d^{-k} f(\frac{a\tau + b}{d}).$$

The origin of this prescription can be understood in several ways. There is the more general background asking for the decomposition of double cosets into cosets for Γ: For instance, for a prime p one has

$$\Gamma \begin{pmatrix} 1 & \\ & p \end{pmatrix} \Gamma = \bigsqcup_{0\leq b<p} \Gamma \begin{pmatrix} 1 & b \\ & p \end{pmatrix}$$

(as exploited in [Sh] p.51ff). But we follow the presentation in [Se] p.98 based on the notion of *correspondences*: Let E be a set and let G_E be the free abelian group generated by E. A *correspondence* on E with integer coefficients is a homomorphism T of G_E to itself. We can describe T by its values on the elements x of E:

$$T(x) = \sum_{y\in E} n_y(x)y, \ n_y(x) \in \mathbf{Z}, \ n_y(x) = 0 \text{ for almost all } y.$$

A **C**-valued function F on E can be extended linearly to a function on G_E, which is again denoted by F. The transform of F by T, denoted TF, is the restriction to E of the function $F \circ T$. With the notation introduced above, it is given by

$$TF(x) = F(T(x)) = \sum_{y\in E} n_y(x)F(y).$$

Let $\mathcal{R}$ be the set of lattices L in **C** and $n \in \mathbf{N}$. We denote by $T(n)$ the correspondence on $\mathcal{R}$ which transforms a lattice to the sum (in $G_\mathcal{R}$) of its sublattices L_n of index n. Thus we have for $L \in \mathcal{R}$

$$T(n)L = \sum_{[L:L_n]} L_n.$$

The sum on the right side is finite. If n is prime, one sees that one has $n + 1$ such sublattices. For the general case one has

Lemma: There is a bijection between the set of matrices

$$M = \begin{pmatrix} a & b \\ 0 & d \end{pmatrix} ; ad = n, a \geq 1, 0 \leq b < d,$$

and the set of lattices L_n of index n in L. For L with basis (τ_1, τ_2) this bijection is given by associating to the matrix M the lattice with basis $\tau'_1 = a\tau_1 + b\tau_2, \tau'_2 = d\tau_2$.

Exercise 9.6: Prove this.

One also uses *homothety* operators R_λ, $\lambda \in \mathbf{C}^*$ defined by

$$R_\lambda L = \lambda L.$$

It makes sense to compose the correspondences $T(n)$ and R_λ, since they are endomorphisms of the abelian group $G_\mathcal{R}$.

Proposition 9.3: The correspondences $T(n)$ and R_λ verify the identities

$$R_\lambda R_\mu = R_{\lambda\mu}, \quad \lambda, \mu \in \mathbf{C}^*,$$
$$R_\lambda T(n) = T(n)R_\lambda, \quad n \in \mathbf{N}, \lambda \in \mathbf{C}^*,$$
$$T(m)T(n) = T(mn), \text{ if } (m,n) = 1,$$
$$T(p^n)T(p) = T(p^{n+1}) + pT(p^{n-1})R_p, \quad p \text{ prime}, \ n \in \mathbf{N}.$$

Proof: We refer to [Se] p.98.

Corollary 1: The $T(p^n)$ are polynomials in $T(p)$ and R_p.
Corollary 2: The algebra generated by the R_λ and the $T(p)$, p prime, is commutative; it contains all the $T(n)$.

For the description of the action of these operators on modular forms of weight k, we associate to f a corresponding function F defined on the set $\mathcal{R}$ of lattices by

$$F(L) := (\tau_1/\tau_2)^{-k} f(\tau_1/\tau_2)$$

if (τ_1, τ_2) is a basis of L. We define $T(n)f$ as the function on $\mathfrak{H}$ associated to the function $n^{k-1}T(n)F$ on $\mathcal{R}$, i.e.

$$T(n)f(\tau) := n^{k-1}T(n)F(L(\tau, 1))$$

where $L(\tau, 1)$ denotes the lattice with basis $(\tau := \tau_1/\tau_2, 1)$ (it can be arranged and is silently understood that the description of the lattices is chosen such that this τ is in $\mathfrak{H}$). Using the Lemma above, we get the formula (9.10) in the definition at the beginning. Certainly, one has to and can verify that this $T(n)$ transforms the spaces of modular forms of weight k into itself (and similarly for cusp forms). We leave this to the reader and only note the relations, which are immediate consequences of Proposition 9.3

(9.11) $T(m)T(n)f = T(mn)f$, if $(m,n) = 1$,

(9.12) $T(p^n)T(p)f = T(p^{n+1})f + p^{k-1}T(p^{n-1})f$, p prime $n \in \mathbf{N}$.

The central point of all this is the action of the Hecke operators expressed by the coefficients of the q–development $f(\tau) = \sum c(n)q^n$:

Proposition 9.4 We have

$$T(n)f(\tau) = \sum_{m \in \mathbf{Z}} \gamma(m)q^m$$

with

(9.13) $$\gamma(m) = \sum_{a \in \mathbf{N}, a|(n,m)} a^{k-1}c(nm/a^2).$$

Using the relations obtained earlier, the proof is straightforward. Again we refer to [Se] p.101. In particular, we come to

$$\gamma(0) = \sigma_{k-1}(n)c(0), \qquad \sigma_k(n) := \sum_{d \in \mathbf{N}, d|n} d^k,$$

$$\gamma(1) = c(n),$$

$$\gamma(m) = c(pm) \qquad\qquad\qquad \text{if } n = p \text{ prime and} \qquad m \not\equiv 0 \mod p,$$

$$= c(pm) + p^{k-1}c(m/p) \qquad\qquad\qquad\qquad\qquad\qquad m \equiv 0 \mod p.$$

Since the Hecke operators commute, one can hope for simultaneous diagonalizability.

Hecke Eigenforms

Let f be a *Hecke eigenform* of weight k, i.e. a modular form of weight k, which is not identically zero and an eigenfunction for all $T(n), n \in \mathbf{N}$,

$$T(n)f(\tau) = \lambda(n)f(\tau).$$

Examples are the Eisenstein series G_k and the discriminant Δ. One has

$$T(n)G_k = \sigma_{k-1}(n)G_k, \ T(n)\Delta = (2\pi)^{12}\tau(n)\Delta, \quad \text{for all } n \in \mathbf{N}.$$

The verification is non-trivial but standard ([Se] p.104).

Our central statement is as follows:

Theorem 9.11: The coefficient $c(1)$ of a Hecke eigenform f is non-zero. If f is *normalized* by $c(1) = 1$, one has

$$\lambda(n) = c(n) \quad \text{for all } n \in \mathbf{N}.$$

Proof: The formula (9.13) in Proposition 9.4 above shows that the coefficient of q in $T(n)f$ is $c(n)$. On the other hand, as f is a $T(n)$–eigenfunction, it is also $\lambda(n)c(1)$. Thus we have $c(n) = \lambda(n)c(1)$. If $c(1)$ were zero, all the $c(n), n > 0$ would be zero, and f would be a constant, which is absurd.

Applying the first statement of the Theorem to the difference of two normalized Hecke eigenforms, we get

Corollary 1: Two modular forms of weight k, $k > 0$, which are both normalized Hecke eigenforms with the same eigenvalues, coincide.

By the second statement of the Theorem the relations (9.11) for the Hecke operators translate immediately to relations for the Fourier coefficients. We get

Corollary 2: For the Fourier coefficients of a normalized Hecke eigenform f of weight k one has the relations

(9.14) $c(m)c(n) = c(mn)$, if $(m,n) = 1$,

(9.15) $c(p^n)c(p) = c(p^{n+1}) + p^{k-1}c(p^{n-1})$, if p is prime and $n \in \mathbf{N}$.

Now, these relations for the Fourier coefficients of a Hecke eigenform $f = \sum c(n)q^n$ translate to the statement that the associated Dirichlet series $L_f(s) = \sum c(n)n^{-s}$ (convergent, as already mentioned, for Re $s > k$) has an Euler product.

Corollary 3: The Dirichlet series L_f of a Hecke eigenform f of weight k has the *Euler product*

$$(9.16) \qquad L_f(s) = \Pi_{p \in P} \frac{1}{1 - c(p)p^{-s} + p^{k-1-2s}}.$$

Proof: From Corollary 2 we know that $n \longmapsto c(n)$ is a multiplicative function, so with $t := p^{-s}$ we can write

$$L_f(s) = \Pi_{p \in P} \left(\sum_{n=0}^{\infty} c(p^n)t^n \right).$$

Hence it is sufficient to show that we have

$$\sum_{n=0}^{\infty} c(p^n)t^n = L_{f,p}(t) \text{ for } L_{f,p}(t) := (1 - c(p)t + p^{k-1}t^2)^{-1}.$$

We compute

$$\psi(t) := \left(\sum_{n=0}^{\infty} c(p^n)t^n \right)(1 - c(p)t + p^{k-1}t^2) =: \sum d(n)t^n$$

and as coefficient of t get $d(1) = c(p) - c(p) = 0$ and, for $n \geq 1$, as coefficient of t^{n+1}

$$d(n+1) = c(p^{n+1}) - c(p)c(p^n) + p^{k-1}c(p^{n-1}) = 0$$

in consequence of the second relation in Corollary 2. I.e., we have $\psi(t) = c(1) = 1$.

Remark 9.16: The statement that L_f has an Euler product as in Corollary 3 is equivalent to the fact that the Fourier coefficients of f fulfill the relations in Corollary 2.

Remark 9.17: Hecke proved that L_f can be analytically continued to a meromorphic function on the whole complex plane, resp. to a holomorphic function if f is a cusp form, and that the *completed L-function*

$$L_f^*(s) := (2\pi)^{-s}\Gamma(s)L_f(s)$$

satisfies the functional equation

$$(9.17) \qquad L_f^*(s) = (-1)^{k/2}L_f^*(k - s).$$

The proof is based on the classical Mellin transform: Let f be a normalized Hecke cusp form of weight k, i.e. we have

$$f(\tau) = \sum_{n=1}^{\infty} c(n)q^n \text{ with } f(-1/\tau) = \tau^k f(\tau).$$

Using the Γ-function

$$\Gamma(s) := \int_0^\infty e^{-t} t^{s-1} dt$$

we determine the *Mellin transform* of f, namely

$$\int_0^\infty f(iy) y^{s-1} dy = \int_0^\infty \left(\sum c(n) e(niy) \right) y^s \frac{dy}{y}$$

$$= \sum c(n) \int_0^\infty e^{-2\pi ny} y^{s-1} dy$$

$$= \sum c(n)(2\pi n)^{-s} \Gamma(s)$$

$$= (2\pi)^{-s} \Gamma(s) L_f(s) = L_f^*(s).$$

In the integral on the left hand side, we put $u = 1/y$ and use the transformation formula for f to get

$$\int_0^\infty f(iy) y^{s-1} dy = -\int_\infty^0 f(i/u) u^{-s} \frac{du}{u}$$

$$= \int_0^\infty f(iu) i^k u^{k-s} \frac{du}{u}$$

$$= i^k L_f^*(k - s).$$

Remark 9.18: The whole theory has been extended to other groups than $\Gamma = \mathrm{SL}(2, \mathbf{Z})$, in particular to $\Gamma_0(N)$ (see for instance [Ge]).

There is also a converse to the statement in Remark 9.17: Every Dirichlet series L, which fulfills a functional equation of the type just treated and has certain regularity and growth properties, comes from a modular form of f of weight k. This was first remarked by Hecke and then refined considerably by Weil in [We1] and is now cited under the heading of the famous *Converse Theorem*.

Remark 9.19: The Hecke operators are hermitian with respect to the *Petersson scalar product* for forms f_1, f_2 of weight k given by

$$< f_1, f_2 > := \int_{\mathcal{F}} f_1(\tau) \overline{f_2(\tau)} y^{k-2} dx dy, \quad \mathcal{F} \text{ as in Remark 9.2,}$$

i.e. one has

$$< T(n) f_1, f_2 > = < f_1, T(n) f_2 > .$$

Arithmetic Modular Forms

Though it does not directly belong to our topic, we can not resist the temptation to complete this section by indicating a concept of arithmetically distinguished modular forms, which has been exploited by Shimura and many others (see [Sh]): We look at modular forms (for the modular group Γ, but this is open to wide generalizations) with integral Fourier coefficients

$$A_k(\Gamma, \mathbf{Z}) := \{ f = \sum c(n) q^n \in A_k(\Gamma); c(n) \in \mathbf{Z} \}.$$

One can show that one can find a **Z**-basis of $A_k(\Gamma, \mathbf{Z})$, which is also a **C**-basis of $A_k(\Gamma)$. Prototypes of these forms are, up to factors, forms we already met with, namely

$$F(\tau) \quad := \quad q \prod_{n=1}^{\infty} (1 - q^n)^{24},$$

$$E_4(\tau) \quad := \quad 1 + 240 \sum_{n=1}^{\infty} \sigma_3(n) q^n,$$

$$E_6(\tau) \quad := \quad 1 - 504 \sum_{n=1}^{\infty} \sigma_5(n) q^n.$$

For $k \in 4\mathbf{N}$ $A_k(\Gamma, \mathbf{Z})$ has as **Z**-basis (see [Se1] p.105)

$$E_4^{\alpha} F^{\beta}, \ \alpha, \beta \in \mathbf{N}_0 \text{ with } \alpha + 3\beta = k/4,$$

and for $k \in 2(2\mathbf{N} + 1)$

$$E_6 E_4^{\alpha} F^{\beta}, \ \alpha, \beta \in \mathbf{N}_0 \text{ with } \alpha + 3\beta = ((k/2) - 3)/2.$$

Our analysis of the action of the Hecke operators above shows that $A_k(\Gamma, \mathbf{Z})$ is stable under all $T(n)$. Hence the coefficients of the characteristic polynomial of $T(n)$ acting on $A_k(\Gamma, \mathbf{Z})$ are integers and one can deduce that the eigenvalues are algebraic integers. There are explicit formulas for the traces of the $T(n)$, the *Eichler-Selberg trace formulas*, starting by [Ei1] and [Sel].

Summary

In this section we looked at L-functions $L(s) = L_f(s)$ coming from modular cusp forms f. Via a functional equation these functions have analytic continuation to the whole complex plane. This functional equation is a consequence of the covariant transformation property of the modular form. The modular form is related to an (infinite-dimensional) automorphic representation of $\mathrm{SL}(2, \mathbf{R})$ (or $\mathrm{GL}(2, \mathbf{R})$). We call such L-functions *automorphic L-functions*. If f is a Hecke eigenform, L_f has an Euler product with *factors of degree 2*, i.e. with

$$(1 - c(p)p^{-s} + p^{k-1-2s})^{-1} = (1 - c(p)t + p^{k-1}t^2)^{-1}, \ t := p^{-s}.$$

There is a general procedure to associate such functions to automorphic representations of other groups leading to Euler products with other degrees. In particular, one has a functional equation and factors of degree 1 in the theory of Hecke's L-series with *grössencharacters* (from the German word "Grössencharakter"). This is fairly easy to be understood ([Ge] p.99f) as a theory of automorphic L-functions for the group $\mathrm{GL}(1)$ if one uses the more functional analytic *idelic* approach initiated by Tate's thesis (reproduced in [CF] or described in [La2] and, provided with the necessary analytic background, in [RV]). Hecke's classic approach is carefully displayed in [Ne] p.515ff. Here we shall indicate the definition of this function later when we have more number theory at hand. For an introduction to more general automorphic L-functions we refer to [Ge] and [GeS].

9.5   Elements of Algebraic Number Theory and Hecke L-Functions

Any book on number theory is good as a background to the following survey, in particular we recommend the books by Lang [La2] and by Neukirch [Ne] (now also in English).

Number Fields and their Galois Groups

Definition 9.5: An *algebraic number field* is a finite extension $\mathbf{K}$ of the field $\mathbf{Q}$ of rational numbers, i.e. a field containing $\mathbf{Q}$, which is a finite-dimensional vector space over $\mathbf{Q}$.

Such a field is obtained by adjoining to $\mathbf{Q}$ roots of polynomials with coefficients in $\mathbf{Q}$. For example, the field

$$\mathbf{Q}(i) := \{a + bi; \, a, b \in \mathbf{Q}\}$$

is obtained by adjoining the roots of the polynomial $x^2 + 1$, denoted as usual by i and $-i$. We obtain a field, which has dimension 2 as a vector space over $\mathbf{Q}$, the field of *Gauss numbers*. This is the most elementary example of an *imaginary quadratic field*, which has the general form $\mathbf{K} = \mathbf{Q}(\sqrt{-d}), d \in \mathbf{N}$, d squarefree.

Similarly, for $N \in \mathbf{N}$, adjoining to $\mathbf{Q}$ a primitive N−th root of unity ζ_N we obtain the N-th *cyclotomic field* $\mathbf{Q}_N := \mathbf{Q}(\zeta_N)$. Its dimension as a vector space over $\mathbf{Q}$ is $\varphi(N)$, the *Euler function* of N, i.e.

$$\varphi(N) := \#\{m \in \mathbf{N}; \, 1 \leq m < N, (m, N) = 1\}.$$

We can embed $\mathbf{Q}(\zeta_N)$ into $\mathbf{C}$ in such a way that $\zeta_N \longmapsto e(1/N)$. But this is not the only possible embedding of $\mathbf{Q}_N$ into $\mathbf{C}$, since we could also send $\zeta_N \longmapsto e(m/N)$ where $(m, N) = 1$.

Suppose now that $\mathbf{K}$ is an algebraic number field and $\mathbf{F}$ a finite extension of $\mathbf{K}$, i.e. another field containing $\mathbf{K}$, which has a finite dimension as a vector space over $\mathbf{K}$. This dimension is called the *degree of* $\mathbf{F}$ *over* $\mathbf{K}$ and denoted by $\deg_{\mathbf{K}} \mathbf{F}$, in particular $\deg \mathbf{F}$ for $\mathbf{K} = \mathbf{Q}$.

Definition 9.6: The group of all field automorphisms σ of $\mathbf{F}$, preserving the field structure and such that $\sigma(x) = x$ for all $x \in \mathbf{K}$, is called the *Galois group of* $\mathbf{F}$ *over* $\mathbf{K}$ and denoted by $\mathrm{Gal}(\mathbf{F}/\mathbf{K})$, in particular $\mathrm{Gal}(\mathbf{F})$ for $\mathbf{K} = \mathbf{Q}$.

Example 9.5: The Galois group $\mathrm{Gal}(\mathbf{Q}_N)$ is naturally identified with the group

$$(\mathbf{Z}/N\mathbf{Z})^\times := \{\bar{n} := n + N\mathbf{Z} \in \mathbf{Z}/N\mathbf{Z}; (n, N) = 1\}$$

with respect to multiplication. The element $\bar{n} \in (\mathbf{Z}/N\mathbf{Z})^\times$ gives rise to the automorphism of $\mathbf{Q}_N$ sending ζ_N to ζ_N^n, and hence ζ_N^m to ζ_N^{mn} for all m. If M divides N, then $\mathbf{Q}_M$ is contained in $\mathbf{Q}_N$, and the corresponding homomorphism of the Galois groups $\mathrm{Gal}(\mathbf{Q}_N) \longrightarrow \mathrm{Gal}(\mathbf{Q}_M)$ coincides under the above identification with the natural surjective homomorphism

$$(\mathbf{Z}/N\mathbf{Z})^\times \longrightarrow (\mathbf{Z}/M\mathbf{Z})^\times; \; \bar{n} \longmapsto n \mod M.$$

A central notion from number theory is the notion of a Galois extension:

Definition 9.7: An algebraic field extension $\mathbf{F}/\mathbf{K}$ is called a *Galois extension* iff the order of the Galois group $G = \mathrm{Gal}(\mathbf{F}/\mathbf{K})$ is equal to the relative degree $\deg_{\mathbf{K}} \mathbf{F}$.

It is a main part of the theory to give different versions of this definition. Then, *Galois theory* consists in establishing for a Galois extension a (contravariant) bijection between subgroups of the Galois group and subfields of $\mathbf{F}$ containing $\mathbf{K}$.

Next, one has *class field theory*, which tries to collect information about a Galois field extension $\mathbf{F}/\mathbf{K}$, in particular the structure of the Galois group, only from the structure of the basefield $\mathbf{K}$. We shall need some of this and have to recall some more notions and facts from algebra:

The Ideal Class Group

Let R be an integral domain, i.e. a commutative ring with unit e and without zero-divisors. A subring $\mathfrak{a}$ in R is an *ideal* in R iff one has $ra \in \mathfrak{a}$ for all $r \in R$ and $a \in \mathfrak{a}$. An ideal $\mathfrak{a}$ is a *principal ideal* iff it can be generated by one element, i.e. is of the form $\mathfrak{a} = aR$ for an element $a \in R$. An ideal $\mathfrak{p}$ in R is a *prime ideal* iff from $ab \in \mathfrak{p}$ one can conclude a or $b \in \mathfrak{p}$, or, equivalently, iff the factor ring $R/\mathfrak{p}$ has no zero-divisors.

For instance, every normalized irreducible polynomial f of degree n in the polynomial ring $R = \mathbf{Q}[x]$ generates a prime ideal $\mathfrak{p} = fR$. This ideal is maximal and the factor ring $R/\mathfrak{p}$ is an algebraic number field of degree n. Every algebraic number field of degree n can be described like this.

If for two ideals we have $\mathfrak{a} \supset \mathfrak{b}$, we say $\mathfrak{a}$ *divides* $\mathfrak{b}$. And $(\mathfrak{a}, \mathfrak{b})$ denotes the greatest common divisor of $\mathfrak{a}$ and $\mathfrak{b}$ (corresponding to the usual notions for integers if $R = \mathbf{Z}$ and $\mathfrak{a} = a\mathbf{Z}, \mathfrak{b} = b\mathbf{Z}$).

From two ideals $\mathfrak{a}$ and $\mathfrak{b}$ one can build new ideals, the *sum* $\mathfrak{a} + \mathfrak{b}$ and the *product* $\mathfrak{a} \cdot \mathfrak{b}$ (the last one consisting of finite sums of products ab, $a \in \mathfrak{a}, b \in \mathfrak{b}$).

In number theory, the main occasion to use these notions is the *maximal order* of an algebraic number field $\mathbf{K}$, i.e. the ring $\mathfrak{o}$ of integral elements of $\mathbf{K}$: an element $a \in \mathbf{K}$ is called *integral* iff it is the root of a normalized polynomial $f \in \mathbf{Z}[x]$ (one has to show that the set of these elements is a ring). The determination of the maximal orders is a fundamental task of number theory. For instance, for the field $\mathbf{K} = \mathbf{Q}(i)$ of Gauss numbers one has the *Gauss integers* $\mathfrak{o} = \mathbf{Z}[i] = \{a + bi; a, b \in \mathbf{Z}\}$ (this is not difficult, try to prove it as **Exercise 9.7**). For the cyclotomic field $\mathbf{K} = \mathbf{Q}(\zeta_N)$ one has also the nice result $\mathfrak{o} = \mathbf{Z}[\zeta_N]$.

Main tools in this context are the maps *norm* and *trace*: For $x \in \mathbf{F}$ one has

$$\lambda_x : \mathbf{F} \longrightarrow \mathbf{F}, \ a \longmapsto xa \quad \text{for all } a \in \mathbf{F},$$

the (additive) trace map

$$Tr_{\mathbf{F}/\mathbf{K}} : \mathbf{F} \longrightarrow \mathbf{K}, \ x \longmapsto \mathrm{Tr}\,\lambda_x$$

and the (multiplicative) norm map

$$N_{\mathbf{F}/\mathbf{K}} : \mathbf{F} \longrightarrow \mathbf{K}, \ x \longmapsto \det \lambda_x.$$

An ideal in $\mathfrak{o}$ is also called an ideal of $\mathbf{K}$ and this notion is extended to the notion of a *fractional ideal* of $\mathbf{K}$: This is an $\mathfrak{o}$-submodule $\mathfrak{a} \neq 0$ of $\mathbf{K}$, which has a common denominator, i.e. there is an element $0 \neq d \in \mathfrak{o}$ with $d\mathfrak{a} \subseteq \mathfrak{o}$.

All this is done with the intention to associate another finite group to an algebraic number field: It is not difficult to see that the set $J_{\mathbf{K}}$ of fractional ideals provided with multiplication as composition has the structure of an abelian group. It is called the *ideal group* of $\mathbf{K}$. The (almost unique) decomposition of an integer in $\mathbf{Z}$ into a product of powers of primes is extended to the statement (see for instance [Ne] p.23):

Every fractional ideal $\mathfrak{a} \in J_{\mathbf{K}}$ has a unique representation as a product of powers of prime ideals

$$\mathfrak{a} = \Pi_{\mathfrak{p}} \, \mathfrak{p}^{\nu_{\mathfrak{p}}}, \ \nu_{\mathfrak{p}} \in \mathbf{Z}, \ \text{almost all } \nu_{\mathfrak{p}} = 0.$$

The principal fractional ideals $P_{\mathbf{K}} := \{(a) = a\mathfrak{o}; \, a \in \mathbf{K}^*\}$ form a subgroup of $J_{\mathbf{K}}$ and, finally, one introduces the factor group

$$Cl_{\mathbf{K}} := J_{\mathbf{K}}/P_{\mathbf{K}}.$$

This comes out as a finite abelian group and is called the *ideal class group* or simply *class group* of $\mathbf{K}$.

Class field theory tries to relate the Galois group of a field extension to the ideal class group of the ground field. An extension $\mathbf{F}/\mathbf{K}$ with abelian Galois group G is called a *Hilbert class field* iff one has

$$G(\mathbf{F}/\mathbf{K}) = Cl_{\mathbf{K}}.$$

Example 9.6: The cyclotomic field $\mathbf{Q}_N$ is a Hilbert class field over $\mathbf{Q}$. We have $Cl_{\mathbf{Q}_N} \simeq (\mathbf{Z}/N\mathbf{Z})^{\times}$.

Dirichlet Characters and Hecke's L-Functions

For $0 \neq m \in \mathbf{Z}$, the classical *Dirichlet character* mod m is a character of the group $(\mathbf{Z}/m\mathbf{Z})^{\times}$

$$\chi : (\mathbf{Z}/m\mathbf{Z})^{\times} \longrightarrow S^1.$$

The character χ is called *primitive* if there is no genuine divisor m' of m such that χ factorizes via a character of $(\mathbf{Z}/m'\mathbf{Z})^{\times}$. Given such a χ, we get a (multiplicative) function for all natural numbers n, which we denote again by χ, by mapping n to $\chi(\bar{n})$ if $(n, m) = 1$ and $\bar{n}$ is the class of n mod m, and to zero if $(n, m) \neq 1$. Hence, to such a primitive character χ, one can associate an L-series

$$L(\chi, s) := \sum_{n=1}^{\infty} \frac{\chi(n)}{n^s}.$$

Obviously, for the trivial character χ^0 mod 1 with $\chi^0(n) = 1$ for all n, we obtain the Riemann Zeta series. Hecke proved that these series have an Euler product (with degree one factors)

$$L(\chi, s) = \Pi_{p \in \mathbf{P}} \frac{1}{1 - \chi(p)p^{-s}},$$

analytic continuation to $\mathbf{C}$ (resp. $\mathbf{C} \backslash \{1\}$ for χ^0), and a functional equation for $s \longmapsto 1-s$. (The proof relies again on the transformation formula of a (generalized) theta series.)

These are examples of Hecke's L-functions, which can be defined more generally: Let $\mathbf{K}$ be an algebraic number field with maximal order $\mathfrak{o}$, $\mathfrak{m} \subset \mathfrak{o}$ an ideal, and $J^{\mathfrak{m}}$ the group of fractional ideals $\mathfrak{a}$ in $\mathbf{K}$, which are prime to $\mathfrak{m}$

$$J^{\mathfrak{m}} := \{\mathfrak{a} \subset \mathbf{K}; \, (\mathfrak{a}, \mathfrak{m}) = 1\}.$$

Let $\chi : J^{\mathfrak{m}} \longrightarrow S^1$ be a character. Then we define an L-series

$$L(\chi, s) := \sum_{\mathfrak{a} \subset \mathfrak{o}} \frac{\chi(\mathfrak{a})}{N(\mathfrak{a})^s}.$$

Here the sum is taken over all ideals $\mathfrak{a}$ in $\mathfrak{o}$, $N(\mathfrak{a})$ denotes the number of elements of the residue class ring $\mathfrak{o}/\mathfrak{a}$, and again we put $\chi(\mathfrak{a}) = 0$ for $(\mathfrak{a}, \mathfrak{m}) \neq 1$. The assumption that χ is a character translates easily to the existence of an Euler product taken over all prime ideals $\mathfrak{p}$ in $\mathfrak{o}$

$$L(\chi, s) := \Pi_{\mathfrak{p}} \frac{1}{1 - \chi(\mathfrak{p})N(\mathfrak{p})^{-s}}.$$

If we take $\chi = \chi^0$ with $\chi^0(\mathfrak{a}) = 1$ for all $\mathfrak{a}$, we get the usual *Dedekind Zeta function* of the number field $\mathbf{K}$

$$L(\chi^0, s) = \sum_{\mathfrak{a} \subset \mathfrak{o}} N(\mathfrak{a})^{-s} =: \zeta_{\mathbf{K}}(s).$$

The question asking for the biggest class of characters, for which one can prove a functional equation (and hence has analytic continuation) leads to Hecke's definition of the notion "Größencharakter":

Definition 9.8: A character $\chi : J^{\mathfrak{m}} \longrightarrow S^1$ is called a *grössencharacter* mod $\mathfrak{m}$ iff there is a pair of characters

$$\chi_f : (\mathfrak{o}/\mathfrak{m})^* \longrightarrow S^1, \; \chi_\infty : \mathbf{R}^* \longrightarrow S^1$$

such that

$$\chi((a)) = \chi_f(a)\chi_\infty(a) \quad \text{for all } a \in \mathfrak{o} \text{ with } ((a), \mathfrak{m}) = 1.$$

We have remarked (in 5.1) that the set of characters of the additive group $\mathbf{R}$ is simply in bijection to $\mathbf{R}$. A sign that multiplicative structure is more intricate than additive structure is again given by the fact that the set of these characters χ_∞ of $\mathbf{R}^*$ is more complicated (for a complete statement see [Ne] p.497).

As already said, a more elegant way to understand these constructions is based on the theory of *adèles* and *idèles* introduced into number theory by Tate and then pursued by many others. Here one has the central notion of a *Hecke character* as a character of the *idel class group*. The relation to the notions presented here is also to be found in [Ne] p.501. There are further refinements using *ray class groups* and their characters but we will stop here and change the topic to get again more concrete examples.

9.6 Arithmetic L-Functions

In classical number theory and in arithmetic geometry there are many occasions to introduce Zeta- and L-functions as Dirichlet series encoding diophantine or arithmetic information. By their nature as Dirichlet series, these series converge in right half planes. Then, as the main task, one has to find a way to extend the function given by such a series to a function on the whole plane, eventually with some poles, and study the values of this function at special, in particular, integral points. This may turn out to be the expression of a mysterious regularity property inherent in the arithmetic or diophantine problem we started with. A way to prove this comes up if this arithmetic L-function can be seen as an automorphic L-function. For instance the proof of the Fermat conjecture by Wiles (and Taylor) incorporates a proof that the L-function of an elliptic curve can be interpreted as an L-function of a modular form. Here we are far from making this sufficiently precise but to give at least some flavour and define the L-functions for number fields Emil Artin discussed around 1923, we borrow heavily from an expository article on the role of representation theory in number theory by Langlands [L].

Two Examples:

We look at *diophantine equations*, i.e. polynomial equations with integral coefficients to which integral solutions are sought. A famous example is the *Fermat equation*

$$x^m + y^m = z^m.$$

For $m = 2$ there are infinitely many solutions, the *Pythagorian triples*, e.g. $(3, 4, 5)$, and for $m > 2$, as Wiles (and Taylor) proved, there are no non-trivial integral solutions. But if there are no integral solutions to a diophantine problem, one can look for solutions mod p, p prime, try to count their number and encode these numbers into a function.

Example 9.7: Probably the simplest example is the equation

(9.18) $$x^2 + 1 = 0.$$

The primes p for which the congruence $x^2 + 1 \equiv 0 \mod p$ can be solved are $2, 5, 13, \ldots$ all of which leave the remainder 1 upon division by 4, whereas primes like $7, 19, 23, \ldots$ that leave the remainder 3 upon division by 4 never do so. In 9.3 we already presented the Legendre symbol

$$\left(\frac{a}{p}\right) = \pm 1 \quad \text{for all } a \in \mathbf{Z}, \text{ with } (p, a) = 1,$$

which indicates whether the congruence $x^2 \equiv a \mod p$ is solvable or not. Having at hand the notion of a Dirichlet character, we see that here we have an example with a multiplicative function χ on $\mathbf{Z}$ given by

$$
\begin{aligned}
\chi(p) &= \left(\tfrac{-1}{p}\right) &= 1 &\quad \text{for } p \equiv 1 \mod 4, \\
&&= -1 &\quad \text{for } p \equiv 3 \mod 4, \\
&= 0 &&\quad \text{for } (p, 4) \neq 1,
\end{aligned}
$$

i.e. with $\chi(1) = 1, \chi(2) = 0, \chi(3) = -1, \chi(4) = 0 \ldots$.

We have

$$L(\chi, s) = \prod_p \left(1 - \frac{\chi(p)}{p^s}\right)^{-1} = \sum_{n=1}^{\infty} \frac{\chi(n)}{n^s} = 1 - \frac{1}{3^s} + \frac{1}{5^s} - \frac{1}{7^s} + \frac{1}{9^s} + \cdots .$$

Like every Dirichlet series, this series defines a function in a right half plane. It is an expression of the mysterious symmetry of our problem that one can bring into play a theta function, which, by its transformation property, produces a function valid for all $s \in \mathbf{C}$.

Now, having prepared some material from algebraic number theory, we can look at the function $L(\chi, s)$ from another point of view: The polynomial $f(x) = x^2 + 1$ defines the Gauss field

$$\mathbf{K} = \mathbf{Q}(i) = \{a + bi, a, b \in \mathbf{Q}\}$$

and this field has the Galois group $G = G(\mathbf{K})$ of automorphisms elementwise fixing $\mathbf{Q}$ consisting just of the identity map and the map φ, uniquely fixed by changing the root $\theta_1 := \sqrt{-1} = i$ into $\theta_2 := -i$. Hence, one has $G \simeq \{\pm 1\}$ and this group has exactly two one-dimensional representations, the trivial one and a representation π with $\pi(\varphi) = -1$. One can use this to define another type of L-function, the Artin L-function. To explain this (following Langlands' article), we go back some steps:

The solvability of the congruence $f(x) = x^2 + 1 \equiv 0 \mod p$ is equivalent to the possibility to factorize $f \mod p$ into a product of two linear factors, e.g.

$$x^2 + 1 \equiv x^2 + 2x + 1 = (x+1)^2 \quad \mod 2$$
$$x^2 + 1 = x^2 - 4 + 5 \equiv (x-2)(x+2) \quad \mod 5.$$

For the primes $p = 7$ and 11 and, in general, for all primes, for which the congruence has no solutions, the polynomial $x^2 + 1$ stays irreducible mod p.

Example 9.8: Another famous example is

(9.19) $$f(x) = x^5 + 10x^3 - 10x^2 + 35x - 18.$$

The polynomial is irreducible mod p for $p = 7, 13, 19, 29, 43, 47, 59, \ldots$ and factorizes into linear factors mod p for $p = 2063, 2213, 2953, 3631, \ldots$.

In general, we can look at a polynomial with integral coefficients

$$f(x) = x^n + a_{n-1}x^{n-1} + \cdots + a_1 x + a_0$$

with roots $\alpha_1, \ldots, \alpha_n$, which we assume to be pairwise different. There will be various relations between these roots with coefficients that are rational,

$$F(\alpha_1, \ldots, \alpha_n) = 0.$$

For example, the roots of $x^3 - 1 = 0$ are $\alpha_1 = 1, \alpha_2 = (-1+\sqrt{-3})/2, \alpha_3 = (-1-\sqrt{-3})/2$ and two of the valid relations for these roots are

$$\alpha_1 = 1, \quad \alpha_2\alpha_3 = \alpha_1.$$

To the equation $f(x) = 0$ we can associate the group of all permutations of its roots that preserve all valid relations. In the last example the sole possibility in addition to the trivial permutation is the permutation that fixes α_1 and interchanges α_2 and α_3. The group G that we get is called the *Galois group of the equation* $f(x) = 0$. It is identical with the Galois group of the algebraic number field **K** defined by the polynomial f. We introduce the *discriminant* Δ of the equation (resp. the number field) as defined by the equation

$$\Delta := \prod_{i \neq j} (\alpha_i - \alpha_j).$$

It is an integer and one of the most important characteristic numbers to be associated to a number field. Now, one can attach to any prime p that does not divide Δ an element $F_p \in G$, called the *Frobenius automorphism*, that determines among other things how the equation factors mod p. More precisely, it is the conjugacy class of F_p within G that is determined (we will indicate below how this is done). To construct an L-series we need a prescription associating numerical information to the primes p. We can come to this by choosing a (finite-dimensional) representation ρ of G and taking the trace or the determinant of the matrix $\rho(F_p)$. Things are particularly easy if we have a one-dimensional representation ρ as for instance for the group G in our first example (9.18) above: We can define an Euler product

$$L(\rho, s) := \prod_{p} (1 - \rho(F_p)p^{-s})^{-1}$$

where the primes dividing Δ are omitted. By closer analysis, one can realize that in the example we get just the Hecke L-function $L(\chi, s)$ if we take for ρ the non-trivial character χ of $G \simeq \{\pm 1\}$. This is another expression of (a special case of) the main theorem of abelian class field theory.

In Chapter 2 we studied very briefly representations ρ of finite groups. Now, here, we find an occasion to expand the subject slightly. Beyond one-dimensional representations, the next possibility is that ρ is a two-dimensional representation, which we may suppose to be unitary. We know that SU(2) is a double cover of SO(3). It is customary to classify finite subgroups of the unitary group by their image in the group of rotations. Taking for an irreducible ρ the finite subgroup to be $\rho(G)$, we obtain dihedral, tetrahedral, octahedral, and icosahedral representations if G has as its image in SU(2) the corresponding group. Dihedral representations can be treated by the classical theory, and also the other representations were treated at the time of Langlands' article with the exception of a complete treatment of the icosahedral case. An example of an equation with an icosahedral representation ρ is given by our equation (9.19). For all primes, which do not divide 800, the *conductor* (related to the discriminant in a way we do not explain here), one can form the matrix $\rho(F_p)$, which has two eigenvalues, λ_p and μ_p. Artin's L-function is then

$$L(\rho, s) := \prod_{p} \frac{1}{1 - \lambda_p p^{-s}} \frac{1}{1 - \mu_p p^{-s}}$$

the primes dividing the conductor being omitted from the product. This product converges for Re $s > 1$ and it is a special case of fhe famous *Artin Conjecture* from 1923 to show that it can be continued to an analytic function for all $s \in$ **C**. For the equation (9.19) this was done in 1970 by J. Buhler. The general case is still open.

Decomposition Theory and Artin's L-Function

To start with, we look again at the example $\mathbf{K} = \mathbf{Q}_N = \mathbf{Q}(\zeta_N)$: Here we have the abelian Galois group $G = \mathrm{Gal}(\mathbf{Q}_N) \simeq (\mathbf{Z}/N\mathbf{Z})^\times$. This isomorphism relates every irreducible representation ρ of G uniquely to a character χ of $(\mathbf{Z}/N\mathbf{Z})^\times$. To each prime number p, which does not divide N, we associate the class $\bar{p} := p \mod N$ in $(\mathbf{Z}/N\mathbf{Z})^\times$ and then the Frobenius automorphism $F_p \in G$ given by $\zeta_N \longmapsto \zeta_N^p$. Now we can define the Artin L-function

$$L(\rho, s) := \prod_{p \nmid N} \frac{1}{1 - \rho(F_p)p^{-s}}$$

and it is clear that it coincides with the Hecke L-function

$$L(\chi, s) := \prod_{p \nmid N} \frac{1}{1 - \chi(p)p^{-s}}.$$

For the general case, we look at an algebraic field extension $\mathbf{F}/\mathbf{K}$ with Galois group $G = \mathrm{Gal}(\mathbf{F}/\mathbf{K})$ and rings of integral elements $\mathfrak{O} \supset \mathfrak{o}$. For a prime ideal $\mathfrak{P}$ in $\mathfrak{O}$ and a prime ideal $\mathfrak{p}$ in $\mathfrak{o}$ we say that $\mathfrak{P}$ *lies over* $\mathfrak{p}$ if $\mathfrak{P}$ contains $\mathfrak{p}$ and hence $\mathfrak{P} \cap \mathfrak{o} = \mathfrak{p}$. We say that $\mathfrak{p}$ *decomposes into the primes* $\mathfrak{P}_i$ if $\mathfrak{p}$ generates in $\mathfrak{O}$ an ideal $(\mathfrak{p})$, which has the decomposition

$$(\mathfrak{p}) = \prod_{i=1}^{h} \mathfrak{P}_i^{e_i}$$

into the primes $\mathfrak{P}_i$ lying over $\mathfrak{p}$ with the *ramification indices* $e_i \in \mathbf{N}$. We denote by $\mathfrak{k}(\mathfrak{P})$ the residue field $\mathfrak{O}/\mathfrak{P}$ and similarly $\mathfrak{k}(\mathfrak{p}) := \mathfrak{o}/\mathfrak{p}$. Both fields are finite fields. If $\mathfrak{P}_i$ lies over $\mathfrak{p}$, $\mathfrak{k}(\mathfrak{P}_i)$ is a finite extension of $\mathfrak{k}(\mathfrak{p})$ and we denote $f_i := \deg_{\mathfrak{k}(\mathfrak{p})} \mathfrak{k}(\mathfrak{P}_i)$. These degrees and the ramification indices are related by the central formula of Hilbert's ramification theory

$$\sum_{i=1}^{h} e_i f_i = n.$$

From now on, we assume that the extension $\mathbf{F}/\mathbf{K}$ is a Galois extension. Then the decomposition simplifies to

$$(\mathfrak{p}) = \left(\prod_{i=1}^{h} \mathfrak{P}_i\right)^e$$

and all degrees f_i coincide $(=: f)$. The Galois group G acts on $\mathfrak{O}$ and moreover even transitively on the primes $\mathfrak{P}_i$ lying over a given $\mathfrak{p} \subset \mathfrak{o}$. We introduce the *decomposition group*

$$G_\mathfrak{P} := \{\sigma \in G; \sigma\mathfrak{P} = \mathfrak{P}\}.$$

The index of the decomposition group in G is equal to the *ramification index*, which is defined as $e = [G : G_\mathfrak{P}]$. One has the two extreme cases:

i) If $G_\mathfrak{P}$ contains only the identity, we say that $\mathfrak{p}$ is *totally decomposed*, and
ii) $G_\mathfrak{P} = G$ where $\mathfrak{p}$ stays *indecomposed*.

Every $\sigma \in G_{\mathfrak{P}}$ induces an automorphism of the residue class field $\ell(\mathfrak{P}) := \mathfrak{O}/\mathfrak{P}$ by

$$a \mod \mathfrak{P} \longmapsto \sigma a \mod \mathfrak{P}$$

Since this automorphism preserves the elements of the subfield $\ell(\mathfrak{p})$, we get a homomorphism

$$\varphi_{\mathfrak{P}} : G_{\mathfrak{P}} \longrightarrow \mathrm{Gal}(\ell(\mathfrak{P})/\ell(\mathfrak{p})),$$

which is surjective. We define the *inertia group* as its kernel, $I_{\mathfrak{P}} := \ker \varphi_{\mathfrak{P}}$. If $\mathfrak{p}$ is unramified, $I_{\mathfrak{P}}$ consists only of the identity, one has $G_{\mathfrak{P}} \simeq \mathrm{Gal}(\ell(\mathfrak{P})/\ell(\mathfrak{p}))$ and, hence, this group may be seen as a subgroup of G. In this case we can find exacly one automorphism $F_{\mathfrak{P}} \in G$ such that

$$F_{\mathfrak{P}} a \equiv a^q \mod \mathfrak{P} \quad \text{for all } a \in \mathfrak{O}$$

with $q := N(\mathfrak{p})$. This is the *Frobenius automorphism*, it generates the cyclic group $G_{\mathfrak{P}}$. If $I_{\mathfrak{P}}$ is not trivial, we take as Frobenius $F_{\mathfrak{P}}$ an element of G such that its image in $\mathrm{Gal}(\ell(\mathfrak{P})/\ell(\mathfrak{p}))$ has again the action $a \mod \mathfrak{P} \longmapsto a^q \mod \mathfrak{P}$.

Now we are ready to define Artin's L-function for this general situation. We take a representation ρ of G in the finite-dimensional vector space V and denote by $V^{\mathfrak{P}}$ the subspace of elements fixed by $\rho(I_{\mathfrak{P}})$. The characteristic polynomial

$$\det(E - \rho(F_{\mathfrak{P}})t; V^{\mathfrak{P}})$$

for the action of $\rho(F_{\mathfrak{P}})$ on $V^{\mathfrak{P}}$ depends only on the prime $\mathfrak{p}$ and not on the prime $\mathfrak{P}$ lying over $\mathfrak{p}$.

Definition 9.9: Let $\mathbf{F}/\mathbf{K}$ be a Galois extension of algebraic number fields and (ρ, V) a representation of its Galois group G. Then Artin's L-series for ρ is defined by

$$L(\rho, s) := L(\mathbf{F}/\mathbf{K}, \rho, s) := \prod_{\mathfrak{p}} \frac{1}{\det(1 - \rho(F_{\mathfrak{P}})N(\mathfrak{p})^{-s}; V^{I_{\mathfrak{P}}})}$$

where the product is taken over all prime ideals $\mathfrak{p}$ in $\mathbf{K}$ (i.e. in $\mathfrak{o}$).

This series converges in the right half plane $\mathrm{Re}\, s > 1$ to an analytic function. For $\mathbf{K} = \mathbf{Q}$ and the trivial one-dimensional representation ρ^0 with $\rho^0(\sigma) = 1$ for all $\sigma \in G$, Artin's L-function is just the Dedekind Zeta-function, one has

$$L(\mathbf{F}/\mathbf{Q}, \rho^0, s) = \zeta_{\mathbf{F}}(s).$$

In [Ne] Kapitel VII, we find a lot of results concerning the Artin L-function. To finish with this topic, we only mention Theorem (10.6) in [Ne], clarifying the relation between Artin's and Hecke's L-function in the abelian case.

Theorem 9.12: Let $\mathbf{F}/\mathbf{K}$ be an abelian extension, $\mathfrak{f}$ the conductor of $\mathbf{F}/\mathbf{K}$, ρ a nontrivial character of $\mathrm{Gal}(\mathbf{F}/\mathbf{K})$ and χ the associated grössencharacter mod $\mathfrak{f}$. Then we have the relation

$$L(\mathbf{F}/\mathbf{K}, \rho, s) = \prod_{\mathfrak{p} \in S} \frac{1}{1 - \rho(F_{\mathfrak{P}})N(\mathfrak{p})^{-s}} \, L(\chi, s), \ \ S := \{\mathfrak{p} \mid \mathfrak{f}, \chi(I_{\mathfrak{P}}) = 1\}.$$

Here we should still explain the notion of a *conductor*, but again we have to refer to [Ne].

L-Functions of Elliptic Curves

Again we follow Langlands' article [L] and look at another diophantine problem: We ask for integral or at least rational solutions of the equation

(9.20) $$y^2 = x^3 + Dx,$$

with $D \in \mathbf{Z}, D \neq 0$, say $D = -1$. This is a special case of an affine equation defining an *elliptic curve* $E = E_D$. More generally, one looks at a smooth projective curve defined by the homogeneous equation

$$y^2 z = x^3 + axz^2 + bz^3, \ a, b \in \mathbf{Q}, \ 4a^3 + 27b^2 \neq 0.$$

A curve like this is distinguished among all smooth curves because it can be given the structure of an abelian group. Elliptic curves now are very popular even outside of number theory because of their importance for cryptography. On the way to define an L-function via Euler factors for the primes p, it seems a good idea to count again the number N_p of solutions of (9.20) mod p (or of the associated homogeneous equation, but we stay here with Langlands in his article). For instance, for $D = -1$ and $p = 5$, we get $N_5 = 7$. We define α_p and β_p by the conditions

$$N_p = p + \alpha_p + \beta_p; \ \beta_p = \bar{\alpha}_p; \ \alpha_p \beta_p = p.$$

(These conditions will define α_p and β_p only if $|N_p - p| \leq 2\sqrt{p}$, but it can be proven that this is true.) Then one should be tempted to define as L-function for the elliptic curve

$$L(E_D, s) = \prod_p \frac{1}{1 - \alpha_p p^{-s}} \frac{1}{1 - \beta_p p^{-s}}$$

omitting perhaps some primes p related to D.

For $D = (6577)^2$, one omits $p = 2$ and 3 and gets a function for any s and not only for Re $s > 3/2$ where the Euler product converges.

At this point it is to be emphasized that the value of an elliptic L-function at $s = 1$ has a particular significance: It is a (special case of a) *conjecture of Birch and Swinnerton-Dyer* that the equation (9.20) must have a solution if $L(E_D, 1) = 0$. This conjecture from 1965 is based on numerical calculations and in general is still open.

The L-functions of elliptic curves are special cases of the *Hasse-Weil Zeta functions* associated to projective varieties over number fields. Starting with work by Hasse, Weil, and Dwork, and up to work by Deligne, Langlands and many others, there are deep theorems and conjectures for these relating them again to automorphic L-functions when the varieties are *Shimura varieties*, i.e. come as homogeneous spaces from a linear group, as for instance (roughly said) the completion of the modular curve $\Gamma \backslash \mathfrak{H} = \Gamma \backslash \mathrm{SL}(2, \mathbf{R})/\mathrm{SO}(2)$. There was a famous conjecture, connected mainly with the names of Taniyama, Shimura, and A. Weil, indicating (again roughly) that each elliptic L-function is an automorphic L-function belonging to some modular form, which has been proved by Wiles and Taylor in the framework of their proof of the Fermat conjecture already mentioned at the beginning of this section.

9.7   Summary and Final Reflections

There are still much more L- and Zeta functions than we mentioned up to now (Dwork once remarked to the author that he had the impression that every mathematician feels obliged to define his own one). As in this text we restricted our treatment to the discussion of real and complex groups, we have to leave aside the *true* story, which happens using the groups $G(\mathbf{Q}_p)$ defined over $\mathbf{Q}_p$, the non-archimedian completions of $\mathbf{Q}$, and the *adelic* groups $G(\mathbb{A})$, which come up as the *restricted direct product* of the groups over all archimedian and non-archimedian completions of $\mathbf{Q}$ and/or as groups of matrices with adelic entries.

Hence we finish our story by giving a rudimentary survey of what Langlands says in [L] concerning the archimedian case, in particular as several notions, which we treated in our chapters 6 and 7, reappear in a language using intuition from elementary particle physics.

On the Way to (Archimedian) Langlands Parameters

We look at representations π of a subgroup G of $\mathrm{GL}(n, \mathbf{R})$, in particular $G = \mathrm{GL}(n, \mathbf{R})$. There are essentially two ways to analyse or construct representations:

   i) the infinitesimal method using the Lie algebra of G, as exploited in Chapter 6,

and

   ii) the method, which came up several times in Chapter 4, 7, and 8 and which starts with the action of G on a manifold M, passes to an associated action on functions living on the manifold, and finally decomposes this action into irreducibles.

If G is compact, as for instance $G = \mathrm{SO}(n)$, one takes $M = G$ and decomposes the right-regular representation on $\mathcal{H} = L^2(G)$. This way, one comes to representations π with square integrable matrix coefficients

$$< \pi(g)v, w >, \ v, w \in \mathcal{H},$$

and discrete parameters characterizing π.

If G is not compact, the situation changes considerably as we saw in our discussion of the example $\mathrm{SL}(2, \mathbf{R})$ in 7.2. Langlands proposes that the system being treated is best compared to a Schrödinger equation for which both asymptotically independent and bound states appear (for instance, in the case of the hydrogen atom, for the electron one has discrete orbits and a continuous spectrum):

Since every element $g \in \mathrm{GL}(n, \mathbf{R})$ can be written as a product $k_1 a k_2$ where k_1 and k_2 are orthogonal matrices and $a := D(\alpha_1, \ldots, \alpha_n)$ a diagonal matrix with positive eigenvalues α_j, the simplest representations should have n freely assignable parameters, and they are analogous to a system of n interacting but asymptotically independent particles to which arbitrary momenta can be assigned. In addition, the presence of the factors k_1 and k_2 may entail the presence of discrete quantum numbers. We exemplified this in 7.2.1 while constructing the principal series representations of $\mathrm{SL}(2, \mathbf{R})$ where we had the purely imaginary "momentum" is and the discrete parameter $\epsilon \in \{0, 1\}$. In the $\mathrm{GL}(n, \mathbf{R})$ case, the general induction procedure from 7.1 proposes to look at first at the group B of superdiagonal matrices

$$
(9.21) \qquad b = \begin{pmatrix} \alpha_1 & * & * & \cdots & * \\ & \alpha_2 & * & \cdots & * \\ & & \alpha_3 & \cdots & * \\ & & & \cdots & \\ & & & & \alpha_n \end{pmatrix}.
$$

Given n real parameters $s_1, \ldots, s_n$ and n numbers $\epsilon_k = \pm 1$ (the supplementary discrete quantum numbers), one introduces the characters

$$
\chi_k : \alpha \longmapsto \operatorname{sgn}(\alpha)^{\epsilon_k} |\alpha|^{is_k}
$$

of $\mathbf{R}^*$, as well as the character χ of B that sends the matrix b to

$$
\prod_{k=1}^{n} \chi_k(\alpha_k) |\alpha_k|^{\delta - k}.
$$

Here we have $\delta = (n+1)/2$ and the supplementary exponent comes from the modular function and makes things unitary: Similar to our treatment in the special case in 7.2 we associate to χ the induced representation $ind_B^G \chi$ given by right translation on the space of functions ϕ on G that satisfy the functional equation

$$
\phi(bg) = \chi(b)\phi(g) \quad \text{for all } b \in B, g \in G.
$$

The representations associated to two sequences of parameters χ_k are equivalent iff one sequence is a permutation of the other.

If one replaces the exponents is_k by arbitrary complex numbers, one gets representations, which in general will be neither irreducible nor unitary. But there is a well-determined process for choosing a specific irreducible factor of the representation that is then taken as the representation associated to the characters $\chi_1, \ldots, \chi_n$. In the intuitive language, this means that one allows complex momenta.

In our study of $SL(2, \mathbf{R})$ in 7.2 we met with the discrete series representations. Keeping this in mind, one can deduce that for $GL(2, \mathbf{R})$ there are representations associated to one continuous parameter (or, if the analogy is pursued, one momentum) and one discrete parameter (or quantum number).

In the treatment of $GL(n, \mathbf{R})$ the case $n = 2$ plays a special role because there is just one algebraic extension of the field $\mathbf{R}$, namely $\mathbf{C}$, and this is of degree 2 over $\mathbf{R}$.

For the group $GL(n, \mathbf{C})$ of complex matrices, the representation theory is simpler. One has no discrete series and therefore no bound states. In Langlands' language, for $\mathbf{K} = \mathbf{R}$ or $\mathbf{C}$ the general irreducible representation of $GL(n, \mathbf{K})$ is analogous to r interacting but asymptotically bound states with $n_1, \ldots, n_r$ particles, $\Sigma n_k = n$ and the n_k being subject to constraints appropriate to the field. To construct the representation, one introduces the group P of matrices of the form (9.21) where n is replaced by r, each α_k is a square n_k–matrix, and the asterisks represent block matrices of the appropriate size. From here, we can again apply the induction process but we have to replace the character χ by the tensor product of (in general infinite-dimensional) representations χ_k of $GL(n, \mathbf{K})$, so that the functions ϕ with their functional equation as above take their values in an infinite-dimensional space. If the momenta are all real, then the representation is irreducible. Otherwise it is again necessary to pass to a specific factor. As before, the order of the χ_k is irrelevant.

After some more consideration, finally one can conclude that the classification of irreducible representations of $GL(n, \mathbf{K})$ has an extreme formal simplicity. For $\mathbf{K} = \mathbf{C}$, to specify representations we specify the n-dimensional representations of $GL(1, \mathbf{C}) = \mathbf{C}^\times$. As Langlands emphasizes, here we compare two very different objects: on one hand, irreducible and in general infinite-dimensional representations of the group $GL(n, \mathbf{C})$, and on the other, finite-dimensional but in general reducible representations of $GL(1, \mathbf{C}) = \mathbf{C}^\times$.

A similar statement for $\mathbf{K} = \mathbf{R}$ requires a group that is not commutative but whose irreducible representations are of degree at most two, in order to accomodate the existence of two particle bound states. The appropriate group is the *Weil group of* $\mathbf{R}$, $W_\mathbf{R}$. It is obtained by adjoining to the group $\mathbf{C}^\times$ an element w such that

$$w^2 = -1, \text{ and } wz = \bar{z}w \quad \text{for all } z \in \mathbf{C}^\times.$$

This is a kind of dihedral group and, in our text, it appeared as an example already in the Chapters 0 and 1. It has as non-trivial irreducible representations the character π_0 given by

$$z \longmapsto z\bar{z}, \ w \longmapsto -1$$

and the two-dimensional representations π_m, for $0 \neq m \in \mathbf{Z}$ given by

$$z \longmapsto \begin{pmatrix} z^m & 0 \\ 0 & \bar{z}^m \end{pmatrix}, \ w \longmapsto \begin{pmatrix} 0 & 1 \\ (-1)^m & 0 \end{pmatrix}.$$

The Weil group $W_\mathbf{C}$ of the complex number field $\mathbf{C}$ is just $\mathbf{C}^\times$. One has the nice result that for $\mathbf{K} = \mathbf{R}$ and $\mathbf{C}$ the irreducible representations of $GL(n, \mathbf{K})$ are parametrized by n-dimensional representations of the Weil group $W_\mathbf{K}$. The Weil group can be attached to many of the fields appearing in number theory. It can be rather complicated because, as Langlands assures, it incorporates many of the deepest facts about the theory of equations known at the present.

To get some more flavour, we add a remark concerning orthogonal subgroups G of $GL(2n + 1, \mathbf{R})$ defined as stability groups of quadratic forms. As we know, there are several types of groups depending on the index of the quadratic form and the representation theory of these will be quite different. But one can get a unified point of view by a parametrization of the representations via homomorphims of the Weil group $W_\mathbf{R}$ into the L-*group* $^L G$ of G. The general definition of the L-group is given in terms of the theory of equations and of root systems. For $G = GL(n, \mathbf{R})$ it is simply $^L G = GL(n, \mathbf{C})$ and for $G = SO(2n + 1)$ it is the symplectic group of $2n \times 2n$-matrices, $^L G = Sp(n, \mathbf{R})$.

The principle that the irreducible representations of the real group G are classified by the (continuous) homomorphisms of $W_\mathbf{R}$ into $^L G$ is valid, but this has consequences. A reducible n-dimensional representation of $W_\mathbf{R}$ is a homomorphism of $W_\mathbf{R}$ into $GL(n, \mathbf{C})$ that is isomorphic to $\mathbf{C}^\times$ and not in the center of $GL(n, \mathbf{C})$. Hence the notion of irreducibility has an obvious analogue for the homomorphisms of $W_\mathbf{R}$ into $^L G$, and it can be verified that the homomorphisms which are irreducible in this sense correspond to representations whose matrix coefficients are square-integrable over the group. If the quadratic form chosen is Euclidean, then G is compact and all representations have this property, so that any homomorphism of $W_{\mathbf{R}}$ into $^L G$ that is not irreducible in this sense will have to be excluded from the list of parameters. A similar but less restrictive condition must also be imposed for the groups with other indices.

It also turns out that the classification provided by $W_{\mathbf{R}}$ and $^L G$ is coarse. Some homomorphisms correspond to several irreducible representations of G. This finally leads to the notion of *L-packets* which indicates representations with the same *L*-functions.

L-Groups and Functoriality

For Langlands the introduction of the *L*-group is an essential step towards his famous principle of functoriality. A homomorphism from a group H to a group G does not provide any way to transfer irreducible representations from one of these groups to another, unless G is abelian, nor are homorphisms between H and G usually related to homomorphisms between $^L H$ and $^L G$. On the other hand, a homomorphism from $^L H$ to $^L G$, by composition,

$$W_{\mathbf{R}} \longrightarrow {}^L H \longrightarrow {}^L G$$

yields a map from the set of parameters for the irreducible representations of H to the irreducible representations of G, and thus implicitly a way to pass from irreducible representations of H to irreducible representations of G. Langlands calls this passage the *principle of functoriality* in the *L*-group. A possibility to realize this principle in general would have deepest consequences.

The Standard *L*-Function

In 9.6 we already insisted on the possibility to associate to a given problem or object an *L*-function via an Euler product by encoding information about the object in such a way that one gets a reasonable "Euler polynomial" in the variable $t = p^{-s}$, which can be put in the denominator for an Euler factor. While constructing representations, say of $\mathrm{GL}(n, \mathbf{K})$, via induction in Chapter 7, the main idea was that at least some representations are fixed by choosing characters for the group H of diagonal matrices $D(\alpha_1, \ldots, \alpha_n)$, which can be understood as distinguishing matrices

$$D(\alpha^{is_1}, \ldots, \alpha^{is_n}), \ s_1, \ldots, s_n \in \mathbf{R}.$$

Even if we still avoid to go into the representation theory of the non-archimedian field $\mathbf{Q}_p$, it should not be too hard to accept that the analogon of the matrix above in the *p*-adic theory is the matrix

$$A(\pi_p) := D(p^{is_1}, \ldots, p^{is_n}), \ s_1, \ldots, s_n \in \mathbf{R}.$$

To this matrix one can attach the function

$$L(\pi_p, s) := \frac{1}{\det(E_n - A(\pi_p)p^{-s})}$$

and then form the Euler product

$$L(\pi, s) := \prod L(\pi_p, s).$$

In this product perhaps some "special" primes are left out or are represented by special factors and one has to find an appropriate "factor at ∞" associated to the archimedian representation. Such a product has the chance to converge for, say, Re $s \geq a + 1$, provided that the eigenvalues of each $A(\pi_p)$ are less than p^a in absolute value. The outcome is called a *standard L-function*. Whoever wants to see that this really happens (and even more...) should be motivated to take a look into the non-archimedian theory, as for instance in [Ge] or [GeS].

L-Groups and Automorphic L-Functions

To finish our fantastic and magical story (the German word *Märchen* appears well in the title of one of Langlands' articles ([L1])), we look at an (automorphic) representation π of a group G (in particular $G = \mathrm{GL}(n)$), take a finite-dimensional representation ρ of the L-group ^LG belonging to G. Then we have again a finite Euler polynomial given by $\det(E - \rho(A(\pi_p)t)$ and one can introduce as *automorphic L-function*

$$L(\pi, \rho, s) := \prod \frac{1}{\det(E - \rho(A(\pi_p))p^{-s})}$$

where again for some primes p we have to put in special factors, which we do not discuss here. In particular, one has to find a *completing* factor belonging to the archimedian theory. The final goal is to generalize what we sketched in the last section by showing that to given π and ρ there is an automorphic representation Π such that

$$A(\Pi_p) = \rho(\pi_p)$$

for almost all p. Then one could have

$$L(\pi, \rho, s) = L(\Pi, s)$$

and use properties of one type of functions to be transfered to the other.

There are many other important topics coming up in this context. We only mention the topic of the *trace formula* but can not touch this here. The only theme whose formulation is within our reach is the following: Given a representation π of a group $G = \mathrm{GL}(n)$ and a representation π' of $G' = \mathrm{GL}(n')$, by the methods from of our text one can construct representations $\pi + \pi'$ of $\mathrm{GL}(n + n')$ and $\pi \otimes \pi'$ of $\mathrm{GL}(nn')$. It is natural to ask for a similar construction for automorphic representations. This is related to functoriality and up to now there are very few results.

We repeat our hope that the vagueness of the presentation of some of the material in the last sections would make the reader curious to know more about all this and to be stimulated to more intense studies of this most fascinating part of mathematics. All the time, we were driven by J. R. Jiménez' words:

> ¡Voz mía, canta, canta;
> que mientras haya algo
> que no hayas dicho tú,
> tú nada has dicho!

Bibliography

[AM]   Abraham, R., Marsden, J.E.: *Foundations of Mechanics*. Benjamin/Cummings, Reading 1978.

[BR]   Barut, A.O., Raczka, R.: *Theory of Group Representations and Applications*. PWN Polish Scientific Publishers Warszawa 1980.

[Ba]   Bargmann, V.: *Irreducible Unitary Representations of the Lorentz Group*. Annals of Math. **48** (1947) 568-640.

[Be]   Berndt, R.: *Einführung in die Symplektische Geometrie*. Vieweg, Braunschweig/Wiesbaden 1998.
         Now also translated: *An Introduction to Symplectic Geometry*. GSM **26**, AMS 2001.

[Be1]   Berndt, R.: *The Heat Equation and Representations of the Jacobi Group*. Contemporary Mathematics **389** (2006) 47 – 68.

[BeS]   Berndt, R. Schmidt, R.: *Elements of the Representation Theory of the Jacobi Group*. PM 163, Birkhäuser, Basel 1998.

[BR]   Brateli, O., Robinson, D.W.: *Operator Algebra and Quantum Statistical Mechanics*. Springer, New York 1979.

[BtD]   Bröcker, T., tom Dieck, T.: *Representations of Compact Lie Groups*. Springer, New York 1985.

[Bu]   Bump, D.: *Automorphic Forms and Representations*. Cambridge University Press, Cambridge 1997.

[CF]   Cassels, J.W.S., Fröhlich, A.: *Algebraic Number Theory*. Thompson, Washington, D.C. 1967.

[CN]   Conway, J.H., Norton, S.P.: *Monstrous Moonshine*. Bull. London Math. Soc. **11** (1979) 308 - 339.

[Co]   Cornwell, J. F.: *Group Theory in Physics*. Academic Press, London 1984.

[Do]   Donley, R. W.: *Irreducible Representations of* SL$(2, \mathbf{R})$. p.51 - 59 in: Representation Theory and Automorphic Forms (Bailey, T.N., Knapp, A.W., eds.), PSPM Vol. **61** AMS 1997.

[Du]   Duflo, M.: *Théorie de Mackey pour les groupes de Lie algébriques*. Acta Math. **149** (1982) 153 - 213.

[Ei]   Eichler, M.: *Einführung in die Theorie der algebraischen Zahlen und Funktio-nen*. Birkhäuser, Basel 1963.

[Ei1]   Eichler, M.: *Einige Anwendungen der Spurformel im Bereich der Modularkor-respondenzen*. Math. Ann. **168** (1967) 128 - 137.

[EZ]   Eichler, M., Zagier, D.: *The Theory of Jacobi Forms*. Birkhäuser, Boston 1985.

[Fi]   Fischer, G.: *Lineare Algebra*. Vieweg, Wiesbaden 2005.

[Fl]   Flicker. Y. Z.: *Automorphic Forms and Shimura Varieties of* PGSp(2). World Scientific, New Jersey 2005.

[Fog]   Fogarty,J.: *Invariant Theory*. Benjamin,New York 1969.

[Fo]   Forster, O.: *Analysis 1*. Vieweg, Wiesbaden 2006.

[FB]   Freitag, E., Busam, R.: *Funktionentheorie*. Springer, Berlin 1993.

[Fr]   Frenkel, E.: *Lectures on the Langlands Program and Conformal Field Theory*. arXiv:hep-th/0512172v1 15 Dec 2005.

[FLM]   Frenkel, E., Lepowski, J., Meurman, A.: *Vertex Operator Algebras and the Monster*. Academic Press, Boston 1988.

[FS]   Fuchs, J., Schweigert, Chr.: *Symmetries, Lie Algebras and Representations*. Cambridge University Press 1997.

[FH]   Fulton, W., Harris, J.: *Representation Theory*. GTM 129, Springer, New York 1991.

[Ge]   Gelbart, S.: *Automorphic Forms on Adele Groups*. Annals of Math. Studies 83, Princeton University Press 1975.

[GeS]   Gelbart, S., Shahidi, F.: *Analytic Properties of Automorphic L-Functions*. Aca-demic Press, Boston 1988.

[Gel]   Gell-Mann, M.: Phys. Rev. **125** 1067-1084, 1962.

[GH]   Griffiths, P., Harris, H.: *Principles of Algebraic Geometry*. Wiley, New York 1978.

[GGP]   Gelfand, I., Graev, M., Pyatetskii-Shapiro, I.: *Representation Theory and Au-tomorphic Functions*. W.B. Saunders, Philadelphia 1963.

[Go]   Goldstein, H.: *Classical Mechanics*. Addison-Wesley, Reading 1980.

[GS]   Guillemin, V., Sternberg, S.: *Symplectic Techniques in Physics*. Cambridge University Press 1984.

[GS1]   Guillemin, V., Sternberg, S.: *Geometric Asymptotics*. Math. Surv. and Mono-graphs 14, AMS 1990.

[Ha]   Halmos, P.: *Measure Theory*. Van Nostrand, New York 1950.

[He]   Hein, W.: *Struktur- und Darstellungstheorie der klassischen Gruppen*. Springer HT Berlin 1990.

[HN]   Hilgert, J., Neeb, K.-H.: *Lie-Gruppen und Lie-Algebren*. Vieweg, Braun-schweig/Wiesbaden 1991.

[HR]   Hewitt, E., Ross, K.A.: *Abstract Harmonic Analysis I*. Springer, Berlin 1963.

[Hu]   Humphreys, J.E.: *Introduction to Lie Algebras and Representation Theory*. GTM 9, Springer New York 1972.

[Ig]   Igusa, J.: *Theta Functions*. Springer, Berlin 1972.

[JL]   Jacquet, H., Langlands, R.P.: *Automorphic Forms on GL(2)*. LNM 114, Springer, New York 1970.

[Ja]   Jacobson, N.: *Lie Algebras*. Interscience, New York 1962.

[Ki]   Kirillov, A.A.: *Elements of the Theory of Representations*. Springer, Berlin 1976.

[Ki1]   Kirillov, A.A.: *Lectures on the Orbit Method*. GSM **64**, AMS 2004.

[Ki2]   Kirillov, A.A.: *Unitary Representations of Nilpotent Lie Groups*. Uspekhi Mat. Nauk **17** (1962), 57 - 110; English transl. in Russian Math. Surveys **17** (1962).

[Ki3]   Kirillov, A.A.: *Merits and Demerits of the Orbit Method*. Bulletin of the AMS **36** (1999) 433 - 488.

[Ko]   Koecher, M.: *Lineare Algebra*. Springer, Berlin 1983.

[Kos]   Kostant, B.: *Quantization and Unitary Representations*. In *Lectures in Modern Analysis* III. (ed. Taam, C.T.) LNM **170**, Springer, Berlin 1970.

[Kn]   Knapp. A.W.: *Representation Theory of Semisimple Groups. An Overview Based on Examples*. Princeton University Press 1986.

[Kn1]   Knapp. A.W.: *Structure Theory of Semisimple Lie Groups*. p.1 - 27 in: Representation Theory and Automorphic Forms (Bailey, T.N., Knapp, A.W., eds.), PSPM Vol. **61**, AMS 1997.

[Kn2]   Knapp. A.W.: *Lie Groups Beyond an Introduction*. PM 140, Birkhäuser, Boston 1996.

[KT]   Knapp, A.W., Trapa, P.E.: *Representations of Semisimple Lie Groups*. p. 7 - 87 in: Representation Theory of Lie Groups (Adams, J., Vogan, D., eds.), IAS/Park City Math. Series **8**, AMS 2000.

[Ku]   Kubota, T.: *Elementary Theory of Eisenstein Series*. Halsted Press, New York 1973.

[La]   Lang, S.: *Algebra*. Addison-Wesley, Reading, Mass. 1965.

[La1]   Lang, S.: SL(2,**R**). Springer, New York 1985.

[La2]   Lang, S.: *Algebraic Number Theory*. Addison - Wesley, Reading, Mass. 1970.

[L]   Langlands, R.P.: *Representation Theory: Its Rise and Its Role in Number Theory*. p. 181-210 in Proceedings of the Gibbs Symposium, Yale University 1989, AMS 1990.

[L1]   Langlands, R.P.: *Automorphic Representations, Shimura Varieties, and Mo-tives. Ein Märchen*. p. 205-246 in PSPM Vol 33, Part 2, AMS 1979.

[LV]   Lion, G., Vergne, M.: *The Weil representation, Maslov index and Theta series*. Birkhäuser, Boston 1980.

[LZ]   Lewis, J., Zagier, D.: *Period functions for Maass wave forms* I. Ann. Math. **153** (2001) 191-253.

[Ma]   Mackey, G.W.: *Unitary Group Representations in Physics, Probability, and Number Theory*. Benjamin/Cummings Publishing Co., Reading, Mass. 1978.

[Ma1]   Mackey, G.W.: *Induced Representations of Locally Compact Groups* I. Ann. of Math. **55** (1952) 101-139.

[Mu]   Mumford, D.: *Tata Lectures on Theta* I,II,III. PM 28, 43, 97, Birkhäuser, Boston 1983, 1984, 1991.

[Na]   Naimark, M.A.: *Linear Representations of the Lorentz Group*. Pergamon Press, London 1964.

[Ne]   Neukirch, J.: *Algebraische Zahlentheorie*. Springer, Berlin 2002.

[RV]   Ramakrishnan, D., Valenza, R.J.: *Fourier Analysis on Number Fields*. GTM 186, Springer, New York 1999.

[Re]   Renouard, P.: *Variétés symplectiques et quantification*. Thèse. Orsay 1969.

[Sch]   Schöneberg, B.: *Elliptic Modular Functions*. Springer, Berlin 1974.

[Se]   Serre, J.P.: *Linear Representations of Finite Groups*. Springer, New York 1977.

[Se1]   Serre, J.P.: *A Course in Arithmetics*. GTM 7, Springer, New York 1973.

[Sel]   Selberg, A.: *Harmonic analysis and discontinuous groups in weakly symmetric riemannian spaces with application to Dirichlet series*. J. Indian Math. Soc. **20** (1956) 47 - 87.

[Sh]   Shimura, G.: *Introduction to the Arithmetic Theory of Automorphic Functions*. Iwanami Shoten and Princeton University Press 1971.

[To]   Torasso, P.: *Méthode des orbites de Kirillov–Duflo et représentations minimales des groupes simples sur un corps local de caractéristique nulle*. Duke Math. J. **90** (1997) 261 – 377.

[vD]   van Dijk, G.: *The irreducible Unitary Representations of* SL(2, **R**). In: Representations of Locally Compact Groups with Applications. ed. Koornwinder, T.H. Mathematisch Centrum Amsterdam 1979.

[Ve]   Vergne, M.: *Geometric Quantization and Equivariant Cohomology*. First European Congress of Mathematics, Vol I, p.249–298. PM 119, Birkhäuser, Basel 1994.

[Ve1]   Vergne, M.: *Quantification géométrique et réduction symplectique*. Séminaire Bourbaki **888** (2001).

[Vo]   Vogan, D.A.: *The Method of Coadjoint Orbits for Real Reductive Groups*. p.179
          – 238 in: Representation Theory of Lie Groups (Adams, J., Vogan, D., eds.),
          IAS/Park City Math. Series **8** AMS 2000.

[Vo1]   Vogan, D.A.: *Associated Varieties and Unipotent Representations*. p.315 – 388
          in: Harmonic Analysis on Reductive Lie Groups (Barker, W., Sally, P., eds.),
          Birkhäuser, Boston 1991.

[Vo2]   Vogan, D.A.: *Cohomology and group representations*. PSPM Vol 61, 219–243
          (1997).

[Wa]   Warner, G.: *Harmonic Analysis on Semi-Simple Lie Groups*. Springer, Berlin
          1972.

[We]   Weil, A.: *Variétés Kählériennes*. Hermann, Paris 1957.

[We1]   Weil, A.: *Über die Bestimmung Dirichletscher Reihen durch ihre Funktionalglei-
          chung*. Math, Ann. **168** 149-156 (1967).

[Wo]   Woodhouse, N.: *Geometric Quantization*. (Second Edition) Oxford University
          Press 1991.

[Ya]   Yang, Y.-H.: *The Method of Orbits for Real Lie Groups*. Kyungpook Math. J.
          42 (2002) 199–272.

Index

Caterina Consani, Matilde Marcolli (Eds.)

Noncommutative Geometry and Number Theory

Where Arithmetic meets Geometry and Physics

2006. viii, 372 pp. with 24 figs. (Aspects of Mathematics E 37, ed. by Diederich, Klas) Hardc. EUR 68,90   ISBN 978-3-8348-0170-8

In recent years, number theory and arithmetic geometry have been enriched by new techniques from noncommutative geometry, operator algebras, dynamical systems, and K-Theory. This volume collects and presents up-to-date research topics in arithmetic and noncommutative geometry and ideas from physics that point to possible new connections between the fields of number theory, algebraic geometry and noncommutative geometry.

The articles collected in this volume present new noncommutative geometry perspectives on classical topics of number theory and arithmetic such as modular forms, class field theory, the theory of reductive p-adic groups, Shimura varieties, the local L-factors of arithmetic varieties. They also show how arithmetic appears naturally in noncommutative geometry and in physics, in the residues of Feynman graphs, in the properties of noncommutative tori, and in the quantum Hall effect.

Abraham-Lincoln-Straße 46
65189 Wiesbaden
Fax 0611.7878-400   Stand 1.1.2007. Änderungen vorbehalten.
www.vieweg.de   Erhältlich im Buchhandel oder im Verlag.

vieweg